Earth and Environmental Science

THE HSC COURSE

Tom Hubble

Chris Huxley

Iain Imlay-Gillespie

CAMBRIDGE
UNIVERSITY PRESS

PUBLISHED BY THE PRESS SYNDICATE OF THE UNIVERSITY OF CAMBRIDGE
The Pitt Building, Trumpington Street, Cambridge, United Kingdom

CAMBRIDGE UNIVERSITY PRESS
The Edinburgh Building, Cambridge CB2 2RU, UK
40 West 20th Street, New York, NY 10011–4211, USA
477 Williamstown Road, Port Melbourne 3207, Australia
Ruiz de Alarcón 13, Madrid 28014, Spain
Dock House, The Waterfront, Cape Town 8001, South Africa

http://www.cambridge.edu.au

All activities in this book have been written with the current safety
regulations in schools in mind. However, it is recommended that
the experiments in this book be carried out in the presence of a
qualified teacher and after consultation with appropriate state
safety regulations for schools.

First published in 2002

Edited by Susan Lee
Artwork and maps by Qiu Lin Yang
Index by Caroline Colton
Design by Mason Design
Printed in Australia by Brown Prior Anderson
Typeface Garamond 10.5 pt *System* QuarkXPress® [MM]

National Library of Australia Cataloguing in Publication data
    Hubble, Tom.
    Earth and environmental science: the HSC course.
    Includes index.
    For senior secondary school students.
    ISBN 0 521 01663 0.
    1. Earth sciences. 2. Environmental sciences. I. Huxley,
    Chris. II. Imlay-Gillespie, Iain. III. Title.
550.712

ISBN 0 521 01663 0

Waiver
The publisher has used its best endeavours to ensure that the URLs for external
websites referred to in this book are correct and active at the time of going to press.
However, the publisher has no responsibility for the websites and can make no
guarantee that a site will remain live or that the content is or will remain appropriate.

Contents

Introduction

In this book we will examine the past, present and possible future of our planet. We begin by looking at the processes that have shaped the continents and the climate through time. The same tectonic processes that shape such things as mountains also affect people and we will examine how and why natural disasters occur and the way science works to reduce the effects of such calamities on people.

An understanding of the interaction of Earth processes and life is developed as we look at the evolution of the atmosphere and key events in the history of life, particularly the role of mass extinctions and how life exploits new environments. Contemporary environmental issues regarding our soils, atmosphere and water are also examined.

Later sections, corresponding to HSC electives, will introduce you to the impact of introduced species, our utilisation of organic and mineral resources as well as the features and processes of the world's oceans.

The text covers a broad spectrum of Earth and Environmental Science issues. It is designed to allow both the school and independent student to follow the NSW HSC course in Earth and Environmental Science. It provides students with regular review activities to test their understanding of the text. In addition, extension activities and a variety of practical exercises will allow students to consolidate their knowledge, understanding and skills. A resource list to assist students in research activities is included at the end of this text. In addition, students will find the 'Effective Research' section in the Preliminary text of use in the HSC course.

Whether you use the book to prepare for the HSC or as a means of learning more about the Earth, we hope that the book will increase your appreciation of our world and your place in it.

Acknowledgments

Cover image Courtesy of Tourism New South Wales; *Fig. 1.1.12* Redrawn fom *Earth's Dynamic Systems* 8/E by Hamblin/Christiansen, © Reprinted by permission of Pearson Education, Inc., Upper Saddle River, NJ; *Fig. 1.1.16* Redrawn from *Geodynamics* by D. Turcotte & G. Schubert, John Wiley & Sons (1982); *Fig. 1.2.1* Redrawn from *The Fundamentals of the Physical Environment*, Briggs et. al, Routledge (1997); *Fig. 1.3.2* Redrawn from *Perspectives of the Earth* by I. F. Clark & B. J. Cook, Australian Academy of Science (1983); *Fig. 1.4.3* Redrawn from *Apocalypse: A Natural History of Global Disasters* by B. McGuire, Cassell (1999); *Fig. 1.4.5* Photograph by D. A. Swanson, Courtesy of the USGS; *Fig. 1.5.2* Redrawn from *Volcanic Eruption and Atmospheric Change* by R. W. Johnson, AGSO (1993); *Fig. 2.2.1 Elements of Palaeontology*, 2nd edn, by Rhona M. Black, University of Cambridge Press (1998); *Fig. 2.2.2* © Queensland Museum; *Fig. 2.2.5* Photo by Bruce Cowell © Queensland Museum; *Figs 2.31, 2.3.2, 2.3.3, 2.3.4 and 2.3.6* Reproduced with kind permission from Roger Buick, University of Washington; *Fig. 2.3.5* Redrawn from *Cradle of Life* by J. William Schopf, Princeton University Press (1999); *Fig. 2.4.5* Redrawn from *Earth through Time* (5th Edition) by H. Levin, Published by Saunders College Publishing; *Fig. 2.4.10* from *Historical Geology, Evolution of Earth and Life through Time* by R. Wicaner & J. S. Monroe, Brooks/Cole (2000); *Figs 2.5.1, 2.5.2 and 2.5.3* Information on the plant fossil record is available in the *Greening of Gondwana*, by Mary E. White, Kangaroo Press (1998); *Fig. 2.5.4* Taken from *Fossil Invertebrates* by Richard S. Boardman, Alan H. Cheethan & Albert J. Rowell, © Blackwell Scientific Publications (1987); *Figs 2.6.2 and 2.6.3* Redrawn from *Scientific American*, Douglas H. Erwin, Scientific American Inc. (July 1996 issue); *Fig. 2.6.4* Courtesy of NASA; *Fig. 2.6.6* Redrawn from *Extinctions Downunder* by Peter Murray, The University of Arizona Press (1984); *Fig. 3.1.1* Natural Resources Conservations Science; *Figs. 3.1.2 and 3.1.3* Redrawn with permission of the NSW Environmental Protection Authority; *Fig. 3.1.4* © NRCS; *Fig. 3.1.5* © USGS; *Fig. 3.2.1* Courtesy of NSW Department of Agriculture; *Fig. 3.2.2* Redrawn from *Australian Academy of Technological Sciences and Engineering Academy Symposium* (November 1999); *Table 3.3.2* NSW Salinity Strategy, NSW Government; *Fig 3.2.3* Courtesy of Fiona Leach, NSW Department of Agriculture; *Fig 3.2.4* © CSIRO Science Images; *Fig. 3.3.3* Courtesy of Malcolm Campbell, Astroloma hortmedia P/L; *Fig. 3.4.1* Courtesy of Tourism New South Wales; *Fig. 3.4.2* Courtesy of the Murray Darling Basin Commission; *Fig. 3.4.4* Melbourne Water; *Fig. 3.5.3* Courtesy of Virginia Naude, Norton Art Conservation, Inc.; *Figs 3.6.1 and 3.6.2* Courtesy of ACT NOWaste; *Figs 3.6.5 and 6.5.1* Michael Dix— Mineral Resources, Tasmania; *Fig. 4.1.2* By permission of the National Library of Australia; *Fig. 4.2.2* © Stephen Doggett; *Figs 4.2.4, 4.4.2 and 6.4.1* News Ltd; *Fig. 4.3.1* Reproduced by permission of CSIRO Australia; *Fig. 4.3.2* Queensland Museum; *Figs 4.5.1 and 4.5.3* Australian Quarantine and Inspection Service (AQIS); *Table 5.1.1* Data reproduced from *Climate Change Science: Current Understanding and Uncertainties* © Australian Academy of Technological Sciences and Engineering; *Fig. 5.1.5* Redrawn from *Introduction to Economic Geology and its Environmental Impact* by A. Evans, Blackwell Scientific (1997); *Fig. 5.2.3* Redrawn from *Basin Analysis Principles and Applications* by P. A. & J. R. Allen, Blackwell Science (1990); *Fig. 6.2.13 and 6.2.14* Redrawn from *Discovery of the Cadia–Ridgeway Gold–Copper Porphyry Deposit* by J. Holliday, C. McMillan & I. Tedder, SMEDG (1999); *Fig. 6.3.1* Image provided from the molybdenum-gold project in south-east Queensland; *Fig. 6.2.14* Redrawn figure © Newcrest Mining Limited (1999); *Figs 7.1.8 and 7.1.9* Courtesy of Ocean Drilling Program; *Figs 7.1.10, 7.1.11, 7.1.14, 7.4.7, 7.5.1 and 7.5.2* Courtesy of National Oceanic and Atmospheric Administration (NOAA); *Figs 7.1.12 and 7.1.15* Redrawn from *Oceanography*, 3rd edn, by M. Grant Gross, Charles E. Merrill Publishing Company (1967); *Fig. 7.2.3* Redrawn from *The Ocean Basins: Their Structure and Evolution*, The Oceanography Course Team, The Open University (1989); *Fig. 7.3.7* Pacific Tsunami Museum.

Every effort has been made to trace and acknowledge copyright. The publisher apologises for any accidental infringement and welcomes information that would rectify any error or omission in subsequent editions.

Tectonic impacts

1

Throughout the Earth's history plate tectonic processes have shaped our planet's surface and caused it to evolve. Continents have grown and been split apart. Oceans have formed and disappeared. Continents move across the surface of the Earth. In the last 500 million years, parts of Australia have moved from the equator towards the South Pole and then north again. During this time, Australia has grown and changed shape.

Tectonic processes have also caused local and global environmental change. As a result, plate tectonics have profoundly affected the evolution of life and the short-term survival of ecosystems. Today, many human communities and associated natural environments are affected by short-term changes caused by tectonic processes, such as volcanic eruptions and earthquakes.

In this section we will examine the nature of the Earth's outer layers and the ways in which mountains and continents are created and evolve. We will look at the way tectonic processes affect the planet's climate and cause a variety of hazards that can destroy or damage communities. We will also investigate the work scientists are doing to reduce the impact of such forces on communities.

CONTENTS

At the end of this chapter you should be able to:

- describe the characteristics of lithospheric plates

- identify the relationship between the general composition of igneous rocks and plate boundary type

- outline the motion of plates and distinguish between the three types of plate boundaries (convergent, divergent and conservative)

- assess current hypotheses used to explain plate motion.

CHARACTERISTICS OF LITHOSPHERIC PLATES AND THEIR MARGINS

Structure of the Earth

Over the last 200 years scientists have studied the Earth's surface and learnt how to investigate its internal structure using seismic waves. By the start of the twentieth century, seismologists recognised four **concentric** layers within the Earth. These consist of an outer pair (the crust and the mantle) and an inner pair (the outer core and the inner core). The crust and mantle are mainly composed of silicate rocks, and the outer core and inner core are predominantly composed of iron and nickel. During the twentieth century, better theories and technological devices led seismologists and other Earth scientists to develop a more detailed model of the Earth that divides the mantle into a number of additional layers.

Today, Earth scientists recognise the following layers of the Earth:

- *Lithosphere.* The outermost cool, rigid part of the Earth. It extends down to an average depth of 100 km and is composed of the crust and a layer of upper mantle that is hard and brittle.

- *Asthenosphere.* A weak layer of upper mantle that flows when forces act on it over a long period of time. This layer is about 250 km thick, but is thin when the lithosphere is thick and thick when the lithosphere is thin.

- *Mesosphere.* The deepest part of the Earth's mantle. It contains material that is rigid enough to transmit seismic waves but plastic enough to allow the material to flow and move by convection.

- *Outer core.* A liquid metal layer containing iron with some nickel and perhaps oxygen and potassium. Convection within the outer core plays an important role in making the Earth's magnetic field.

- *Inner core.* A solid metal core consisting mainly of iron and nickel.

A comparison of older and more recent views on the structure of the Earth is shown in Figure 1.1.1. Note that the mantle forms much of the lithosphere, and all of both the asthenosphere and the mesosphere.

Older model

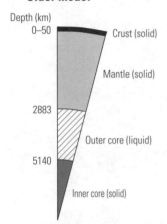

Depth (km)
0–50 — Crust (solid)
Mantle (solid)
2883
Outer core (liquid)
5140
Inner core (solid)

More recent model

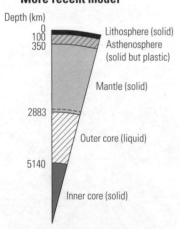

Depth (km)
0
100
350 — Lithosphere (solid)
Asthenosphere (solid but plastic)
Mantle (solid)
2883
Outer core (liquid)
5140
Inner core (solid)

Figure 1.1.1 Older and more recent models of the Earth's structure.

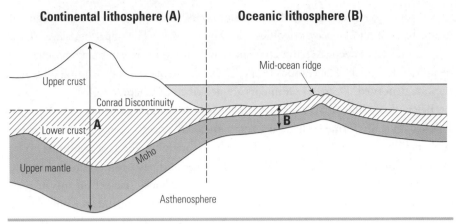

Figure 1.1.2 The structure of the lithosphere. (*Note:* This diagram is not drawn to scale.)

THE LITHOSPHERE

The concept of the lithosphere (a word meaning 'the rocky sphere') has developed over the last 100 years. The lithosphere is cool, solid and rigid. It is approximately 100 km thick and consists of an upper part called the crust and a lower part that is made of mantle rocks. The lower part has a mineral and chemical composition very similar to that of the underlying, softer asthenospheric and mesospheric mantle. (See Figure 1.1.2.) The lithosphere rides on top of the asthenosphere and is broken up into a number of lithospheric plates.

The lithosphere can be divided into two types—oceanic or continental—depending on the nature of the upper layer of crustal material. (See Figure 1.1.3.) Oceanic crust consists of basaltic materials and oceanic sediments, while continental crust is dominated by granitic materials, metamorphic rocks and sedimentary rocks. Note that oceanic crust can be incorporated into continental crust in certain circumstances.

Oceanic crust and lithosphere

Although oceanic crust is difficult to sample, we probably know more about it than continental crust. Scientists have built up a model of oceanic crust using studies of seismic, gravity and heat flow measurements, together with studies of **ophiolites**. A diagram of the model is shown in Figure 1.1.3.

concentric
a set of curved surfaces or lines that have a common centre

ophiolites
rocks found on land that have an origin as oceanic crust

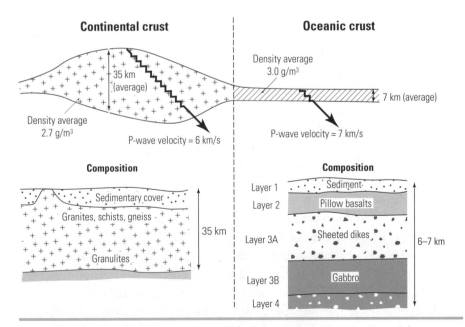

Figure 1.1.3 A comparison of continental and oceanic crust. (*Note:* This diagram is not drawn to scale.)

Oceanic crust is composed of four major layers, which were originally recognised on the basis of seismic wave velocities. (See Figure 1.1.3.) The uppermost layer (layer 1) is composed of deep-sea sediment. The sediment layer varies in thickness and may be up to 4 km thick. The composition of these sediments varies depending on where they are deposited. Deep-sea sediments are described in Chapter 7.6.

Below the sediment layer is a layer of **pillow basalts** and basaltic breccias (layer 2). Pillow basalts are formed when lava enters the cold ocean water. The outer surface cools rapidly and forms a glassy skin, or chilled margin. The lava continues to be extruded through cracks in the chilled margin like toothpaste being squeezed from a tube. The basalt forms a layer resembling elongated sacks stacked on top of each other. Basaltic dikes (also known as dolerite dikes) sometimes intrude the pillow layer from below. The basalts and breccias form a layer 1–2.5 km thick.

The third layer usually consists of two parts. Below the pillow basalts is a 2 km thick sheeted dike complex (layer 3A). The dikes are about a metre wide and are composed of relatively fine crystals, indicating that they cooled relatively quickly. The outermost margins of these dikes cool even more quickly than their interiors and dikes intrude each other.

Below the sheeted dikes is a layer of massive gabbro and layered gabbro (layer 3B). These gabbros have virtually the same mineral and chemical composition as the layers above them but are given a different name because they are coarser grained and have different textures. They form from batches of the same liquid magma that makes up the upper layers. The gabbros form on the bottom and sides of magma chambers located about 2–3 km beneath the crest of the mid-ocean ridge. Sometimes crystals rich in iron and magnesium minerals form in batches that sink to the bottom of the magma chamber, giving these gabbros a layered appearance when seen exposed in outcrops onshore.

Layer 4 of oceanic crust is a thin layer of upper mantle material that has become attached to the layer 3B gabbros. This layer of mantle is unusual in that its mineral grains are generally orientated parallel to one another in a particular direction, giving the rock a laminated appearance. This characteristic indicates that the layer 4 rocks have been stretched and have flowed during their formation. Consequently, these rocks are often called **tectonites** to indicate the involvement of plate motion in their formation. When oceanic crust first forms at a mid-ocean ridge, layer 4 is usually 5–7 km thick but can be thicker than 10 km.

After its initial formation, oceanic crust moves away from the mid-ocean ridge and cools. As the crust cools, so does the mantle below it. As the mantle cools it becomes attached to the base of the layer 4 tectonites. This process changes the oceanic crust into oceanic lithosphere. Over time, this process continues, causing the thickness of the oceanic lithosphere to increase with age. Because the oceanic lithosphere is cooling it also becomes more dense and 'floats' lower in the mantle. This is why the ocean becomes deeper as the distance from the mid-ocean ridge increases.

Continental crust and lithosphere

The continental crust is both older and more complex than the oceanic crust. The surface of the continents is better studied than the floor of the oceans, but what occurs deep within the continents is not well known. We do know that the continental crust contains the oldest pieces of the Earth's crust. While the oceanic crust is no more than about 180 million years (**Ma**) old, several areas of the continental crust in Western Australia's Pilbara and Yilgarn areas contain rocks over 3800 Ma old as well as some zircon mineral grains dated to be 4200 Ma old.

The complexity of the continental crust is a function of its great age. Periods of plate **extension** and collision have produced processes that have distilled components

of the mantle and given the continental crust a very different composition from that of the oceanic crust. (See Table 1.1.1, page 6.) It is possible to identify four components that make up the majority of continental crust:

- continental **shields** that contain Pre-Cambrian igneous and metamorphic rocks
- continental **platforms** consisting of a relatively thin layer of sedimentary rocks overlaying a shield made of Pre-Cambrian basement
- relatively young **fold mountains** that contain older **metasediments** and young volcanic and intrusive igneous rocks
- continental rifts and continental rift margins.

The fourth component generally contains one or more of the other three components of crust listed above, but the continental crust that occurs in rifts and rift margins is somewhat different because it has been stretched. The rift margins form many continental edges and are the sites of continental break-up due to sea-floor spreading. The stretching of continental crust does not always lead to the formation of a new ocean. Instead, it can create a rifted continental valley that fills with material shed from the crust on either side.

Seismic studies have established that these four components of crust have different thicknesses. (See Figures 1.1.4 and 1.1.5, page 6.) The crust is thinnest at continental rifts, where the lithosphere is being stretched, and thickest within fold mountains. Seismic studies have also established that the continental crust consists of three recognisable layers:

- The uppermost layer consists of unmetamorphosed sedimentary and igneous rocks at the surface. With depth, the rocks change to medium-grade metasedimentary rocks. Seismic waves vary in speed in this layer but **P-waves** usually travel at less than 5.7 km/s.
- The second layer consists of metamorphic rocks with a composition similar to the igneous rock granodiorite. This layer is separated from the layer below it by a seismic boundary called the Conrad Discontinuity. (See Figure 1.1.2, page 3.)
- The lower crust is thought to consist of rocks called **granulites**. Granulites form under high-pressure and high-temperature conditions. The granulites have a composition that is **silica**-rich in the upper part of the lower crust, grading into more silica-poor rocks at the base of the crust.

Like the oceanic crust, the base of the continental crust is recognised as a seismic boundary: the Mohorovicic Discontinuity (or Moho for short).

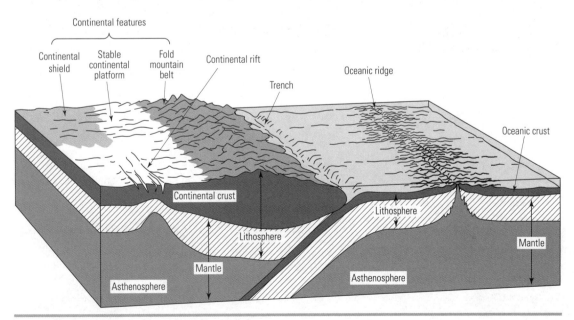

Figure 1.1.4 Major features of the Earth's continental crust.

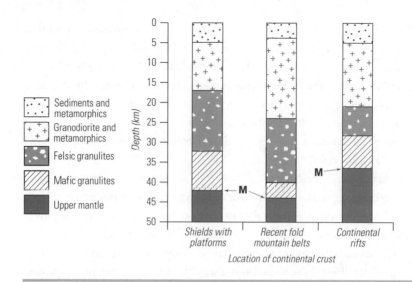

Figure 1.1.5 Variations in the vertical structure of the continental crust. Note that these layers are recognised on the basis of seismic behaviour. The Moho is marked M.

Lithospheric mantle

Below the crust lies the lithospheric upper mantle. Like the crust, this material is hard and rigid. It is firmly attached to the crust so that it moves with the overlying continental crust as a single unit. Continental lithospheric mantle is, on average, about 80 km thick and is mostly composed of a rock type called **peridotite**, with lesser amounts of a rock called eclogite. Peridotite is an **ultramafic** rock composed of 65% olivine, with lesser amounts of pyroxenes and garnet. The eclogite is similar in composition to basalt, but is denser than basalt because it has been metamorphosed under high temperature and pressure. We know about these peridotites and eclogites because slices of them are found uplifted in fold mountains. Nodules of both these rock types have also been found in basalts that were generated deeper in the mantle. As these lavas made their way to the surface they ripped pieces of eclogite and peridotite off the sides of the tubes and pipes they flowed through. The lower boundary of the lithospheric mantle is the asthenosphere.

Table 1.1.1 The chemical composition of the Earth's crust*

	Oceanic crust				Continental crust			
	Layer 1	Layer 2 (see Figure 1.1.3, page 3)	Layer 3	Total oceanic	Sediments and metamorphics	Granodiorite	Granulites	Total continental
Mass (%)	3.0	16.0	81.0	100.0	8.0	44.0	48.0	100.0
Element-oxide								
Silicon (SiO_2)	50.2	50.2	49.9	50.0	56.0	64.2	54.4	58.9
Aluminium (Al_2O_3)	14.0	16.0	17.3	17.0	16.8	16.3	16.1	16.2
Iron (FeO)	6.9	9.2	10.2	9.9	6.1	5.8	10.6	8.1
Magnesium (MgO)	3.7	6.9	8.4	8.0	3.5	2.6	6.3	4.5
Calcium (CaO)	20.6	13.2	9.9	10.8	12.7	4.6	8.5	7.1
Titanium (TiO_2)	0.7	1.2	1.5	1.4	0.8	0.6	1.0	0.8
Sodium (Na_2O)	1.4	2.2	2.7	2.6	1.8	3.4	2.8	3.0
Potassium (K_2O)	2.5	1.1	0.1	0.3	2.3	2.5	0.3	1.4

*The data are expressed as a percentage of the total amount of the eight most common element-oxides. Notice that volatiles, such as water and carbon dioxide, are missing. This makes the top layer for each type of crust an approximation.

Continental and oceanic crust both have a lower density than mantle and they effectively float on it. While thicker than the oceanic crust, the continental crust is less dense (2.7 tonnes/m³ compared with 3.2 tonnes/m³) and it floats higher than the oceanic crust. Changes in a plate's thickness and average temperature as well as in the forces acting on the plate can all cause it to rise or sink. You can think of a lithospheric plate as behaving a bit like a giant sheet of ice floating on water.

The asthenosphere and plate motion

The asthenosphere plays a very important role in the motion of plates because it is plastic. Although the asthenosphere is not liquid, it can flow very slowly when forces are applied to it. The forces on plates can result in plate motion because the asthenosphere allows the lithosphere to move over it.

The lithosphere also plays a role in what is called **isostatic adjustment**. If material is removed from part of a plate, as when erosion removes material from a mountain, the plate rises vertically. It is like a cargo ship floating higher in the water when its cargo is unloaded.

isostatic adjustment
movement that balances weight forces and buoyancy forces

definition

REVIEW ACTIVITIES

1
List the layers of the Earth and briefly describe the characteristics of each layer.

2
Describe, using a diagram, the structure of the lithosphere.

3
Compare the features of the oceanic and continental lithosphere.

4
Contrast the age of continental and oceanic crust.

5
List the main components of the continental crust.

EXTENSION ACTIVITIES

6
Why is it incorrect to say that the term 'lithosphere' is another name for the crust?

7
Why is basalt the most common volcanic rock on Earth?

PLATE BOUNDARIES

The lithosphere is something like a cracked eggshell. It is broken up into a number of parts called lithospheric plates. An important characteristic of lithospheric plates is the nature of their margins, or boundaries. Plate boundaries are commonly described in terms of the relative motion of the plates on either side of the boundary.

Types of plate boundaries

There are three types of plate boundaries. They are divergent boundaries, convergent boundaries and **conservative** boundaries.

DIVERGENT BOUNDARIES

Divergent boundaries are sometimes called **constructive** or accreting boundaries. (See Figure 1.1.6, page 8.) Except for the active rift zones of Iceland, western North America and Africa, all the Earth's divergent boundaries are submerged beneath the ocean. Divergent boundaries form where two plates move away from one another or when a continental plate is pulled apart. The boundary is characterised by forces that produce normal faulting and volcanic activity. The volcanic activity occurs quietly along the central rift zone. Earthquakes also originate along the central rift at shallow depths.

conservative
in relation to plates, it describes a boundary where crust is neither created or subducted

constructive
in relation to plates, it describes a boundary where new ocean crust is formed

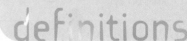
definitions

Oceanic divergent boundary

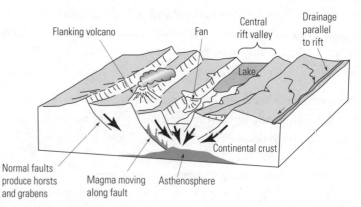

Graben — Fault scarp — Linear zone of volcanic activity — Axial rift — Abyssal hill — Normal faults — Mantle — Magma chamber — Oceanic crust

Continental rift

Flanking volcano — Fan — Central rift valley — Drainage parallel to rift — Lake — Continental crust — Normal faults produce horsts and grabens — Magma moving along fault — Asthenosphere

Figure 1.1.6 Characteristics of a divergent boundary.

It is at divergent boundaries that new ocean crust is produced and the age of the crust at such boundaries is very young (0–10 Ma). The rough topography of divergent boundaries includes an ocean ridge, which may have a central rift valley. The segments of a divergent boundary are usually offset from one another by comparatively short transform faults, which really are a special type of conservative boundary. Divergent boundaries form some of the longest continuous mountain ranges on Earth. Their structure will be examined in more detail in Chapter 1.2.

CONVERGENT BOUNDARIES

Convergent boundaries are sometimes called **destructive**, or consuming, boundaries because oceanic crust is often removed from the Earth's surface and descends into the mantle at these sites. (See Figure 1.1.7.) Here, two plates approach each other and one plate is usually forced under the other. Such boundaries are characterised by subduction zones consisting of a trench and **volcanic island arc** or a trench and continental arc and mountain belt if at least one of the plates is oceanic; or by the construction of a mountain range when both plates are continental. In zones where an oceanic plate is dragged down deep into the mantle, volcanic activity is produced in a chain of volcanoes, or volcanic arc, that develops above the downgoing plate. Earthquakes are triggered along the upper surface of the plate as it descends. The inclined zone of earthquakes descends up to 700 km into the mantle. This zone is commonly called the **Benioff zone** after one of the scientists who first described it. Convergent boundaries with active Benioff zones are also referred to as sites of B-subduction.

Geophysical signatures of convergent zones include the Benioff zone and a relatively low heat flow from within the Earth near the trench but a relatively high heat flow at the volcanic arc. Examples of subduction zones include the Andes Mountains and the Japanese islands, where oceanic lithosphere is subducted beneath continental lithosphere. The Tongan islands are an example of a volcanic island arc that has formed due to the subduction of oceanic lithosphere beneath other oceanic lithosphere.

If two continental lithospheric plates converge and collide with each other, mountain ranges such as the Himalayas and the Zagros Mountains are formed. Before the continental masses can collide, an ocean is subducted under one or both of the

definitions

destructive
in relation to plates, it describes a boundary where crust is being subducted into the mantle

volcanic island arc
an arc of volcanic activity marked by a chain of volcanic islands. It lies parallel to an ocean trench.

Benioff zone
an inclined area of earthquake activity dipping into the mantle away from a trench

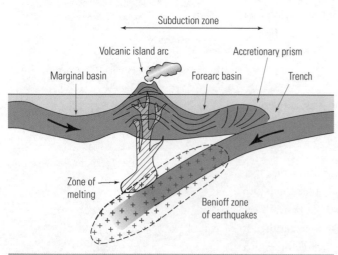

Subduction zone

Volcanic island arc — Accretionary prism — Marginal basin — Forearc basin — Trench — Zone of melting — Benioff zone of earthquakes

Figure 1.1.7 Characteristics of a convergent boundary.

continents. After the oceanic crust is subducted, one of the continental plates is thrust under the other plate. However, the low density of the crust stops it from being subducted. This situation is sometimes referred to as A-subduction. The 'A' refers to 'Alpine-type subduction', or 'Ampferer subduction' after the scientist who first suggested that such a situation might exist. More detail on the features of mountains appears in Chapter 1.2.

CONSERVATIVE BOUNDARIES

A conservative boundary occurs where two plates slide past each other and move parallel to their boundary. This motion usually forms either a strike-slip fault or a transform fault. (See Figure 1.1.8.) Lithosphere at these boundaries is neither created nor destroyed, which is why they are called conservative boundaries, but the shearing involved creates characteristic topographies, such as cliffs, ridges and troughs. The direction of the motion of the two plates along these boundaries is rarely perfectly parallel and this gives rise to restricted zones of compression and extension. Compression can cause folding, thrust faulting and mountain building, while extension can give rise to small pull-apart basins as well as some igneous activity.

Most conservative boundaries occur in the ocean, but notable conservative continental boundaries include the Alpine Fault of New Zealand (where slight compression has produced mountain ranges), the San Andreas Fault of California and the Dead Sea Fault in Israel (where slight extension has produced a pull-apart basin).

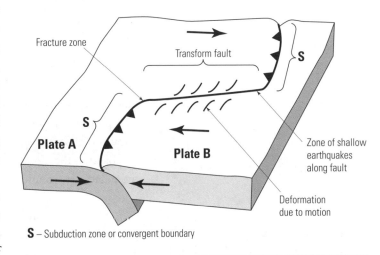

S – Subduction zone or convergent boundary

Figure 1.1.8 Features of a transform fault. (*Note:* One side of the fault may be higher than the other.)

Igneous rocks formed at plate boundaries

The nature of the magmas that are erupted as lava at the surface or intruded into the subsurface depends on three factors. First, the composition of the source is very important. The source is usually either mantle rocks or lower continental crust rocks, but source materials can also be a combination of both mantle and crustal rocks. A second important factor is the way in which the source melts. The source rocks rarely melt completely. Instead, they partially melt and the amount of melting determines the composition of the rocks produced. Thirdly, processes that occur in the magma chamber can modify the initial composition of the magma. This mostly involves the removal of low-silica minerals, which tends to increase the total amount of silica present in the magma.

The igneous activity that occurs on Earth is classified or described in many different ways. We will classify igneous activity using a simplified scheme that focuses on the composition of the volcanic rocks and their tectonic setting. There are two broad types of vulcanism and related intrusions:
- intra-plate vulcanism, which occurs in the interior of a plate
- plate boundary vulcanism, which occurs at both divergent and convergent boundaries.

In this chapter we will focus on plate boundary or active-margin vulcanism and the igneous rocks that occur in these locations because this is where the vast majority of the world's igneous activity occurs. Nevertheless, intra-plate vulcanism is important because it has formed Hawaii, Iceland and many of the volcanoes that formed in Eastern Australia.

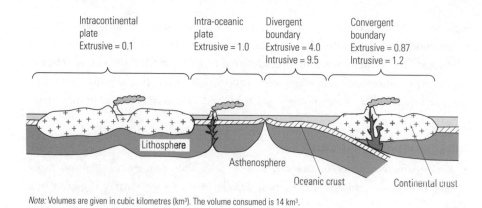

Intracontinental plate
Extrusive = 0.1

Intra-oceanic plate
Extrusive = 1.0

Divergent boundary
Extrusive = 4.0
Intrusive = 9.5

Convergent boundary
Extrusive = 0.87
Intrusive = 1.2

Lithosphere

Asthenosphere

Oceanic crust

Continental crust

Note: Volumes are given in cubic kilometres (km³). The volume consumed is 14 km³.

Figure 1.1.9 Rates of annual magma production.

IGNEOUS ROCKS FORMED AT DIVERGENT PLATE BOUNDARIES

The greatest volume of igneous rock formed on Earth is erupted or intruded at divergent boundaries. (See Figure 1.1.9.) These rocks are almost entirely basaltic in composition. Figure 1.1.10 shows the structure of a mid-ocean ridge and the locations of some of the processes discussed here.

Basalt is the most common volcanic igneous rock on Earth. It is found in all plate boundary environments and also within plates. Most basaltic magma forms when mantle peridotites partially melt. This forms a magma with a relatively low silica content (about 50%) that also contains aluminium oxide (about 15%), calcium oxide (about 10%), iron oxide (about 10%), magnesium oxide (about 10%) and sodium or potassium oxides (about 5%). If this magma is erupted at the surface it cools quickly to produce a fine-grained, dark-grey or black volcanic rock. Basalts may contain small gas holes called vesicles and occasional crystals of olivine, pyroxene and plagioclase feldspars that are set in a fine-grained matrix of glassy material. The crystals can usually be observed with a hand lens.

When basaltic magma is intruded into the deep subsurface it cools slowly and forms a rock called gabbro, which is composed entirely of large (3 mm) crystal grains of the same minerals as those found in basalt. Basaltic magma intruded at intermediate depths forms a rock called dolerite, which is also known as microgabbro because it is just like a gabbro only the crystals are smaller (1 mm).

The low silica content of the basaltic igneous rocks formed at constructive boundaries means that the liquid magma flows easily and gases can escape from the cooling magma relatively easily. As a result, the eruptions of such magmas are relatively quiet and gentle. They tend not to produce much ash or many explosions. Such eruptions generally occur when magma streams out of elongated fractures that run along the length of a mid-ocean ridge's rift valley. Figure 1.1.10 shows the characteristic layers of igneous material produced within the oceanic crust. Note that the age of the rocks increases as the distance from the ridge increases.

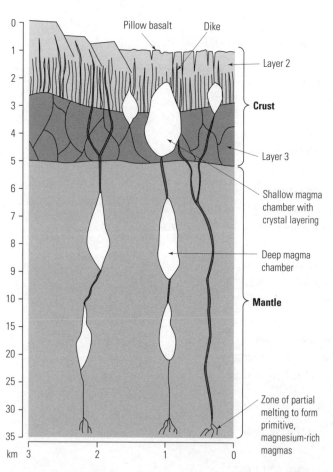

Pillow basalt   Dike

Layer 2

Crust

Layer 3

Shallow magma chamber with crystal layering

Deep magma chamber

Mantle

Zone of partial melting to form primitive, magnesium-rich magmas

Figure 1.1.10 The origin of the igneous rocks erupted at a mid-ocean ridge.

The origin of the magma formed at a divergent boundary lies in the mantle below the boundary. Convection causes mantle material to rise towards the surface, and as it rises the pressure on the material decreases. (See Figure 1.1.11.) The upper mantle is composed of a material called peridotite. The melting temperature of peridotite depends on both temperature and pressure. It begins to melt at 1300 °C and a depth of 30–40 km below the surface. When the mantle material begins to melt, small droplets of magma with a basalt composition are produced. The droplets join together to form larger bodies of magma, which rise towards the surface and feed the long, linear chamber below the mid-ocean ridge crest.

IGNEOUS ROCKS FORMED AT CONVERGENT PLATE BOUNDARIES

Igneous rocks formed at convergent boundaries show far more diversity than those formed at divergent boundaries. The volcanic rocks formed at convergent boundaries do include basalts, but other volcanic rocks called andesite and rhyolite tend to be more common. Intrusive bodies called **plutons** have similar compositions to the volcanic rocks and form when the magmas are intruded into the crust rather than erupted at the surface.

Igneous activity at convergent boundaries occurs at both subduction zones and continental collision zones. The origin of the magmas in subduction zones is somewhat different from the origin of magmas in continental collision zones, and characteristic igneous rocks are formed in each of these two convergent environments.

The volcanic rock that is characteristic of convergent boundaries where subduction of oceanic lithosphere is occurring is **andesite**. Andesite is classified as an intermediate igneous rock because it contains more silica (about 58%) than basaltic rocks but less silica than rhyolitic and granitic rocks (about 65%). Andesites are medium-grey rocks with an easily recognised spotted appearance resulting from their texture. This texture is formed by abundant crystals of rectangular prisms of plagioclase and lesser amounts of dark minerals, such as pyroxenes or hornblende, set in a matrix of microscopic mineral grains. The matrix is referred to as **aphanitic**, which means that the crystals are too small to be identified with the naked eye. The larger crystals are called **phenocrysts** and a rock such as andesite, which contains phenocrysts in a finer crystal matrix, is referred to as having a **porphyritic** texture.

It is thought that much andesite magma is produced above subduction zones when basaltic magma, generated in the mantle above the subducted plate, gets trapped at the base of the overlaying plate. This basaltic magma originates hundreds of kilometres below the Earth's surface, above the top of the plate descending into the mantle. Water contained in altered oceanic crust in the descending plate escapes into the overlaying mantle material. This water also carries some dissolved materials. As it rises into the mantle above the descending plate it triggers partial melting, forming basalts. (See Figure 1.1.11.) These basalts rise to a level within the overlaying plate where they become trapped. Low-silica minerals crystallise here and are removed from the basaltic magma while it is trapped. This alters the remaining magma's composition to a less dense andesitic one, which then rises to the surface where it can erupt. Some andesitic magmas are thought to be generated when other batches of basaltic magma mix with granitic magma generated by partial melting of the base of the crust in the overlaying plate. The mixing of low-silica basaltic magma with high-silica granitic magma produces the andesites of intermediate composition.

The composition of the magma may change as it cools in the crust. The change is called magmatic **differentiation** and may be caused by one or more processes. (See Figure 1.1.12, page 12.) **Assimilation** occurs when pieces of rock surrounding the magma melt into the magma. Magma mixing is a process in which two magmas of different compositions come together and form a magma with a new composition. Fractional crystallisation occurs when some of the mineral components are removed from the liquid as the magma cools. Minerals that crystallise lock up atoms and alter the composition of the liquid. The first minerals to form are metal oxides and

plutons
masses of igneous rock formed below the Earth's surface

phenocryst
a crystal that is large compared to those surrounding it

andesite
a fine-grained igneous volcanic rock crystallised from magmas formed at subduction zones

porphyritic
a rock texture in which some crystals are significantly larger than others

aphanitic
a rock texture in which crystals are too small to be identified without a microscope

differentiation
in relation to magmas, this is the separation of a magma into parts with different compositions

assimilation
the process in which a rock melts and becomes part of a magma

definitions

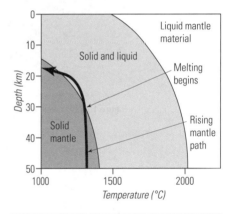

Figure 1.1.11 Decompressional melting.

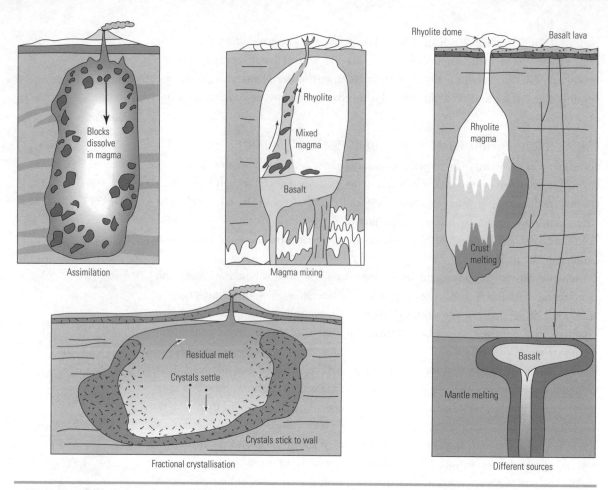

Figure 1.1.12 Differentiation of magmas.

silicates, such as olivine, that are rich in metals. Minerals with a low melting point, such as quartz and potassium-rich feldspars, remain in the liquid part of the magma. If the liquid erupts at the surface, the volcanic rocks produced are richer in quartz than volcanic rocks formed in oceanic crust.

The eruption of andesites

Andesitic magma contains more silica than basaltic magma and is more **viscous**, which is why andesite volcanoes tend to be cone shaped. As a result of the higher viscosity, gases within the magma cannot easily escape while the molten material cools. The trapped gases finally escape violently, shattering the cooling material into ash. Such eruptions are called explosive eruptions and occur through central vents. They produce ash flows and viscous lavas, which form composite volcanoes. Other features formed include lava domes and collapsed calderas.

The debris formed in the blasts of andesite eruptions may consist of rocks ranging in size from enormous boulders bigger than a house down to very fine-grained ash. These materials can become what are called **pyroclastic** rocks. The fragments, or clasts, may be cemented together by precipitated minerals some time after the fragments are laid down. Alternatively, the clasts may be welded together by sticky glass fragments as the hot materials are deposited.

Rhyolites

Rhyolite is another volcanic rock formed at convergent boundaries. Rhyolite is very rich in silica (more than 65%) and is classified as a felsic rock. Rhyolite magmas tend to form from the partial melting of the lower continental crust of the overriding plate. When the magma cools it forms fine-grained rocks that are light-grey, pink and cream in colour. These rocks usually contain visible crystals of potassium-rich feldspars and quartz and sometimes crystals of dark minerals, such as hornblende or biotite mica.

The higher silica content of rhyolite magma makes it extremely viscous. This means that rhyolite lavas do not flow very easily and so the rhyolite lava eruptions tend to form domes above a vent rather than extensive lava flows. Rhyolite eruptions are usually much more violent than andesite eruptions. Extensive sheets of igneous material with a rhyolitic composition are quite common throughout the Western USA and in the ancient convergent zone rocks found in New South Wales. These pyroclastic rhyolites formed during giant catastrophic eruptions when high-level granite magma chambers literally blew their roofs off or were breached from the surface.

OTHER IGNEOUS ROCKS FORMED BY CONTINENT-CONTINENT CONVERGENCE

In continental collision zones, where one piece of continental crust is thrust under another, mantle-derived magmas are rare. However, the heat and pressure deep in the crust is enough to melt some of the deeply buried metamorphosed sedimentary rocks. Partial melting produces magmas very rich in silica and aluminium. The magma crystallises close to where it forms and rarely causes eruptions at the surface. The granites that result are rich in minerals such as muscovite mica and quartz and contain characteristic minerals such as tourmaline and cordierite. Some geologists refer to these granites as S-type granites because of their origin. Examples of environments where this is happening today include the Himalayas, where continental lithosphere is thrust under another plate of continental lithosphere. Contraction and compressional forces dominate such environments. Melting of the crust in both the lower and upper of the two converging lithospheric plates occurs.

Australian examples

In New South Wales, igneous rocks formed in a zone of ocean–ocean convergence are the volcanics and small intrusions formed during the Ordovician in the Molong Volcanic Arc. Young island arcs are structurally simple and the crust below them is about 20 km thick. Comparable **modern** examples of such young arcs are the Tonga-Kermadec Volcanic Arc to the north of New Zealand and the Aleutian Volcanic Arc between Alaska and Kamchatka.

The volcanic rocks and granites found throughout most of New South Wales, especially those of the Lachlan and New England Fold Belts, formed in both ocean–continent and continent–continent convergent zones. A modern example of an ocean–continent convergent zone is the Andes in South America.

SUMMARY OF IGNEOUS ROCKS

The main volcanic igneous rocks erupted on Earth were described in detail above. Coarse-grained rocks with similar compositions exist. (See Table 1.1.2.) They are derived from the partial melting of mantle and crustal rocks. Basaltic rocks

viscous
the property of a fluid that describes its resistance to flow. A very viscous fluid does not flow easily.

pyroclastic
fragmented rock material formed in a volcanic eruption

rhyolite
a fine-grained volcanic igneous rock with the same composition as granite

modern
something forming at present

definitions

Table 1.1.2 Types of igneous rocks and their occurrence in the crust

Location in which the rock crystallises	Composition			
	Extremely poor in silica	Relatively poor in silica	Intermediate silica content	Relatively rich in silica
At the surface	Komatiite	Basalt	Andesite	Rhyolite
At shallow depths (tens to hundreds of m below the surface)	–	Dolerite	–	–
At intermediate depths (a few km below the surface)	Peridotite	Gabbro	Diorite	Granite

characterise divergent boundaries, while andesites and rhyolites characterise convergent boundaries. Coarse-grained intrusive rocks with equivalent compositions (such as gabbros, diorites and granites) form large intrusions at depth in the various plate boundaries.

There is a clear relationship between igneous composition and plate boundary type. The plate boundary's nature determines the origin of the magmas that form igneous rocks there. Mantle-derived magmas give rise to the basaltic-composition rocks that form the oceanic crust. Water released from descending plates causes partial melting of mantle material, which is further modified in the crust. This produces rocks with a characteristic andesitic composition, but a wide range of other rock types also arise. Such rocks form an essential part of the continental crust.

The high silica content of igneous rock formed at destructive boundaries has important implications for the formation of continents. The lower density of rocks such as granite, diorite and andesite means they float in the mantle and do not subduct. The crust that is composed of these materials is the raw material from which continental crust is formed.

Explosive vulcanism is characterised by eruptions that produce large amounts of ash and other fragments. The fragmented volcanic material forms what are known as volcaniclastic rocks: rocks formed from volcanic fragments. Pyroclastic rocks are classified according to particle size and composition. Compositional terms include 'vitric' (refers to glass fragments), 'lithic' (refers to rock fragments) and 'crystal' (refers to the broken crystals produced in eruptions). Ash is the name given to particles less than 2 mm in diameter. Rocks formed from ash are called tuff. Ash has a similar size to clay, silt and sand. Particles the size of granules and pebbles (2–64 mm) are called lapilli. Larger particles the size of cobbles and boulders, are called blocks and bombs, respectively. A rock consisting of blocks cemented together is called a volcanic breccia. It is rare to find rocks formed only from lapilli or blocks. Smaller fragments fill in the spaces between the larger particles and the resulting rocks are named according to both the size types present. Tuff-breccias and lapilli-tuff are rock types found in volcanic areas. Sometimes the ash deposited is so hot that the fragments weld together to form welded ash-flow tuffs, or ignimbrites. Such rocks are evidence of extremely hot ash clouds. In Chapter 1.4 we will look at the events that produce such materials and the damage this causes.

REVIEW ACTIVITIES

1
List the three types of plate boundaries.

2
Construct a table to list and describe the features of divergent boundaries.

3
Construct a table to compare the features of different convergent boundaries.

4
Classify the types of volcanic rocks found at different types of plate boundaries.

5
Explain what is meant by explosive vulcanism.

6
Summarise the five environments in which igneous rocks are produced at convergent boundaries.

EXTENSION ACTIVITIES

7
What would it mean if you found andesitic lavas and tuffs in an area?

8
Is it correct to say that magmas formed at divergent boundaries originate in the mantle and magmas formed at convergent boundaries originate in the crust? Explain your answer.

THE MOTION OF PLATES: CHARACTERISTICS

Why we believe plates move

Over the last thirty years, Earth scientists have found a number of ways to measure the motion of plates. The motion can be measured relative to another plate or relative to the mantle. A summary of plate motions is shown in Figure 1.1.13 and Table 1.1.3 (page 16). Some of the sources that provide information about the motion of plates are described below.

SATELLITES

Satellites can be used to directly measure the relative motion between plates. Prior to the 1990s such surveying was done by traditional surveying methods using lasers, Such methods, while accurate, took a long time and could not be used across large oceans. Two methods developed during the last two decades are satellite laser ranging and satellite radio positioning.

Satellite laser ranging involves measuring the time it takes a pulse of laser light to travel to a satellite and back. The time is converted to a distance using the speed of light. When two stations on different plates simultaneously track a satellite in this way they can determine their position relative to each other. Repeating such methods over a number of years allows the relative motion of the two stations to be determined.

Satellite radio positioning makes use of the Global Positioning System (GPS). This system relies on a number of satellites orbiting the Earth that broadcast signals. A GPS receiver on the Earth's surface uses signals from a number of satellites to work out its position in three dimensions. Results of GPS measurements are quite accurate and can be made easily. Repeated measurements over time allow plate motions and plate stretching (extension) to be determined.

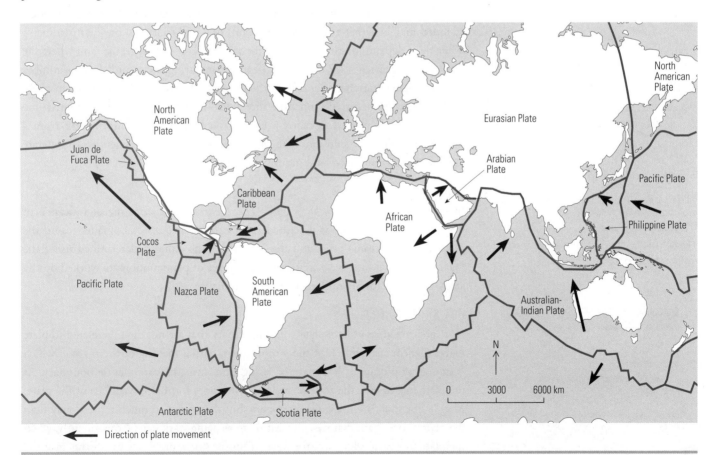

Direction of plate movement

Figure 1.1.13 Relative directions of plate motion.

Plate	Total plate area (millions of km²)	Continental area (millions of km²)	Average velocity (cm/year)	Effective length (hundreds of km)		
				Ridge	Trench	Transform fault
Pacific	108.0	0.0	8.0	119	113	180
African	79.0	31.0	2.1	58	9	119
Eurasian	69.0	51.0	0.7	35	–	56
Australian-Indian	60.0	15.0	6.1	108	83	125
North American	60.0	36.0	1.1	86	10	122
Antarctic	59.0	15.0	1.7	17	–	131
South American	41.0	20.0	1.3	71	3	107
Nazca	15.0	0.0	7.6	54	52	48
Philippine	5.4	0.0	6.4	–	30	32
Arabian	4.9	4.4	4.2	27	–	36
Caribbean	3.8	0.0	2.4	–	–	44
Cocos	2.9	0.0	8.6	29	25	16

Table 1.1.3 Plate geometries and motions

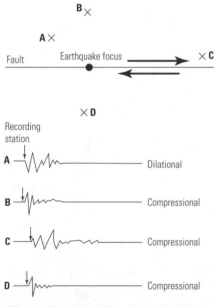

Different stations receive different first motions (shown by arrows) due to their location relative to fault motion

Figure 1.1.14 First motion at seismometers around an earthquake origin.

fault plane
the surface along which rocks are moved in a fault

seismometer
an instrument that records earthquake, or other seismic, waves

intra-plate within a plate

EARTHQUAKES

Earthquakes are caused by movement along **fault planes**. The fault planes may lie along convergent, divergent or conservative boundaries. Earthquakes may also occur within a plate. When an earthquake occurs, energy in the form of waves moves away from the site where the motion occurred in all directions. The waves are called seismic waves and their motion is recorded by instruments called **seismometers**.

When the seismic waves of an earthquake arrive at a seismometer the first motion recorded may be either a push (compression) or a pull (dilation) on the instrument. Which it is depends on the location of the seismometer relative to the fault where the earthquake occurred. (See Figure 1.1.14.) By comparing the records from a number of seismometer stations, scientists can determine the orientation of the fault where the earthquake occurred and the direction of movement that generated the earthquake. This process is called a fault-plane solution. The information from a number of fault-plane solutions along a plate boundary can help scientists to generate a picture of current forces and motions along the boundary.

MAGNETIC REVERSALS

In *Earth and Environmental Science: The Preliminary Course* the significance of magnetic field reversals was described. The sea-floor magnetic anomalies record the growth of the oceanic crust and the age of the anomalies is determined using the magnetic reversal time scale. Rates and directions of plate motion are worked out by analysing the maps of sea-floor magnetic anomalies.

HOT SPOT TRACES

Hot spots are areas of **intra-plate** volcanic activity that are not directly related to plate margin igneous activity. Hot spots result from partial melting of the mantle, which is caused by plumes of hot material rising from near the core mantle boundary. As the plate moves over the zone of melting, volcanoes form at the surface of the plate. Mantle hot spots are not quite stationary, but they move so much more slowly than the plate above them that we can treat them as being stationary. Consequently, when the plate moves an older volcano away from the hot spot, a new one forms over the hot spot. With time this leads to the formation of a chain of volcanoes that are progressively older as their distance from the hot spot increases. (See Figures 1.1.15 and 1.1.16.)

The distribution of volcanoes generated by hot spots, together with radiometric dating of their ages, allows scientists to calculate the direction and rate of a plate's motion. It is also possible to work out the relative motion of other plates once the motion of one plate is established.

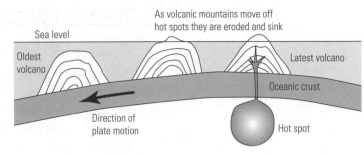

Figure 1.1.15 How a seamount chain forms over a hot spot.

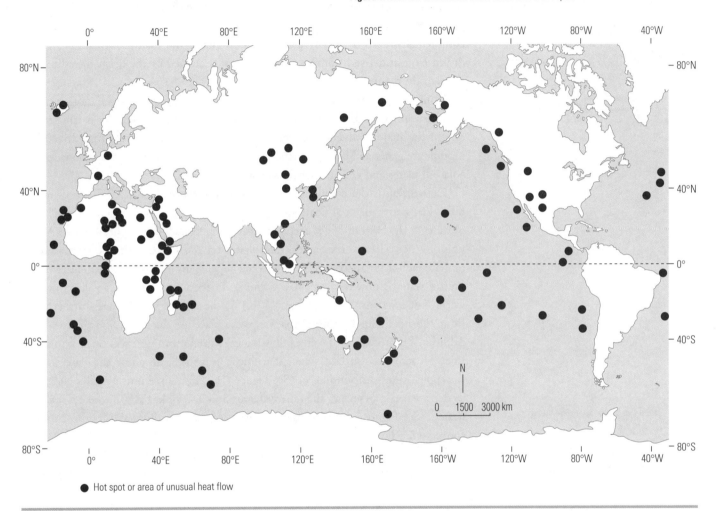

● Hot spot or area of unusual heat flow

Figure 1.1.16 Global distribution of hot spots.

REVIEW
ACTIVITIES

EXTENSION
ACTIVITY

1
Summarise the methods used to measure plate motion.

2
Draw a diagram to show the history of a piece of oceanic crust from the time it forms until it remelts.

3
Three volcanic islands form a straight line across a plate. Their ages are 120.3 Ma, 86.2 Ma and 1.2 Ma. They are arranged in order from oldest to youngest and the distance of the oldest from the youngest is 1350 km. Calculate the speed at which the plate is passing over the hot spot and comment on whether your answer is a reasonable value.

THE MOTION OF PLATES: MECHANISMS

Convection

Why do plates move? No-one is completely sure, but the geophysicists who work on this problem are getting close to an agreed answer. So far we have described how plates move and the effects on the lithosphere when plates interact with each other. However, to fully understand plate tectonics, scientists need to understand the mechanisms that cause plates to move. Part of the reason that Alfred Wegener had so much trouble convincing scientists that continental drift occurred was because he could not supply a feasible mechanism for his wandering continents. In 1928, two years before Wegener died, the British scientist Arthur Holmes proposed that continental drift is driven by **convection** currents in the mantle, which are in turn driven by heat derived from the core and heat generated by the decay of radioactive elements in the mantle.

While the current scientific view of how convection may drive the motion of the plates is somewhat different from Holmes's original idea, convection is still seen as one of the major influences on plate motion. Other likely forces that drive plate motion are generated by the Earth's gravity acting on density contrasts between the adjacent plates and within a particular plate.

Convection is the circulation of fluids resulting from density differences in the fluid. Less dense material floats and more dense material sinks. Imagine a small unit of water at the bottom of a saucepan that is being heated. (See Figure 1.1.17.) As the unit of water is heated it expands. (See Figure 1.1.17a.) It contains the same amount of material but the space it fills has become larger. This means that the density of the unit of water is now less than it was and it experiences a force that will propel the warm unit upwards. (See Figure 1.1.17b.) As the unit of water moves further away from the heat source and approaches the surface of the water it will lose heat. Additional heated units of water following the initial warm unit will cause it to move sideways across the surface, where it will lose heat to the air above the water and to the cooler water surrounding it. (See Figure 1.1.17c.) As the unit of water cools and its density becomes greater than the water around it, it sinks back to where it was first heated. (See Figure 1.1.17d.) The circulation of water in this way is referred to as a convection cell.

In the Earth, convection occurs very slowly within the solid mantle. During short intervals of time (days) the mantle is solid, but over longer periods of tens or hundreds of years the mantle material can creep and behave like a fluid. Convection within the Earth is also different from the simple model just described because convection occurs in three dimensions. Under the appropriate conditions, convection occurs in hexagonal cells with warm material rising, or upwelling, in the centre and moving downwards, or downwelling, at the border of the cell. It was once thought that the horizontal motion of convection carried a plate along like a raft being carried along by the current in a river. The situation, as we will see, is more complex, with the plates interacting with the convecting mantle.

The fate of plates

Plate motion occurs because of forces acting on the plate. These forces can cause motion of the plate and also deform it and change its shape. Before we examine the nature of these forces we will look in general terms at the fate of continental and oceanic plates.

Consider first the history of a piece of oceanic crust. (See Figure 1.1.18.) The crust forms at the top of a mid-ocean ridge. Heat in the upwelling mantle material provides the energy to elevate the ridge above the surrounding crust. The new crust

convection
the process in which heat is transferred by the motion of material in a fluid

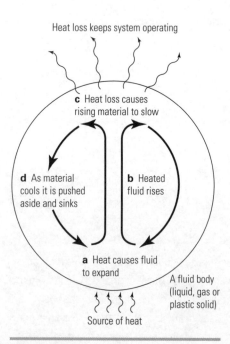

Heat loss keeps system operating

c Heat loss causes rising material to slow

d As material cools it is pushed aside and sinks

b Heated fluid rises

a Heat causes fluid to expand

A fluid body (liquid, gas or plastic solid)

Source of heat

Figure 1.1.17 A simple convection model.

is colder than the mantle below it, but because it is made of less dense basaltic material it floats. As the plate ages and moves away from the ridge, it cools and thickens. The thickening occurs because mantle material freezes onto the base of the plate. Gravitational force acts on the plate and it sinks further into the mantle.

The elevation of the oceanic crust at a mid-ocean ridge is much higher than the crust 200 km from the ridge. A force due to the difference in elevation is like the force on a book resting on a tilted desk. The gravitational force, or weight, of the book will cause it to slide down the desk if the desk and book do not have too much friction between them. The book is like the oceanic plate and the desk is like the upper surface of the mesosphere. The boundary between the book and desk represents the weak asthenosphere that allows the plate to move over the underlying mantle.

There comes a time when the density of the oceanic plate exceeds the density of the underlying mantle. If the plate breaks free of surrounding plate material it will sink into the mantle and form a subduction zone. Once it begins to sink into the hotter mantle, the density of the plate will cause the plate to continue to sink towards the core-mantle boundary through the less dense mantle. As the descending slab sinks it probably pulls more of the plate behind it.

Continental lithosphere behaves differently from oceanic lithosphere. Continental crust is less dense than oceanic crust because of its more silica-rich composition. Because it is about 20% less dense than mantle, continental crust does not subduct. However, if an upwelling cell forms under a piece of continental crust it may cause the lithosphere to bulge and thin. The resulting tilt of the crust may also cause forces that pull the lithosphere down and assist in rifting of the continent.

The forces acting on plates

Forces can assist or oppose the motion of plates. Figure 1.1.19 shows the location of the forces discussed in this chapter. The preceding description of the history of a plate contains clues to the forces that drive plates. There are four possible forces:

- *Slab-pull.* The force acting on a plate due to the weight of a subducting plate. As part of the dense plate sinks it pulls the rest of the plate behind it.
- *Ridge-push.* The weight of the lithosphere on the inclined surface of the asthenosphere causes the plate to slip away from the mid-ocean ridge. This is also referred to as gravitational sliding.
- *Basal-drag.* The convecting mantle may pull the overlaying lithosphere along with it. This is also referred to as shear traction.
- *Trench suction.* A force that pulls the overriding plate at a subduction zone towards the trench. This may be due to gravitational sliding because the plate is being pulled downwards by the subducting plate. Alternatively, it may be due to flow in the mantle caused by the subducting lithosphere affecting a convection cell.

Slab-pull, ridge-push and trench suction can be thought of as forces acting at the plate edges, while basal-drag forces act over the whole lower surface of the plate. We may expect that continental lithosphere, which extends further into the mantle, will be more affected by basal-drag than oceanic lithosphere. On the other hand, slab-pull and ridge-push forces would have a strong affect on oceanic lithosphere.

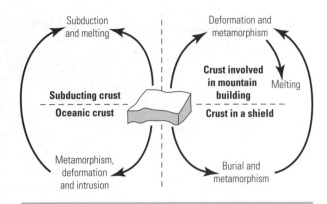

Figure 1.1.18 The fate of a piece of crust.

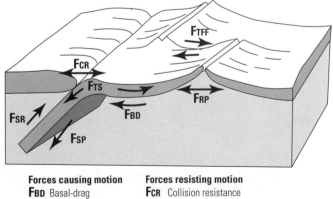

Figure 1.1.19 The forces acting on a plate.

There are also forces that act on the plates that oppose motion. This is to be expected because without such forces the plates would accelerate. Forces that would resist the motion of a plate include:

- *Collision resistance.* This occurs where two plates come into contact at a subduction zone.
- *Transform fault friction.* The pressure between plates along a transform fault will resist motion of the plates.
- *Basal-drag.* This may resist motion if convection is moving in the opposite direction from plate motion.
- *Slab resistance.* As the subducting plate sinks into the mantle, friction between the mantle and lithosphere will oppose plate motion.

The speed at which a plate moves depends on the size of the driving forces. If the characteristics of a plate, such as types of plate motion and plate area, are compared it is possible to analyse the relative importance of the driving forces. Such an analysis is the subject of the practical exercise on page 23.

Analysis of the available evidence indicates that slab-pull and ridge-push are the most important forces driving the plates. It has also been found that the 'rough' base of lithosphere under continents provides a better surface for basal-drag to act on than the base of lithosphere under the oceans.

REVIEW ACTIVITIES

EXTENSION ACTIVITY

1

Explain what convection is and its importance to plate motion.

2

Draw a diagram to summarise the forces acting on a plate.

3

Summarise the most important forces that cause plates to move.

4

Draw up a table to summarise the topography, igneous rocks and earthquake activity found at the three types of plate boundaries.

5

It is thought that the temperature of lavas erupted 4 billion years ago was much higher than the temperatures of lavas erupted today. Explain how higher temperatures in the mantle would affect convection and the rate of plate motion.

SUMMARY

The Earth's surface is composed of lithospheric plates composed of crust and a rigid layer of the mantle.

Continental and oceanic lithosphere differ in terms of their thickness, composition, structure and age.

There is a relationship between the composition of igneous rocks and the tectonic environments in which they form.

Basalts are characteristic of magmas derived from the mantle and occur at divergent boundaries and at hot spot volcanoes.

Andesites are characteristic of subduction zones and convergent boundaries.

Rhyolites are volcanic rocks that are very rich in quartz and form in continents or in mature island arcs.

Plates interact in three basic ways: they move together (convergent boundaries), move apart (divergent boundaries) and move parallel to one another (conservative boundaries).

Different types of plate boundaries have different patterns of volcanic activity, earthquake activity and rock structures.

A plate's motion depends on both the forces that move it and the forces that resist motion.

Slab-pull and ridge-push are the most important forces propelling plates.

Tectonic locations and environments

In this exercise you will use information from a world map and atlas to examine the environments found at different tectonic boundaries. A suitable map for this exercise is the world map of volcanoes, earthquakes, impact craters and plate tectonics in *This Dynamic Planet* (Simpkin, T., et al, US Geological Survey, Washington, 1994). On the basis of the information you gather you will then make some predictions about the importance of plate tectonics in producing environments.

ACTIVITIES

Divergent boundaries

1

Locate the following features on your map or in an atlas:

a the boundary between the African and South American Plates

b the boundary between the Australian and Antarctic Plates

c the boundary between the Pacific and Nazca Plates.

2

Record the average elevation of each plate boundary and describe the geographical features found near the boundary.

3

Summarise the similar features found at these boundaries.

Convergent boundaries

4

Locate the following features on your map or in an atlas:

a the boundary between the Pacific and Philippine Plates

b the Himalayas

c the boundary between the South American and Nazca Plates.

5

Describe the geographical features of each region. Include the width and elevation of mountains, rivers and their paths, and earthquake and volcanic activity.

6

Use an atlas to determine the types of vegetation found in each region.

7

Summarise the similar features found in these regions.

8

Contrast the features of these three regions.

Drainage patterns

9

Compare the river drainage pattern along the margins of the Himalayas with the drainage pattern along the margin of the African Rift Valley in Eastern Africa.

Predictions

Use the information from earlier parts of this exercise to complete the following activities.

10

Predict the type of plate tectonic process you would expect to find in an ocean basin.

11

How would the depth of an ocean basin be related to the age of the crust there?

12

Predict how the features formed along a continental margin due to plate convergence would be similar to, and different from, the features formed at a convergent margin between two pieces of oceanic crust.

13

Predict the sort of river drainage pattern you would expect to find at a continental margin formed at a divergent boundary.

Summary

This exercise is a very simple introduction to a powerful Earth science technique: using modern environments to understand the types of environments that have formed in the past.

14

What environments appear to be characteristic of each type of tectonic boundary?

15

Summarise how drainage patterns appear to be influenced by plate tectonics.

PRACTICAL EXERCISE
Comparing hypotheses of plate motion

In this exercise you will examine the relative importance of three forces that may drive the motion of plates. You will analyse the data provided in order to determine the relative importance of ridge-push, basal-drag and slab-pull as forces that drive plates. Table 1.1.3 (page 16) shows the characteristics and motions of the twelve major plates of the Earth.

Procedure

Record your calculations in tables. This exercise benefits from the use of a spreadsheet program, but if you do not have access to a computer then a calculator, ruler and pencil will do the job. Write down answers to the questions in the exercise as you come to them.

ACTIVITIES

Preliminary task

1

Copy Figure 1.1.19 (page 19) and label the forces that act on the lithospheric plates shown. Mark the driving forces in red and the forces resisting motion in blue.

Lithosphere type and the speed of the plate

2

Predict whether a plate composed mainly of oceanic crust will move faster than a plate made mainly of continental lithosphere. Write down your prediction and the reasons for it. You will use the information in Table 1.1.3 to test your prediction.

3

Determine the relative percentage of continental or oceanic lithosphere. To do this, assume that the continental area reflects the amount of continental lithosphere. Calculate the percentage of the plate covered by the continental lithosphere for the plates. Remember, the percentage of the plate that is crust is calcu-lated by dividing the continent area by the plate area and multiplying your answer by 100.

4

Create two groups of plates:
a plates with less than 15% continental lithosphere
b plates with more than 20% continental lithosphere.

5

Calculate the average velocity of each group of plates and compare the averages.

6

Do the averages support your prediction? Explain your reasons. List other factors that may affect the rate of plate motion.

Area and motion of a plate

7

The size of a basal-drag force will depend on the area with which the convecting mantle is in contact. You would expect that the bigger the surface of a plate, the greater the basal-drag on it will be and the faster the plate will move. A graph of average plate velocity against plate area can be used to test this idea. What would you expect to see on a graph if the area of a plate does affect the plate's average velocity?

8

Graph average plate velocity against plate area for the twelve plates. The plates are in descending order of size in the table. Does your graph show a clear relationship between plate area and average velocity?

9

Repeat the procedure using only plates that are mainly oceanic lithosphere. Does this graph show a clear relationship between plate area and average velocity? Which plate seems different from the others? Look at Figure 1.1.13 (page 15) and see if there is anything about the plate that may suggest a reason for the difference.

Slab-pull and ridge-push

10

Two significant forces that may affect plate motion are slab-pull and ridge-push. Slab-pull is related to the amount of trench present on the edge of the plate in the direction in which the plate is moving. Ridge-push, on the other hand, will depend on the length of ridge on the side of the plate opposite to the direction in which the plate is moving. These lengths are called the effective lengths of ridge or trench in Table 1.1.3. Assess the significance of each force with a graph:
a a graph of average plate velocity against effective ridge length (a measure of ridge-push)
b a graph of average plate velocity against effective trench length (a measure of slab-pull).
Does one of the graphs show a clearer relationship than the other?

Discussion and conclusions

11

Assess the three forces studied here and suggest which shows the closest relationship with plate velocity.

12

Write a summary of what your analysis has shown. Can you identify any other factors in Table 1.1.3 or Figure 1.1.19 that seem to affect the average velocity of the plates?

1.2 Mountain building

At the end of this chapter you should be able to:

describe mountains formed at:
- ocean–ocean boundaries
- ocean–continent boundaries
- continent–continent boundaries

in terms of general rock types and structures including folding and/or faulting.

OVERVIEW OF MOUNTAINS

In everyday use, the term 'mountain' is used to describe a feature of the Earth's surface that rises high above the surrounding surface. Volcanoes built from the eruption of lava can form an individual mountain and chains of separated individual mountains, but most long ranges of mountains have other origins. Some mountain chains are unseen because they lie beneath the surface of the ocean, but they still conform to the everyday definition of a mountain.

To Earth scientists, mountains and mountain ranges are relatively high-standing parts of the Earth's oceanic or continental crust. The rocks within mountain ranges are nearly always a combination of igneous, sedimentary and metamorphic types. They generally show evidence of deformation in the form of large-scale structures (such as folds and faults) as well as small-scale (millimetres) recrystallisation and regrowth of the individual crystals that make up the rock. The large-scale structures may be kilometres or tens of kilometres across, while the rock fabric characteristics are more likely to be apparent at scales of millimetres to tens of metres. The summits of young mountains stand several thousand metres or more above the average elevation of the surrounding crust, but they tend to be continuously eroded. This leads to the exposure of the deep interior of the mountain range and reduces the elevation of the mountain summits.

The characteristic structures of mountain ranges are a consequence of the different types of motion that occur at the three main types of plate boundaries. At these sites either compressional forces or heating lead to the uplift of high-standing crust along with the deformation, igneous activity and metamorphism that characterise mountain building. While the majority of mountain ranges have formed at convergent or divergent boundaries, very small components of compression on conservative margins can also lead to uplift, as has happened along the Alpine Fault in New Zealand's South Island.

Figure 1.2.1 shows the location of the ancient Palaeozoic and Mesozoic–Cainozoic mountain belts. These ancient mountains are called fold mountains belts because they contain extensive areas of strongly folded rocks. Many of these belts now show relatively low relief because they stopped being uplifted when the plate convergence that formed them ceased. They have been worn down by erosion since they stopped being uplifted. As a result, the internal structure of the mountains, which might have formed 3 km below the original surface, is exposed, allowing us to study it. The folding and faulting found in these belts represent the effects of compressional forces that are typical of convergent boundaries.

Mountain ranges that form at divergent margins are found in the ocean and at some continental edges that were the site of continental break-up prior to the commencement of sea-floor formation and sea-floor spreading.

The ocean's mountain ranges are the mid-ocean ridges where high heat flow beneath the ridge causes expansion of the ocean crust around the ridge. This uplift is accompanied by stretching, which can produce tensional faults in the crest of the mid-ocean ridge.

The mountain ranges that occur on the edges of some continents form when heat from the developing ocean-ridge system rifts the continent into two pieces just before sea-floor spreading commences. This heat causes thermal expansion of the crust around the zone of break-up, while processes associated with the creation of the spreading ridge thickens some of the crust on either side of it. These two processes cause uplift and produce tensional faults in the uplifted continental edge. The mountain ranges and coastal escarpment along the Eastern Australian seaboard formed in this way, as did those found on the North American Atlantic coast.

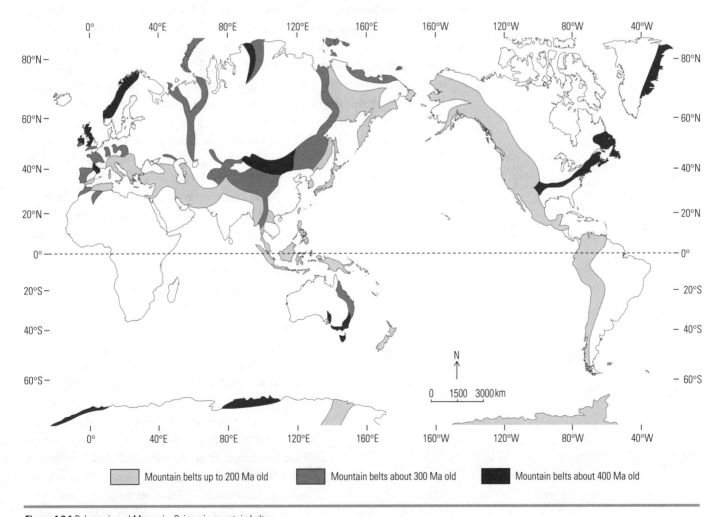

Figure 1.2.1 Palaeozoic and Mesozoic–Cainozoic mountain belts.

1
Compare the everyday definition of a mountain with that used by scientists.

2
Summarise the locations of mountain belts of Mesozoic–Cainozoic age from Figure 1.2.1 (page 25).

3
Explain why both compressional and extensional structures exist in fold mountains.

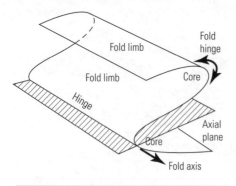

Figure 1.2.2 The parts of a fold.

STRUCTURES FOUND WITHIN MOUNTAINS

Folds

The compressional forces that build mountains commonly cause the rocks within them to deform by bending. This process is called folding. Folds come in many sizes, but they all have the effect of shortening the width, and increasing the thickness, of the body of rock in which they form. Folding that occurs under conditions of high heat and pressure normally occurs at the same time as the rocks are metamorphosed. Under certain conditions the rock may be stretched and flow.

Two simple types of folds are called synforms and antiforms. (See Figure 1.2.3.) Synforms consist of rock layers that have been bent into a U shape. The rock layers in the fold have been bent so that the sides slope down into the bend of the fold. Antiforms have the opposite shape and are folds in which the sides of the fold have been bent up. The sides of the fold slope up into the bend of the fold. This gives them a shape like an upside-down U. The bend in the fold is called the fold hinge, while the straight layers on either side of the hinge are called the fold limbs. (See Figure 1.2.2.) If we know that the rock layers that form a synform or antiform are the right way up then we can use the two terms 'syncline' and 'anticline' to give each fold a more accurate geological name. The rock layers in a fold are right way up if younger rocks form the top layers of the fold and older rocks form the bottom layers of the fold.

Many other terms are used to describe the folds that are found in mountain belts. Some terms you may come across if you study folds in your local area are 'asymmetrical', 'overturned', 'tight' and 'recumbent'. An asymmetrical fold is not symmetrical. Rather, it is a fold that has one limb inclined at a different angle from the other limb. An overturned fold is one in which the rock layers have been turned completely upside down so that the oldest layers are at the top and the youngest layers are at the bottom. If the two limbs of a fold are parallel to each other we say that the fold is isoclinal, a word that means 'inclined in the same way'. A fold is said to be tight if its limbs are nearly parallel. If a fold's two limbs are horizontal or nearly horizontal then we say that the fold is recumbent, a word that means 'lying down'. Asymmetric folds, tight folds, isoclinal folds, overturned folds and recumbent folds are more typically found in the highly deformed rocks of mountain belts formed by continental convergence.

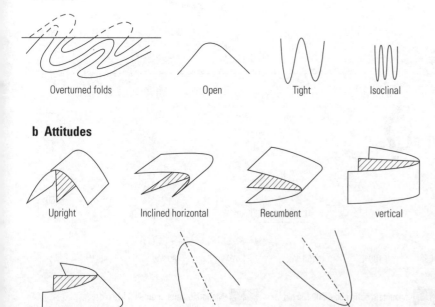

Figure 1.2.3 Forms and attitudes of folds.

Folds can often be recognised on aerial photographs and on geological maps because they produce curved outcrop patterns. The shape of the fold outcrop pattern will be affected by the tightness of the folds. A tight fold is one in which the angle between the limbs of a fold is small (less than 30°). An open fold is one in which the angle between the arms is large (70–120°). A gentle fold is one in which the angle between the limbs is very large (greater than 120°). Folded rocks can be extremely complex, with the rocks being refolded during several events. Such complex folds are usually exposed by erosion in the centre of ancient mountain belts.

Faults

If a set of rocks is brittle or particularly strong the rocks will resist folding and instead the forces exerted on them during mountain building may cause them to break. This breaking, or fracturing, results in a block of rock on one side of the fracture moving relative to the block of rock on the other side of the fracture. The structure that results from this process is called a fault.

Faults are described in terms of the way that the blocks of rock on either side of the fracture, or fault plane, have moved in relation to each other. (See Figure 1.2.4.) Three main types of faults are recognised. These are normal faults, reverse faults and strike-slip faults.

Tensional forces that stretch and pull apart the crust produce most normal faults. These faults are commonly found at divergent boundaries, such as mid-ocean ridges, and at sites of continental rifting. The fracture that forms a normal fault is usually steeply inclined (60–90°) and the block of rock above the fault plane ('the hanging wall') moves downwards relative to the block of rock below the fault plane ('the footwall'). Normal faults often occur as groups of faults that are aligned roughly parallel to one another. This can produce a set of associated fault blocks that are positioned slightly differently from one another. A fault block that has risen above its neighbours is called a horst, while a fault block that has dropped down between its neighbours is called a graben.

Reverse faults are caused by compressional forces that shorten the length of the crust. In a reverse fault the hanging wall block moves upwards relative to the footwall block. Reverse faults in which the fault plane is steep (more than 45°) are simply referred to as reverse faults, but compressional faults with low-angle fault planes are called thrust faults. Thrust faults are also characteristic features of the fold mountain belts that commonly form at convergent boundaries. Movement along a thrust fault can transport an enormous slab of crust several tens, or even hundreds, of kilometres in area and emplace older rocks over the top of younger ones. In strongly folded mountain belts, thrust faults often separate groups of folds, which forms a structure called a fold nappe.

Strike-slip faults are those with vertical or almost vertical fault planes. The displacement of the rock blocks that occurs on them is horizontal. These faults are named because of the way their movement occurs. A horizontal line that lies on a planar geological structure is known as its strike direction, while the distance of movement along a fault is known as its slip. So, fault movement that occurs along the strike direction is strike-slip. All the movement on these faults occurs along their strike. Hence the name strike-slip fault.

Strike-slip faults generally occur at locations where the plates move past each other. In mountain ranges, strike-slip faults can occur where parts of the crust are

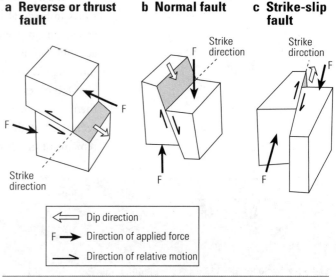

a Reverse or thrust fault   **b Normal fault**   **c Strike-slip fault**

⇐	Dip direction
F ➞	Direction of applied force
➞	Direction of relative motion

Figure 1.2.4 Faulting.

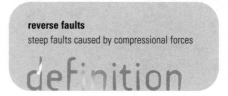

reverse faults
steep faults caused by compressional forces

definition

being compressed or stretched at different rates. We will examine a special group of strike-slip faults that separate the discrete segments of mid-ocean ridge later in this chapter. These special faults are called transform faults.

Intrusions and ophiolites

Two other structural elements that are commonly found in fold mountain belts are intrusions and ophiolites.

The intrusions found in mountain belts are named according to their size and shape. They are batholiths, dikes and sills. Batholiths are extensive bodies of igneous rock made up of a number of large, bulbous bodies of solidified magma called plutons and smaller bodies of solidified magma called stocks. Dikes are relatively thin but extensive vertical, planar sheets of igneous rock and sills are horizontal, planar bodies of igneous rock.

Most igneous rocks found in mountain belts were initially generated by partial melting of the lower crust or upper mantle. Granites and granodiorites are common in mountain belts and occur over wide areas in the large batholiths. Some granites form from the melting of sedimentary rocks that are buried deep within mountain belts. Other granitic rocks are thought to originate by partial melting of igneous rocks incorporated into the lower crust and uppermost mantle above subduction zones in areas of both ocean-ocean and ocean-continent convergence. The magma generated in these settings moves towards the surface through fractures and accumulates as giant fluid masses that are eventually injected up pipes and channels into magma chambers nearer the surface.

Ophiolites are sequences of igneous rocks characterised by gabbros at the base and then a sequence of sheeted dikes, pillow basalts and rocks formed from oceanic sediments at the top. Such sequences represent parts of oceanic crust and are the remains of pieces that are incorporated into mountains at collision zones.

REVIEW ACTIVITIES

1

Explain why some rocks fold and others break when compressional forces act on them.

2

Sketch the basic types of folds and label them.

3

Identify fault types that are formed by compression and those caused by extensional forces. Present your answer in a table.

EXTENSION ACTIVITY

4

Contrast the structure of the two mountain types formed when oceanic crust is subducted.

MOUNTAINS FORMED AT PLATE BOUNDARIES

Ocean-ocean boundaries

The mountains formed at plate boundaries involving oceanic lithosphere are of two types: those occurring at divergent boundaries and those forming at convergent boundaries. Mountains formed at divergent boundaries include mid-ocean ridges and mountain ranges that form at continental rifts. Where two lithospheric plates composed of oceanic crust converge, one plate descends under the other in a process called subduction. The area of subduction is characterised by a deep-sea trench and the formation of a curved island arc. The island arc is a curved ridge composed mainly of volcanic materials. Separate volcanoes are spread fairly evenly along the length of the arc. The Pacific Ocean contains a number of such island arc-trench systems, including the Tonga-Kermadec Arc system (located due east of Australia) and the Mariana Arc system (located between the Philippines and Japan).

MID-OCEAN RIDGES

Oceanic ridges are the largest mountain systems on Earth. If we could drain the oceanic water away we would see that these ridges cover a little more than 20% of the Earth's surface. Comparing this area with the 29% of the surface of our planet that is dry land allows us to appreciate the vast extent of this mountain system.

The oceanic mountain ranges form as broad, linear swells that may be as much as 1000 km wide. They usually rise about 3 km above the abyssal plains that flank the ridge. Mid-ocean ridges show the simplest structure of the mountains we will study. The raw material of the mountains is the basaltic crust that is formed at the divergent boundary. (Review the description of the structure of oceanic lithosphere given in Chapter 1.1.) The features of a mid-ocean ridge are shown in Figure 1.2.5.

Such ridge systems begin when hot, buoyant material in the mantle causes the lithosphere above it to form a dome. The cold lithosphere is stretched and cracks, creating faulted blocks that form linear hills and valleys running parallel to the axis of the ridge. The axis of the ridge is often marked by a rift valley. The presence of a rift valley is related to the rate at which new ocean crust is being formed. Slow-spreading ridges tend to have a central rift valley and rugged topography. Fast-spreading ridges, on the other hand, have a lower relief, appear to be relatively

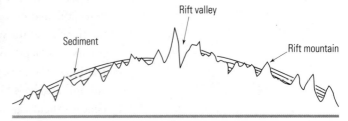

Figure 1.2.5 The structure of a mid-ocean ridge.

smooth and lack a central rift valley. The depth of the highest part of a mid-ocean ridge varies along its length, with the highest parts being about 2.5 km below sea level.

Our planet's mid-ocean ridges are thought to be almost constantly active and their submarine volcanic vents and fissures erupt on a regular basis. The plate divergence that occurs there also causes many shallow earthquakes. The volcanic activity involves the eruption of basaltic lava along and near the ridge axis. The lava forms structures called pillows when the hot molten rock forms a skin as it enters the much colder sea water. Heated water from within the ridge also escapes and carries with it dissolved metals and sulfur compounds. These metals are also deposited in vast sheets of sediment that may become valuable mineral deposits.

Many earthquakes that occur in this setting are caused by movements along normal faults that are related to rift valley structures found at the summit of mid-ocean ridges. These ridge axis earthquakes are generally restricted to a narrow band about 15 km wide. The earthquakes occur at relatively shallow depths, with few earthquakes occurring at a depth greater than 10 km. Other earthquakes that occur in these locations are related to faults that separate mid-ocean ridge segments from one another. These structures are called transform faults. Sediment accumulates on the ocean crust well away from the zone of shallow earthquakes and does not seem to be disturbed by the seismic activity.

Sediment thickness and elevation of the sea floor are both related to the age of the ocean crust. Sediment from the surface of the ocean sinks to the sea floor as a constant rain of material. As the crust moves away from the ridge the thickness of sediment increases. The high elevation of the ridge axis is due to low-density, high-temperature mantle material that buoys up the ridge. As newly formed crust is pushed away from the ridge it cools and becomes more dense. This causes the lithosphere to sink a little deeper into the asthenosphere. The depth of the sea floor below the ridge summit is inversely proportional to the square of its age. This means that in the first 2 Ma a piece of ocean crust will lose about 500 m of elevation, and about 1.5 km by the time the lithosphere has an age of 20 Ma.

The rocks of the mid-ocean ridge also show signs of metamorphism. The hot fluids that circulate through fractures in the crust alter the rocks they pass through. The altered basalts are referred to as **metabasalts** or greenstones. Their green colour is due to the formation of metamorphic minerals, such as chlorite and serpentine.

VOLCANIC ISLAND ARCS AND SUBDUCTION ZONES

The simplest type of convergence in terms of the structures produced occurs when two pieces of oceanic lithosphere converge. One plate is thrust under the other to form a subduction zone with the characteristic Benioff zone of earthquake activity. As such systems age they change both in terms of structure and the types of materials formed within them.

When one plate is thrust under the other to form a subduction zone, a characteristic set of topographic features are formed. The features are shown in Figure 1.2.6. The distribution of the features is affected by the angle at which the subducted plate enters the mantle. Shallow dipping plates produce relatively large distances between the trench and the volcanic arc because the melting that produces magma for the arc occurs at a depth of about 80 km below the surface. Young volcanic arcs have a relatively simple structure and have up to 20 km of crust beneath them. Older volcanic arcs have built on the materials scaped up and erupted before, and the crust beneath them can be 20–30 km thick. As the crust thickens, the elevation of the mountains rise.

The parts of an ocean–ocean subduction zone are described below.

Outer swell and trench

The outer swell is a raised part of the subducting plate. It bulges upwards due to the flexing of the plate and can sometimes form islands above the surface. The trench marks the surface boundary between the two plates. Trenches are usually 50–100 km wide and have an asymmetrical V shape. They can be as deep as 11 km below sea level, which is about 6–8 km deeper than the sea floor. The steeper side of the trench is that closest to the volcanic arc.

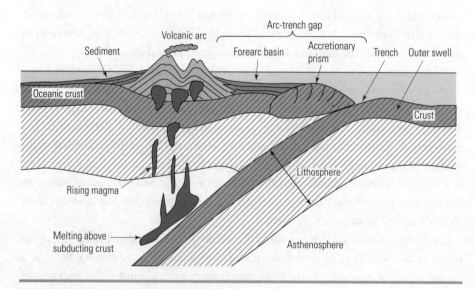

Figure 1.2.6 A cross-section of an ocean–ocean convergence boundary.

Accretionary prism

The accretionary prism is a structure formed on the volcanic arc side of the trench. The prism forms when sediments that have been deposited in the trench (**flysch**), together with deep-sea sediments and crust from the subducting plate, are scraped off and attached to the overriding plate. (See Figure 1.2.7.) As the accretionary prism grows, new material is added near the trench and older material is forced towards the volcanic arc. The growing thickness of the accretionary prism may help to buoy up the overriding plate.

The structure of the prism consists of blocks of material separated by thrust faults. Within the blocks the material is strongly folded and deformed. The blocks of material, termed thrust blocks, become more steeply inclined as time passes. The material in an accretionary prism is sometimes called a **melange**. While the material at the surface may be loosely compacted, the material deep in the prism is strongly metamorphosed. Over time the prism grows in size and thickness. The base of the prism may eventually reach the asthenosphere and the top of the prism may form islands at the surface.

Forearc and back-arc basins

Depositional basins may form on either side of the volcanic arc. The forearc basin is found between the volcanic arc and the accretionary prism. It may overlay older parts of the prism and receive sediment from both the prism and the volcanic arc. On the other side of the volcanic arc is the back-arc basin. This basin also receives sediment from the volcanic arc. Under some conditions, crustal thinning beneath the back-arc basin can give rise to sea-floor spreading.

Volcanic arc

The volcanic arc is formed from erupted material that originates from melting above the subducting plate. Volcanoes form at relatively regular intervals along the arc (50–70 km) and at a uniform distance from the trench. (See Figure 1.2.8.) Around the edges of the volcanic arc coral reefs may form in the relatively shallow waters if they are the right temperature. The volcanoes may

flysch
sediment deposited in a trench or a thick sequence of marine shales and turbidites

melange
a chaotic mixture of sediments

definitions

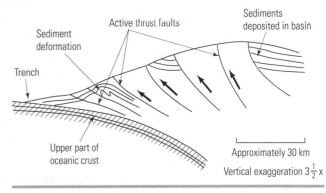

Figure 1.2.7 The structure of an accretionary prism. Note the inclination of thrusts away from the trench.

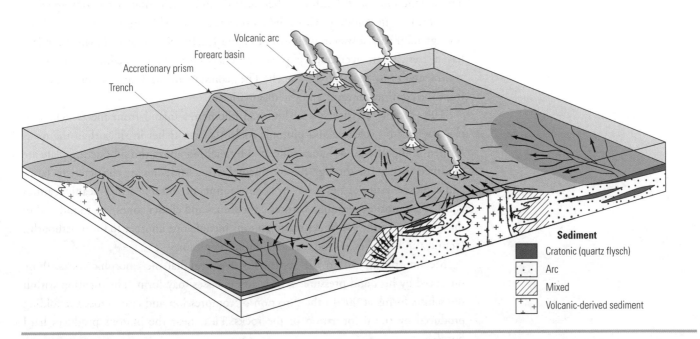

Figure 1.2.8 A volcanic arc-trench system.

rise 2 km above sea level, and form larger islands when weathering and erosion produce sediments on their flanks.

In young arcs the volcanic rocks erupted are basalts, but as the arc ages andesites become common. Andesitic magmas produce explosive eruptions with large amounts of ash. In relatively old arcs erosion may expose plutonic rocks, which were once the magma chambers of the volcanoes. Such rocks are generally granodiorite.

Metamorphic activity

Ocean–ocean subduction zones show characteristic patterns of metamorphism. (See Figure 1.2.9.) Close to the trench, conditions of high pressure and relatively low temperatures are found. These conditions give rise to characteristic metamorphic rocks. The rocks are fine-grained schists and slates that contain the blue amphibole mineral glaucophane. The rocks as a group are referred to as blueschists. Around the volcanic arc itself pressures are relatively low but, due to the igneous activity, temperatures are high. In this area, characteristic high-temperature, low-pressure metamorphic rocks form. These include hornfels that form near the magma bodies and metamorphic rocks called greenschists. These zones of metamorphism form characteristic metamorphic belts parallel to the trench.

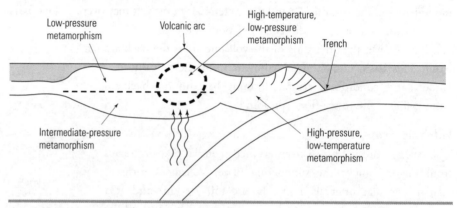

Figure 1.2.9 Metamorphism within an island-arc system.

Ocean–continent boundaries

The subduction zones that form when oceanic and continental plates converge have some features in common with subduction zones formed from oceanic lithosphere. A diagram of such a system is shown in Figure 1.2.10. Note that this plate boundary has a trench, an accretionary prism and a forearc basin. Sediment from the continental crust may fill the trench. The sediment in the forearc basin is derived from the continental crust and the magmatic arc.

The volcanic arc lies several hundred kilometres inland from the trench. It is characterised by volcanic and plutonic rocks that are richer in silica than the rocks formed in island arcs. Magmas generated by melting above the subducting plate are modified as they rise through the continental crust. Andesites are characteristic lavas, and explosive eruptions give rise to large amounts of volcanic ash. Volcanic features (such as composite volcanoes, fissures, lava domes and cinder cones) are found within the arc. Beneath the volcanoes many plutons intrude one another to form batholiths. Granites and diorites are the rocks found in the plutons.

Metamorphism within the system involves regional metamorphic rocks: those produced by heat and pressure. Schists and gneisses may form. The foliation within the schists forms at 90° to the direction of compression and cuts across the folding produced by the deformation of the rocks. Heat near the plutons produces hard hornfels.

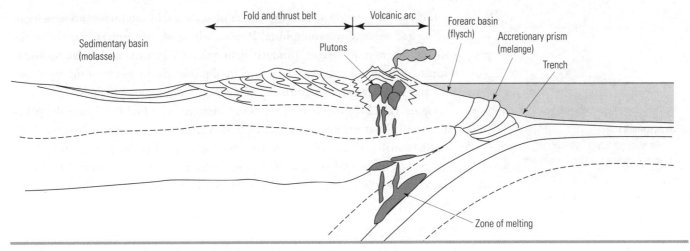

Figure 1.2.10 Fold mountain structure due to ocean–continent collision.

FOLD AND THRUST BELT

Landwards of the volcanic arc, ocean–continent mountains contain fold and thrust belts. The thrust faults dip towards the volcanic arc and the folds are overturned away from the arc. Both these features indicate that pressure from the direction of subduction has produced the structures. The thrust faults dip less steeply with depth. This leads to a situation where material between the thrust faults—the thrust sheets—are moved over the ones below. This can lead to older rock being thrust over younger ones. As the distance from the centre of the fold mountain increases the temperatures are less and the thrusts move relatively undeformed sedimentary layers.

SEDIMENTARY BASIN AND MOLASSE

The material that accumulates in forearc basins consists of muds and **turbidites**. The muds and turbidites are deposited in marine conditions and are collectively known as flysch deposits. Landwards of the magmatic arc, sediments are also deposited in basins running parallel to the arc. The sediments are derived from the land and are called **molasse**. Molasse sediments consist of coarse sands and silts that show little modification due to transport. As the fold mountain ages, the flysch of the forearc basin may be covered by molasse deposits from the range.

Continent–continent boundaries

The convergence of two continental plates consumes the ocean between them. As it does so it folds and deforms the oceanic sediments and these materials become part of the mountain range formed. The features of such a mountain range are shown in Figure 1.2.10.

The development of these mountain ranges can be considered in three stages (see Figure 1.2.11, page 34):

- *Stage 1.* As the continents approach one another, the oceanic crust between them is consumed at a subduction zone. The accretionary prism, forearc basin and any volcanic arcs from the subducting crust build up and are deformed.
- *Stage 2.* When the ocean finally closes, the continental crust of the continent with the passive (non-subducting) margin attempts to subduct under the other continental plate. Its buoyancy prevents it from subducting, and so the sediments and underlaying crust become detached along a thrust.
- *Stage 3.* The crust below the thrust continues to move forwards and forces the detached sheet, and sediments in front of it, upwards. As this occurs the wedge of sediments between the two pieces of continental crust is deformed into a fold and thrust belt. At the same time weathering and erosion produce molasse sediments, which are deposited in the basins on either side of the fold mountains.

turbidites
characteristically sorted sediments deposited from undersea avalanches of sediment called turbidity currents

molasse
thick sequences of sediment deposited by rapid erosion of newly formed mountains

definitions

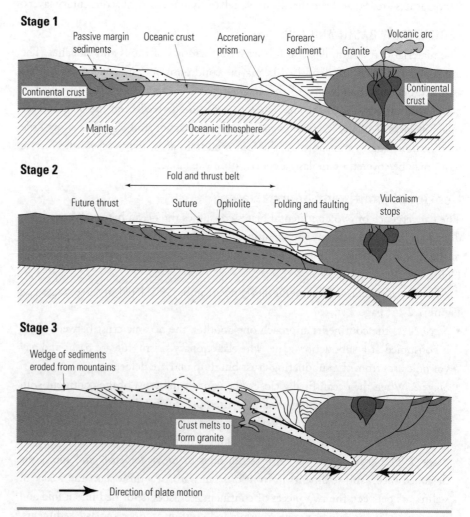

Intense deformation within the centre of such a fold mountain produces tight folding and regional metamorphism. Partial melting of sedimentary material in the core of the mountain may produce light-coloured granites called **leucogranites**. Thrust faults on either side of the mountain dip towards the centre of the mountain belt, and folds are often overturned. Nappes occur and so do ophiolites.

Figure 1.2.12 shows the distribution of materials in a fold belt where the plates have parallel, but irregular, margins. Note that when the margins are irregular both compressional and extensional features can be formed. Rifts such as the Rhine Graben in Europe and the Lake Baikal structure in Siberia were formed by events related to the plate convergence that built the European Alps and the Himalayas, respectively.

CONTINENTAL RIFTING

Continental rifts occur when a divergent plate margin begins to form within a continent. Like mid-ocean ridges, continental rifts are characterised by vulcanism and shallow earthquakes along the rift axis. Like mid-ocean ridges, continental rifts are composed of normal faults and fault blocks. Continental rifts differ from mid-ocean ridges in terms of the composition of the volcanic products produced. While basalts are commonly erupted so are more viscous rhyolites. This is known as bimodal vulcanism because two types of lavas are erupted.

The arching and pulling apart of the crust in continental rifts (see Figure 1.1.6, page 8) is similar to the process that occurs in oceanic crust. Within the rift, linear uplifted blocks (horsts) form mountain ranges and down-dropped blocks form

leucogranite
a granite composed of light-coloured minerals, such as orthoclase and quartz

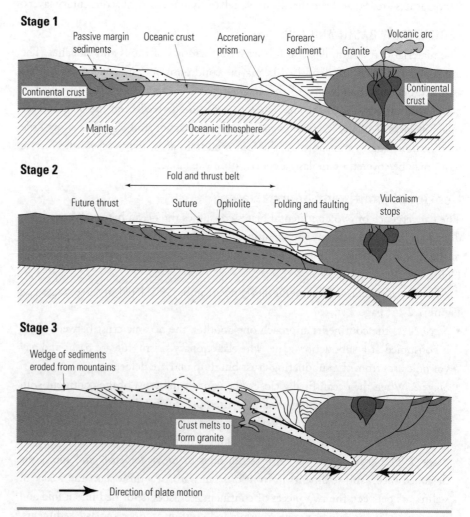

Figure 1.2.11 Stages in a continent–continent collision.

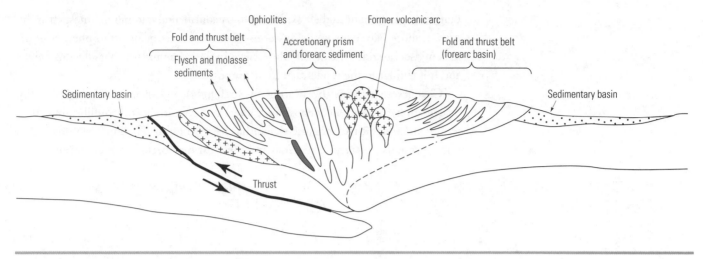

Figure 1.2.12 The structure of a continent–continent collision zone.

grabens or rift valleys. In the Western USA the Basin and Range area is an example of present-day continental rifting.

ISOSTASY AND MOUNTAINS

In order to understand more fully why thickening of the crust produces mountains it helps to know about the concept of **isostasy**. Isostasy is a state of equilibrium, or balance, between the forces on the lithosphere that make it sink and the forces that tend to make it float. When these forces are balanced the lithosphere is steady, but when one force is greater than the other the lithosphere sinks or rises. The Himalayas are not in equilibrium because they continue to increase in height every year.

The idea of isostasy was developed in the nineteenth century to explain differences between the measured and calculated values for the gravitational attraction of large mountains. The mountains appeared to contain less mass than expected. Two models were proposed to explain the mass difference, and they are known by the names of their authors. (See Figure 1.2.13.)

The Airy model says that parts of the lithosphere are of the same density but float at different heights because of their varying thicknesses. Where the lithosphere is thickest we find the highest part of the lithosphere. As a result of this model we may expect the base of the lithosphere to resemble the inverted shape of the surface.

The Pratt model describes parts of the lithosphere floating to the same depth but having different altitudes because of differences in the density of parts of the lithosphere. Less dense lithosphere floats higher than more dense lithosphere.

In terms of what you have read about types of tectonic mountains you already appreciate that real mountains vary in both density and thickness. Both Airy and

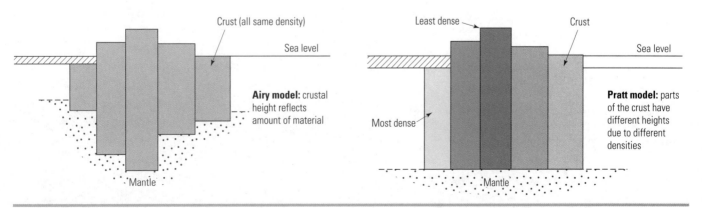

Figure 1.2.13 Isostasy models.

Pratt were partly right in their explanations. A continental fold mountain, such as the Alps, is more like Airy's model in terms of the thickness of the lithosphere beneath it. A mid-ocean ridge has density differences in the upper mantle that affect its height and it is similar to the model that Pratt proposed.

Changes to density (by heating or stretching) and changes to thickness (by deformation, erosion or deposition) alter the conditions of equilibrium. As a mountain erodes it rises because weight has been removed. The intrusion of granites into crust adds low-density material and the crust may become more buoyant.

REVIEW
ACTIVITIES

1

Sketch the features of a mid-ocean ridge.

2

Explain how the elevation of part of a mid-ocean ridge is related to its age.

3

Summarise the parts of an ocean–ocean convergent boundary.

4

Summarise the parts of a fold mountain formed at an ocean–continent boundary.

5

Compare the rock types formed in three different types of tectonic mountains.

6

Explain the role of isostasy in forming mountains.

7

Explain what the terms 'turbidite', 'flysch, 'molasse' and 'melange' mean.

8

Summarise the stages in the formation of a mountain range when continents collide. Present your summary as a flow chart.

9

Copy Table 1.2.1 into your notebook and complete it.

EXTENSION
ACTIVITIES

10

Describe the geological features you would find as you walked across the Himalayas into Tibet.

11

Assess the statement: 'Heat and gravity are the origins of mountains'.

12

Write two definitions of 'fold mountain'. Contrast the definitions and explain why multiple definitions of some terms occur in science.

SUMMARY

At ocean–ocean plate boundaries, mountains form due to convergent and divergent motion.

At convergent boundaries in oceans, island arcs form from the erupted material produced by subduction. Andesite is a characteristic rock from this environment.

At divergent boundaries, mountains form along the mid-ocean ridge and are composed of basaltic material. Normal faulting is a characteristic structure of such mountains.

At ocean–continent boundaries, mountains such as the Andes form. They are characterised by a volcanic arc and fold and thrust belts. Andesites are characteristic volcanic rocks of this environment.

Continent–continent boundaries produce fold mountains. These mountains contain folding, thrusting and regional metamorphism. They also contain the remains of oceanic rocks from the ocean consumed during continental collision.

PRACTICAL EXERCISE
Comparing convergent and divergent boundary mountain ranges

In this exercise you will compare the mountain ranges formed at a mid-ocean ridge with those formed at a convergent boundary.

Procedure

You may begin this exercise using information from this text. However, in order to complete the exercise, you need to use information from at least four sources. These sources can be supplied by your teacher or be those you have identified through your own research.

Collate your information using the headings in Table 1.2.1. You may research all five areas or a selection decided on by you or your teacher.

Ensure you record the sources of your information. Be conscious of any conflicting information you read and try to resolve any conflicts you find.

ACTIVITIES

1

Draw a diagram to compare the features of a mountain chain formed at a divergent boundary with one formed at a convergent boundary.

2

Write a summary of the differences and similarities between the mountain chain formed at a divergent boundary and that formed at a convergent boundary.

Table 1.2.1 Comparison of mountain ranges

Mountain range	Topographic features	Age of rocks within the mountain (Ma)	Volcanic activity (past and present)	Structures and their distribution	Metamorphic products and their distribution	Sedimentary deposition: types and location
Mid-Atlantic Ridge						
East African Rift						
Himalayas						
Andes Mountains						
Tongan islands						

1.3 How continents evolve

At the end of this chapter you should be able to:

- outline how the Australian continent has grown over geological time as a result of plate tectonic processes

- summarise the plate tectonic super-cycle.

THE PLATE TECTONIC SUPER-CYCLE

The term 'tectonic super-cycle', or the supercontinent-cycle model, was proposed by Nance, Worsley and Moody of Ohio University during the 1980s. They were not the first scientists to suggest that tectonic events show some regularity. One way of seeing how a cycle works in plate tectonics is the Wilson Cycle.

The Wilson Cycle, or Wilsonian Cycle, was proposed by Dewey and Burke in 1974 and named after J. Tuzo Wilson, who was the first to suggest that the Atlantic Ocean may have once closed and then opened again. The Wilson Cycle recognises that there have been cycles of ocean creation and destruction during Earth's history. Periodically, oceans open and close. As a result of this process, material is added to continental margins and the continents evolve.

John Veevers of Macquarie University in Sydney proposed a model for a tectonic-climatic super-cycle in 1990. Veevers proposed that the Earth oscillates between a situation where there is a single supercontinent and ocean to a situation where there are many continents and oceans. John Veevers recognised that there is relationship between climate and tectonics. Supercontinents occur at times when the climate is cold: an icehouse situation. When there are many continents and several oceans the global climate is generally warm, that is, there is a greenhouse climate. The whole process takes about 400 Ma to complete one cycle.

Supercontinent cycle models involve the continents coming together to form a single supercontinent. The last supercontinent to form was Pangaea, which formed some 300 Ma before the present. The name given to the super ocean that surrounded Pangaea was Panthalassa. (See Table 1.3.1.)

The importance of a supercontinent model is that it suggests that the continents evolve over time. When continents collide to form a supercontinent they form mountain ranges, which incorporate and modify material from the ocean being consumed. It was the ages of old, eroded mountain belts that led the US scientists to develop the idea of a cycle. As well as explaining perceived periodicity in mountain building, the model suggests how sea level may change over geological time. John Veevers's model is, in addition, based on a variety of stratigraphic information and ages supplied by radiometric dating.

The stages in Veevers's supercontinent cycle are illustrated in Figure 1.3.1 and are described below.

The cycle arbitrarily starts as a supercontinent begins to break up. The amount of carbon dioxide emitted from mid-ocean ridges is small and helps to develop a cold

Table 1.3.1 Supercontinents and their oceans

Age of assembly (Ma ago)	Supercontinent	Ocean
320	Pangaea	Panthalassa (formed Atlantic and Indian Oceans on break-up)
720	Sturtia	– (formed Palaeo-Pacific Ocean and Iapetus Ocean on break-up)
1120	Rodinia	– (formed Mozambique and Pan-African Oceans on break-up)

climate. It is possible that the ocean currents affected by the supercontinent may also affect climate. The sea level is low. The slowly spreading mid-ocean ridges displace less water than at other times. The build-up of heat under the supercontinent causes it to be elevated higher than it otherwise would be. There is less continental shelf than at other times because much of it has been consumed by plate convergence. This stage exists for about 20 Ma.

In the next stage, the heat build-up under the supercontinent causes the crust to sag. Crust of orogenic, or recent mountain-building, origin is prone to rifting and it is within this material that the boundaries between subsequent continents begin to form. During the last cycle this stage existed from the Late Carboniferous to the Mid-Triassic (about 70 Ma).

In the third stage, rifting between the new continents leads to supercontinent break-up. This lasts for about 70 Ma. During the last cycle this stage occurred from the Mid-Triassic to the Mid-Jurassic.

In the fourth stage, sea-floor spreading rates are very rapid and the sea level achieves a maximum height. The reasons for the change in sea level involve a number of factors. The rapidly spreading ridges displace more water and much of the heat that buoyed up the supercontinent is gone. High levels of carbon dioxide also raise the temperature that moves ice in glaciers to water in the oceans. In the last supercycle this stage lasted from the Mid-Jurassic through to the Mid-Cretaceous (75 Ma).

The fifth stage is marked by slower spreading rates in the oceans. This is due to the removal of heat in earlier stages. Cold oceanic crust founders and subduction brings continents back together again. What we see today in the Western Pacific is characteristic of this stage. As part of the latest supercycle this stage has been going on for the last 85 Ma, since the Late Cretaceous. It is likely that this stage will continue for another 80 Ma. During this stage, the sea level decreases, mountain building leads to crustal shortening and the continent is thermally uplifted. A stable supercontinent is formed and stage 1 begins again. In Chapter 1.5 we will see how the changes that occur in the supercontinent cycle not only alter the Earth's continents but climate too.

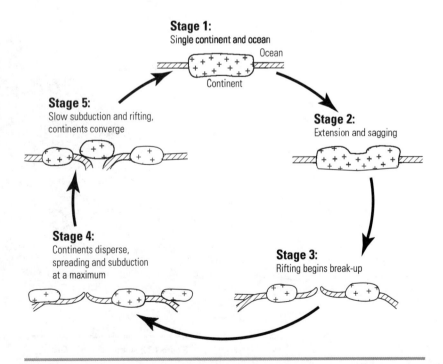

Figure 1.3.1 Stages in the supercontinent cycle. (A summary of J. J. Veevers's supercontinent cycle.)

1

Summarise the stages in the supercontinent cycle.

2

Describe how the supercontinent cycle is related to warm and cool periods of global climate.

3

Explain why sea-floor spreading is more rapid in stage 4 of the cycle than in stage 5.

4

Using the stages of the last supercontinent cycle, match the events in the creation of the Australian continent to the stages of the cycle.

THE EVOLUTION OF AUSTRALIA THROUGH TIME

Australia or, rather, the Australian continent, has grown over time. We have already seen how plate tectonics can form fold belts at the edge of plates and where continental plates converge. The process by which such tectonic products are turned into stable parts of continental lithosphere is called cratonisation. The product of mountain building is called an orogen.

Cratonisation involves the folding and metamorphism of material, its subsequent faulting and intrusion by granitic plutons. Only when it is eroded and beginning to acquire a covering of flat lying sediments is the material said to be cratonised. One way of measuring the age at which an area became fully cratonised is to measure the age of the oldest sedimentary rocks deposited over the crust.

Figure 1.3.2 shows that different parts of Australia are of different ages. The oldest parts of Australia are the Pilbara and Yilgarn Cratons in Western Australia, which were cratonised before 2500 Ma ago. Note that most of Eastern Australia is less than 500 Ma old. In general terms Australia has grown from west to east as we view it now and it continues to grow to our north in Papua New Guinea.

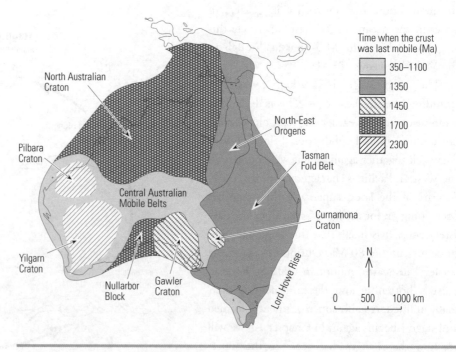

Figure 1.3.2 Structural units of Australia.

We will examine the evolution of the Australian continent in three parts:

- during the Archaean (prior to 2500 Ma)
- during the Proterozoic (between 2500 Ma and 545 Ma);
- during the Phanerozoic (from 545 Ma to the present).

The Archaean

Tectonic processes during the Archaean were very different from those we can observe in the world today. There was a great deal more heat in the mantle and the high heat flow to the surface would have caused processes to operate in different ways from those we know. Convection would have been more rapid and perhaps chaotic. The lithosphere would have been much thinner and more molten material was probably present within the Earth.

Material formed during the Archaean tends to be of two types. One type consists of belts of highly altered metamorphic rocks called granulite-gneisses. They resemble the granites formed deep within continent–ocean subduction zones, such as the Andes, and if plate tectonics gave rise to the granulite-gneisses it may have been in such an environment.

Greenstone belts consist of three parts that show low-pressure metamorphism. They are intruded by igneous rocks called tonalites and granodiorites. The lowermost part of the greenstones contains ultramafic and mafic rocks that erupted at very high temperatures. These rocks are called komatiites and they often show pillow-like structures, such as the basalts that erupt onto the sea floor. Above the komatiites is found a layer of andesites and related volcanic rocks. Chemical evidence suggests they are like the rocks erupted in island arcs. The uppermost layer consists of clastic sedimentary rocks (such as greywacke, chert and sandstone) and banded-iron formations.

One model for the formation of greenstone belts sees them forming in back-arc basins. (See Figure 1.3.3.) The volcanics erupt through thinned continental lithosphere and are covered by sediments from the flanking volcanic arc and continent. The rocks are subsequently folded and intruded by granites when small continents collide.

During the Late Archaean the small cratons came together to form a single supercontinent. This has been deduced from apparent polar wandering curves for the Archaean cratons and the evidence suggests that the supercontinent remained as a single entity throughout most of the Proterozoic. It is unlikely it was rigid and that compression and rifting occurred within it.

The Proterozoic

Between 2500 Ma and 545 Ma fold belts between the Archaean and Early Proterozoic cratons underwent a number of periods of deformation. The Central Australian Mobile Belt surrounded the Yilgarn, Pilbara and Gawler Cratons and experienced periods of deformation throughout the Proterozoic. To the north, the North Australian Craton evolved in what are now northern Western Australia and the Northern Territory. This part of the crust was cratonised progressively from west to east and was fully cratonised by about 1400 Ma.

By about 900 Ma the Central Australian Craton became cratonised, with north–south thrust faulting and silica-rich igneous rocks being emplaced in Central Australia. Within earlier

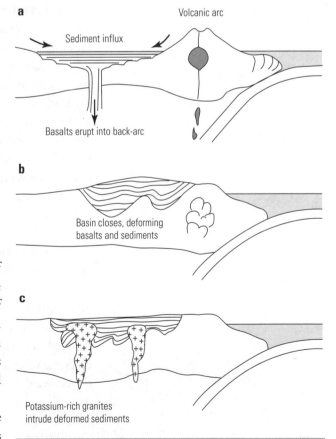

Figure 1.3.3 Formation of greenstone belts.

geosyncline
the thick accumulation of sediments formed near mountain belts. The term predates plate tectonics by 100 years.

orogeny
episode of mountain building

definitions

cratonised areas, basins and rifts developed. To the east of the Gawler Craton, the Adelaide **Geosyncline** (a rift basin) formed. To the east of the Kimberley Block, the Victoria River Basin also formed. It was during this time that the supercontinent Rodinia formed (1120–850 Ma). Another supercontinent, named Sturtia, assembled at 720 Ma and broke up at 560 Ma. It is during this time that the Ediacaran fauna were being preserved in the sediments of the Adelaide Geosyncline.

The Phanerozoic

The development of Eastern Australia during the Palaeozoic is marked by a series of eastward jumps by subduction zones. During the Early Cambrian, the sea level rose and then fell. Volcanic island arcs ran through what is now Western Victoria in the Early Cambrian, but were accreted to the coast during the mountain-building episode called the Dalamerian **Orogeny**.

During the Ordovician the subduction zone was centred in what is now New South Wales. Part of the accretionary wedge is preserved in rocks at Narooma on the South Coast and the island arc, called the Molong Volcanic Arc, was situated in what is now the centre of the state. Compression during the Benambran Orogeny formed the Benambran Highlands in the centre of the state.

During the Silurian and Devonian the area of subduction moved east again. In the New England area and extending northwards into Central Queensland, from Newcastle to Mackay, volcanic arcs were active. The New England Volcanic Arc was located further west than the arc in Queensland. In the Late Silurian and Early Devonian, crustal extension created rifts and volcanic activity.

In the Mid-Devonian another orogeny, the Tabberabberan Orogeny, helped to cratonise crust that had already formed and the volcanic arc was reorganised yet again. During the Early Carboniferous the Kanimblan Orogeny occurred, which saw the Kanmantoo Fold Belt and the Lachlan Fold Belt both form new cratons.

From the Late Carboniferous to the Triassic, periods of crustal extension and contraction occurred. Note that during this time the supercontinent Pangaea was forming. During the Triassic the New England Fold Belt was completed and the Sydney-Gunnedah Basin was deformed by thrusting from the Eastern Fold Belt. By the end of the Triassic all of what is now Eastern Australia was consolidated as dry land.

Two further events are worthy of note in describing the evolution of the Australian continent. The first is the development of our passive continental margins from the Early Cretaceous to the present. Sea-floor spreading began in the Argo area of northern Western Australia and moved down the western edge of the continent. During the Cretaceous, rifting along the southern margin of the continent occurred and the formation of the Tasman Sea began. By about 60 Ma ago, in the Paleocene, the Tasman Sea had effectively ceased spreading but the Southern Ocean was growing and Australia was moving northwards. The changes in ocean circulation south of Australia, and elsewhere, led to changes in climate.

The second significant event occurred as a result of Australia's northward movement. Papua New Guinea began to form in the Early Cretaceous as the Pacific Plate was subducted under the Australian-Indian Plate. In the last 5 Ma what is now the Highland area of Papua New Guinea has risen some 6 km. In relatively recent times the volcanic arc has moved northwards to the Bismarck Sea. This is similar to the eastward movement of the volcanic arcs during the development of Eastern Australia.

1

Explain what the term 'cratonisation' means.

2

Describe the features of the Yilgarn Craton.

3

Summarise the three stages in the development of the Australian continent.

4

Outline how the Australian continent has grown over geological time as a result of tectonic processes. In your answer compare the formation of the greenstone belts in Western Australia with the fold belts of Eastern Australia.

SUMMARY

The continental parts of the Earth combined to form a single supercontinent.

The coming together of continents and their subsequent dispersal forms a repeating cycle known as the plate tectonic super-cycle.

There is a relationship between the stages of the plate tectonic super-cycle, global sea level and climate.

Supercontinents occur when there are icehouse conditions. Times of continental fragmentation produce periods of warm climate.

The Australian continent has grown over time by the process of cratonisation. The continent continues to grow as Australia moves northwards in the mountains of Papua New Guinea.

Australia has grown around cratons such as the Yilgarn and Pilbara Cratons, which contain some of the oldest rocks on Earth.

The evolution of Australia can be summarised in three parts: during the Archaean, Proterozoic and Phanerozoic.

Eastern Australia has grown from west to east since the Cambrian.

PRACTICAL EXERCISE
An analysis of a geological or tectonic map

The aim of this exercise is to develop a better understanding of the age and evolution of parts of Australia.

For this exercise, use some of the references listed for Section 1 on page 387.

ACTIVITIES

1

On a map of Australia draw a straight line from Perth to Sydney.

2

Use Figure 1.3.2 (page 40) to identify the structural units you would pass over as you move in a straight line from Perth to Sydney. List the units in a table.

(*Note:* you are not recording sedimentary units, but the units making up the craton.)

3

Use the resources available to you to identify the age, composition and environment of formation for each unit.

4

Do the units show increasing age towards the west? Would this be obvious if the line was drawn through the centre of Australia?

5

What are the major environments in which the materials making up Australia were formed?

1.4 Natural disasters and tectonics

OUTCOMES

At the end of this chapter you should be able to:

- predict where earthquakes and volcanoes are currently likely to occur based on the plate tectonic model

- describe methods used for the prediction of volcanic eruptions and earthquakes

- describe the general physical, chemical and biotic characteristics of a volcanic region and explain why people would inhabit such regions despite the risk

- describe hazards associated with earthquakes, including ground motion, tsunamis and collapse of structures

- describe hazards associated with volcanoes, including poisonous gas emissions, ash flows, lahars and lava flows and examine the impact of these hazards on the environment, on people and on other living things

- justify continued research into reliable prediction of volcanic activity and earthquakes

- describe and explain the impacts of shock waves (earthquakes) on natural and built environments with reference to specific examples

- distinguish between plate margin and intra-plate earthquakes with reference to the origins of specific earthquakes recorded on the Australian continent.

PREDICTING EARTHQUAKES AND VOLCANIC OCCURRENCES USING PLATE TECTONICS

Natural hazards are those natural events that endanger life and property. Droughts, cyclones and landslides are natural hazards that occur all too regularly in Australia, but for many people living around the rim of the Pacific it is natural hazards due to tectonic processes that pose long-term threats.

Plate tectonics and natural hazards

Figure 1.4.1 reveals that most active volcanoes and earthquakes are not randomly distributed. You should understand by now that most earthquakes and volcanic eruptions occur along plate margins. Hot spot volcanoes also produce eruptions and earthquakes within the plates, but even these events are predicted by the theory.

The theory of plate tectonics allows us to confidently predict that the majority of the world's future earthquakes and volcanic eruptions will occur at or near the plate boundaries. Convergent ocean–ocean and ocean–continent plate boundaries produce both widespread vulcanism and many earthquakes. Areas of the planet where continent plates are converging and have formed fold mountains also experience many earthquakes and occasional vulcanism. For example, there are many volcanoes and frequent earthquakes in Japan and along the length of the Andes and the Tongan Trench. Earthquakes regularly occur in the Himalayas and the European Alps as the

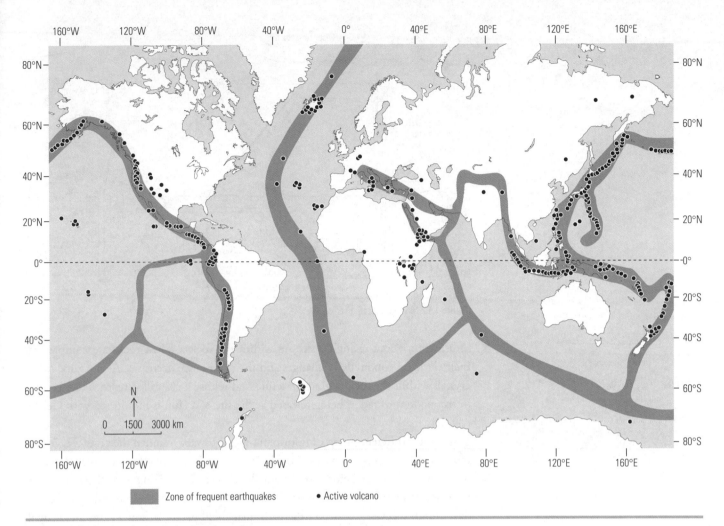

Figure 1.4.1 Earthquake epicentres and active volcano sites.

Key: Zone of frequent earthquakes • Active volcano

crust there adjusts to the compressional forces resulting from plate convergence. The plate tectonic theory also explains why conservative plate margins and transform faults generate earthquakes when plates slide past each other. The historical earthquakes that have occurred on the mid-ocean ridge transform faults, the Dead Sea Fault in the Middle East and the Alpine Fault in New Zealand are examples of conservative margin earthquakes. Divergent margins are also expected sites of much earthquake activity because rifting produces movement on the normal faults on the rift valleys of mid-ocean ridges.

While we can predict the probable occurrence of earthquakes and volcanoes in terms of plate tectonics it is much harder to predict exactly where and when such events will happen. We do not know enough about the history of most areas to be able to test ideas so that we can predict the exact timing of the earthquakes that may occur at a given location.

Monitoring seismic and volcanic activity

It is possible, however, to monitor areas that are likely to produce natural hazards, and this is done where the risk to human life is thought to be high and where the resources are available.

Earthquakes are monitored in a number of ways that aim to measure changes in the stresses around a fault. (See Figure 1.4.2, page 46.) Seismometers detect small movements along faults, and the information allows us to determine how the fault moves. Strain gauges measure bending within rocks. Tilt meters measure flexing of the surface, and measurements of ground water levels and temperature provide

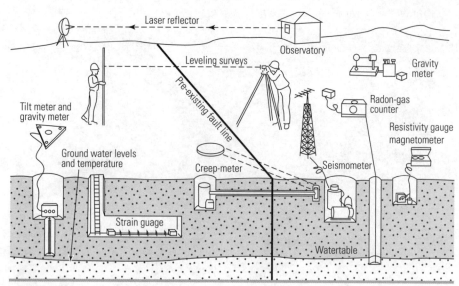

Figure 1.4.2 Monitoring earthquakes.

information about changing stresses within the rocks. When rocks are compressed their electrical conductivity changes and their resistance decreases. At the same time, crystals within the rock begin to fracture and release a gas called **radon**. Fluctuations in the amount of radon being released at the site of a fault is monitored with Geiger counters.

All these different pieces of information tend to vary in much the same way in the months or years prior to an earthquake, but there is no obvious pattern that occurs in the days or hours before an earthquake strikes. This means that these methods warn scientists that an earthquake is due, but do not allow them to know the exact day, month or even year when the earthquake will occur. In some cases a fault will develop enough stress to produce an earthquake, but the earthquake may not occur until several years after the necessary stress is present. In Parkfield, USA, the San Andreas Fault has been studied for over twenty years with a vast array of sensors. It is the most studied area in the world and yet it is not possible to predict the specific time of large earthquakes on this fault with any confidence.

Like earthquakes, volcanoes are monitored in a variety of ways. Some of the useful evidence comes from direct observation of such changes as new or larger cracks in the ground. Changes in steaming from vents and the death or damage of plants on the sides of the volcano also provide information about changes. Historical observations can also be used to build up an eruption history. Unlike the speed and unpredictability of an earthquake, volcanoes tend to take some time to wake up and unleash a major eruption. For instance, the catastrophic Mt Saint Helens lateral blasts of 18 May 1980, when half the volcano disintegrated and exploded, was preceded by three months of ever increasing activity. This allowed the evacuation of most of the population from the area before the blast.

Seismometers, gravity and accurate surveying are important physical methods used to monitor a volcano. (See Figure 1.4.3.) An array of portable seismometers can be used to detect the position of minor tremors in the volcano's magma chamber and build a picture of magma rising in the volcano. Gravity meters and tilt meters are also used on the surface of the volcano to measure the ascent of the magma. Lasers can be used in surveys to measure changes in the shape of the volcano. The injection of magma into a near surface magma chamber beneath the volcano usually causes it to swell or bulge slightly as pressure builds up in the chamber before the eruption.

Chemical monitoring of gases also supplies information about imminent eruptions. As the magma starts to rise in the volcano the composition of the gases

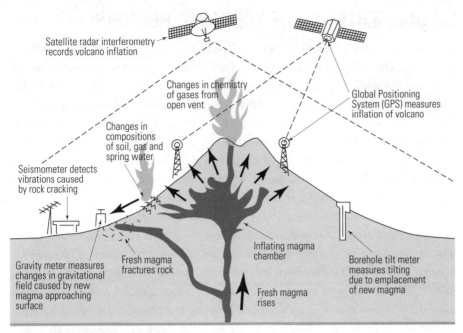

Satellite radar interferometry records volcano inflation

Global Positioning System (GPS) measures inflation of volcano

Changes in chemistry of gases from open vent

Changes in compositions of soil, gas and spring water

Seismometer detects vibrations caused by rock cracking

Gravity meter measures changes in gravitational field caused by new magma approaching surface

Fresh magma fractures rock

Inflating magma chamber

Fresh magma rises

Borehole tilt meter measures tilting due to emplacement of new magma

Figure 1.4.3 Monitoring volcanic activity.

from it change. Sulfur dioxide, carbon dioxide and hydrogen are three gases that are sampled and measured, often by data loggers that do not require people to be present.

Monitoring volcanoes to give early warning can be a very hazardous business. In 1993 the Galeras Volcano in South America erupted while scientists were working on it, killing a number of them. While such work is dangerous, it is important. The information gathered by these studies has allowed vulcanologists to understand the causes and progress of some eruptions well enough to order evacuations of areas likely to be affected by volcanic eruptions. For example, many tens of thousands of lives were saved when vulcanologists ordered a general evacuation of the area around Mt Pinatubo just twenty-four hours before its major eruption in 1983. Many large and dangerous eruptions occur in the volcanic islands above subduction zones. Eruptions on islands in the Indonesian volcanic arc have accounted for about two-thirds of recorded deaths by volcanic activity during the last 300 years. So, monitoring of the volcanoes is vitally important for ensuring the safety of the people who live around them.

REVIEW ACTIVITIES

1

Predict the nature of earthquakes and volcanic activity you would expect to find at a convergent boundary.

2

Outline the methods used to monitor and predict earthquake activity.

3

Outline the methods used to monitor volcanic eruptions.

EXTENSION ACTIVITY

4

Outline the resources you would request if you were given the job of monitoring an active volcano. Justify your choices.

CHARACTERISTICS OF VOLCANIC REGIONS

The characteristics of volcanic regions depend not only on the type of volcano present but also on the volcano's location. One of the few places where people live on a mid-ocean ridge is Iceland. Iceland rises above the sea because a hot spot lies under the mid-ocean ridge. It is a cold place that is thinly vegetated and sparsely populated. Iceland was originally settled during a short warm period early in the second millennium and the locals have adapted to their island. Eruptions here are comparatively gentle and are rarely lethal because the basaltic lavas that are erupted are quite fluid. But even these gently erupting basalts can cause damage. This is especially so when the magma is erupted into the glaciers and sheets of ice for which the island is named. If there is enough erupting lava, enormous amounts of ice can be melted and the resulting floods cause lots of damage. Such floods are called jokullups and can be much larger than any flood caused by rainfall. The one that occurred during 2000 in Southern Iceland buried enormous areas of the coastal plain under metres of sediment and destroyed several bridges. Expulsion of volcanic gases from the many vents in Iceland has killed livestock and people. This mostly happens when carbon dioxide gas, which is denser than air, flows into low-lying areas and displaces the atmosphere, causing asphyxiation or poisoning. Sulfur-oxide can also cause asphyxia or poisoning in much the same way.

Composite volcanoes are characteristic of convergent boundaries and their eruptive style shapes the areas around them. Eruptions give rise to large volumes of ash, which can bury vegetation and easily cause the collapse of built structures, such as houses. When Mt St Helens erupted in 1980 after being dormant for 123 years, ash falls, ash flows and **lahars** destroyed the vegetation over an area of 800 km². Plant successions and animal colonisation occurred reasonably rapidly and the volcano continues to grow.

The elevation of volcanoes can produce particular environments, cooler than nearby areas at lower altitudes. Volcanoes can also draw in clouds and rain, which makes the flanks better watered and possibly produces a rain shadow elsewhere. The rich soils of the volcano's flanks attract people there to farm and harvest trees. The mineral deposits that occur in such environments also attract mining. Volcanoes also draw tourists because of their scenery and island cultures or because they provide good ski slopes.

For many people the rewards to be found living near a volcano outweigh the risk. If a volcano erupts once every 200 years, possibly one generation in six will live through the event. Flora and fauna regenerate quite quickly after eruptions, memories are short and over time people forget how destructive their volcano actually is.

REVIEW ACTIVITIES

1
Describe how the physical, chemical and biological characteristics of a volcanic island would be different from a non-volcanic island.

2
Describe five of the hazards associated with volcanoes.

EXTENSION ACTIVITY

3
Make a diagram to show the potential hazards around a composite volcano.

HAZARDS ASSOCIATED WITH EARTHQUAKES AND VOLCANOES

Earthquake hazards

Earthquake hazards include violent ground motion, tsunamis, landslides and the collapse of built structures. In all cases the hazards are due to ground motions induced by the passage of earthquake waves. An earthquake wave produces a variety of wave types. (See Figure 1.4.4.) The fastest waves (P-waves and S-waves) cause motion in the direction of the wave and at right angles to it. Surface waves have a variety of motions and it is generally these waves that produce the greatest damage to structures and natural systems because they produce the largest ground motions.

The passage of an earthquake can mobilise a slope, causing a debris avalanche, a large landslide or a small landslip. Even the slip may be enough to damage the roots of plants and cause them to die. Such unstable areas can be recognised by their tilted and dead trees and understorey of young plants.

The sudden collapse of coastal cliffs into the sea, the triggering of underwater landslides, and motion of the sea floor are common consequences of earthquakes. Any and all of these events can trigger a **tsunami**. A tsunami travels across the ocean very rapidly, and as it approaches a shoreline its height increases as its wavelength shortens. A tsunami wave that is 40 cm high in the middle of the deep ocean can develop into a wave that is as much as 30 m high when it surges onto coastal land or up a tidal waterway. The force of a surging wall of water produced by a large tsunami can demolish buildings and trees on and behind the shoreline. They commonly affect sites up to 1 km inland and flood vegetation and gardens. Many people and animals have drowned and been washed out to sea by these waves. If a tsunami damages sand dunes behind beaches it can lead to changes in local geography.

Human structures rarely survive the large earthquakes. The use of rigid, brittle materials, such as stone and unreinforced cement, means that some structures will fail when the blocks they are built from disintegrate due to the vibration. Materials such

> **tsunami**
> a water wave with a very long wavelength
>
> definition

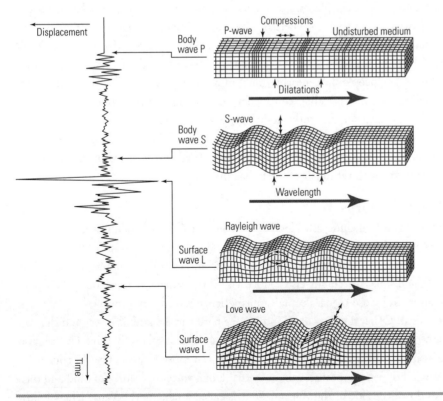

Figure 1.4.4 Earthquake waves.

as cement and glass are very strong when compressed but fracture easily when they are stretched. The oscillation of the ground by surface waves can cause such materials to fail. In some areas buildings and roads are placed on unconsolidated materials, such as sand or mud. Earthquake waves are often amplified in these settings, causing much more violent ground shaking than on nearby hard rock. Earthquakes can cause deposits of unconsolidated sediments to wobble like jelly. More rarely these materials become like liquid. Structures built on soft soils often collapse due to the amplified ground-shake and sink or subside into them if they liquefy. Unreinforced masonry buildings are at greatest risk in an earthquake and, because they are cheap to build, the poor people who inhabit them are often the victims of earthquakes.

INTRA-PLATE EARTHQUAKES

You may think that living on the Australian continent and well away from a plate boundary makes us completely safe from earthquakes. While it is true that we live in a relatively earthquake-free environment, intra-plate earthquakes do occur in Australia. The 1989 Newcastle earthquake killed sixteen people when buildings and awnings collapsed. As a consequence, the Australian building codes were reviewed and upgraded.

The causes of intra-plate earthquakes are not completely understood but are thought to occur along already existing faults that are slowly adjusting to changes in the local stress regime. The stress regime within a plate changes as the plate drifts and is stretched or compressed. Similarly, as the lithospheric plate moves over hotter or colder asthenosphere the heat flow regime changes. The stresses within the plate alter due to heating and expansion of the lithospheric rocks or cooling and contraction of the lithosphere. Uplift and subsidence of the crust due to the erosion or deposition of material can also change the stress regime acting within the plate. So these are the three effects that probably cause intra-plate earthquakes.

Intra-plate earthquakes also differ from plate margin earthquakes in that they are much less frequent events. This means that the large, very destructive earthquakes that occur quite frequently at the convergent margins are relatively rare in an intra-plate location. For example, an earthquake like the Newcastle earthquake may occur once or twice a century somewhere in New South Wales, whereas an earthquake this size occurs perhaps once every couple of years along Turkey's active Anatolian Fault system. Intra-plate earthquakes can occur at a variety of depths and cause a great deal of damage if people live nearby. The important difference is that intra-plate earthquakes occur in stable continental or oceanic crust while plate margin earthquakes occur in currently active plate boundaries. Currently, the Simpson Desert is probably the most seismically active part of Australia. This is due to the release of forces stored within the rocks of this folded and thrusted block, which has been active throughout much of Australia's history.

Volcanic hazards

Volcanic hazards include ash clouds (see Figure 1.4.5), ash fallout, lahars, floods, debris avalanches, direct blasts, tsunamis and volcanic gas emissions.

Ash clouds can suffocate animals and cause an aircraft's engine to be choked by the ash and fail. Ash clouds also reduce visibility. When the ash lands it can bury vegetation and close roads. People can experience respiratory problems as they kick up the dust. Often the dust turns into a very hard surface after it is wet and this can inhibit germination of plants even when the ash layer is relatively thin. The ash also reflects more light than other surfaces and this can cause local changes in climate.

When hot water mobilises ash on the flank of a volcano, a slurry of mud and rock moves down the side of the volcano. This is called a lahar and because lahars follow

Figure 1.4.5 Mt St Helens' eruption in 1980 was explosive and produced enormous amounts of volcanic ash.

creek lines they cause many casualties in populated areas around the base of volcanoes. The damming of rivers by lahars and avalanches produces hazards when the dam fails and the water floods down the river.

Failure of material on the side of a volcano may produce a debris avalanche. Such avalanches can involve tens of cubic kilometres of material and can travel many kilometres from their point of origin. When debris avalanches descend rapidly into the sea they can initiate volcanic tsunamis, which can affect nearby islands. The scar of such an avalanche strips soil away and slows revegetation. Erosion of the scar can produce gullying and further erosion.

Volcanic gas emissions can be poisonous. Volcanoes produce sulfur dioxide and carbon dioxide, but carbon dioxide is perhaps more dangerous. Carbon dioxide is heavier than air and it will settle in low areas near the volcano. This proves lethal to any animals and humans who inadvertently enter the depressions. Volcanic lakes can contain substantial amounts of dissolved carbon dioxide. Earth tremors can release the gas, poisoning people and animals living by the side of the lake.

REVIEW ACTIVITIES

1
Explain how a tsunami is caused.

2
Explain why surface waves can cause damage to rigid structures.

3
Explain why it is dangerous to build structures on unconsolidated ground.

4
Explain how intra-plate earthquakes are different from plate margin earthquakes.

5
Justify continued research into reliable prediction of volcanic activity and earthquakes.

EXTENSION ACTIVITIES

6
Discuss how the hazard an earthquake poses to a person is related to the distance the person is from the epicentre.

7
Explain what is needed to study the recurrence of natural hazards in an area.

8
Make a series of diagrams to show the life of a tsunami.

9
Create a web page on natural hazards. Present it as a webquest with questions and links to web pages where the answers are found.

SUMMARY

Earthquakes and volcanoes occur along plate boundaries and at hot spots.

Volcanic eruptions can be predicted imprecisely using changes in seismic activity, volcano shape, gas emissions and other factors.

Volcanic regions are often fertile and have particular biological communities.

Hazards due to volcanoes include poisonous gas emissions, ash flows, lahars and lava flows. Each hazard occurs in a particular pattern and poses a risk for varying periods of time.

Hazards due to earthquakes include damage to structures by ground movement and liquefaction, as well as tsunamis and landslides.

Australia is subject to earthquakes that occur within the plate. They arise from stresses and forces originally produced during the formation of Australia and adjustment to stresses within the plate today.

Research into natural hazards is justified by the effect that such events have on human populations. By early warning of such events we may improve the quality of people's lives.

PRACTICAL EXERCISE
Measuring crustal movement

In this exercise you will gather information about the technologies used to measure crustal movements at collision boundaries.

ACTIVITIES

1
Review the material in Chapter 1.1 that describes some of the methods used to measure plate motion. Also review the information in this chapter that details the methods used to study plate movements and volcanic activity.

2
Make a list of key terms and also organisations from which relevant information may be obtained.

3
Carry out a search of information on the Internet and in your library. On the basis of your results, make a list of the technologies used to measure crustal motion.

4
If you are working alone, you may wish to summarise the features of the techniques you found. If you are working with others in the class, you may be able to work as a cooperative group teaching each other about the technology each of you researched.

PRACTICAL EXERCISE
Plotting natural disasters

IIn this exercise you will plot the occurrence of natural disasters around the world.

ACTIVITIES

1
Use the Geoscience Australia website <http://www.agso.gov.au> to locate information about earthquakes. Search for other sites using appropriate search terms.

2
On a world map, mark on the location of the natural disasters you have compiled.

3
How do these sites relate to:
a convergent boundaries
b divergent boundaries?

4
Summarise the relationship between sites of natural disasters and plate tectonic environments.

PRACTICAL EXERCISE
A natural disaster case study

Task description

You are to collect information from various sources (for example, newspapers, TV and the Internet) of a *recent* natural disaster that involved tectonic (earthquake, volcano and/or tsunami) activity. The natural disaster you select must be one that occurred in the past twelve months.

You will present this information in the form of a diagrammatic poster or a PowerPoint presentation to the class. Your presentation will include information on the:

- cause of the disaster
- damage caused
- cost to the economy and deaths that resulted
- technology used to measure the impact on the local environment (if applicable), for example, the size of the earthquake, and the gases released
- accuracy of details of information provided by the media.

Resources required

You must keep a logbook of your research to validate your research work.

Sample marking criteria

This assumes the task will be marked out of 20.

An excellent performance (15–20 marks) will reflect your ability to:

- demonstrate a high level of proficiency in identifying, gathering and processing relevant information
- present detailed information about each of the issues listed in the task description
- show a high level of proficiency in selecting appropriate text types, media for presentation and acknowledgment of sources
- demonstrate a high level of proficiency in analysing the information collected.

A satisfactory performance (8–14 marks) will reflect your ability to:

- demonstrate proficiency in identifying, gathering and processing relevant information
- present detailed information for most of the issues listed in the task description
- show a satisfactory level of proficiency in selecting appropriate text types, media for presentation and acknowledgment of sources
- demonstrate some proficiency in analysing some of the information collected.

An unsatisfactory portfolio mark (7 marks or less) will reflect your inability to fulfil a substantial number of the 'satisfactory' criteria listed above.

1.5 Climate and tectonics

greenhouse gases
gases that increase the heat retained in the atmosphere

aerosol
a small droplet of liquid that is suspended in air

troposphere
layer of the atmosphere in which clouds form

definitions

EFFECTS OF VOLCANIC ACTIVITY ON GLOBAL AND LOCAL CLIMATES

In order to predict the effect of explosive eruptions on climates we need to know the following: what there is in an eruption that can affect climate, how that material affects climate and how long such materials are active.

Composite volcanoes produce andesitic magmas as well as ash and gases. Such magmas are viscous and contain large amounts of dissolved gas. When the magma reaches the Earth's surface, sudden loss of pressure leads the gas to form bubbles that try to escape from the magma. As the gas escapes it breaks the magma into small fragments that cool in the air to form ash. The ash, together with volcanic gases, is then carried into the atmosphere.

The gases in the erupting column consist of water, carbon dioxide and other gases, such as sulfur oxides. Carbon dioxide and water vapour are both **greenhouse gases**. This means that they absorb heat and, by doing so, raise the air temperature. Sulfur dioxide reacts with water to form sulfuric acid and this material forms a cloud of small droplets called an **aerosol**. The droplets may form around a very small piece of ash or remain as a simple acid droplet.

Aerosols reflect and scatter incoming light and stop it reaching the Earth's surface. The presence of aerosols in the atmosphere produces effects such as rings around stars and vivid, colourful sunsets. The aerosols are heated by infra-red radiation from the surface. By blocking incoming light, aerosols reduce surface heating and, in turn, surface temperatures. Ash can also block incoming radiation. The amount of cooling produced by ash or aerosols depends largely on how big an area the cloud covers and how long the volcanic materials stay in the atmosphere.

The type of eruption that is most successful in placing ash or aerosols into the atmosphere is called a Plinian eruption. (See Figure 1.5.1.) Such eruptions are very explosive and can propel material 30 km or more into the atmosphere. If the material is ejected high enough it may reach the stratosphere. Here there are no clouds for the material to be washed out of the atmosphere by rain, and the material may stay in the atmosphere for months or years, rather than days in the **troposphere**.

When the eruption cloud reaches high altitudes, winds may carry it around the Earth. This can occur quite quickly. When the Hudson Volcano in Chile erupted in 1991, the eruption cloud took only eight days to circle the Earth. The cloud can also

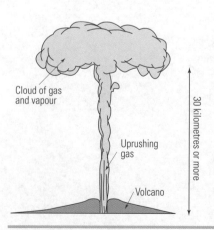

Cloud of gas and vapour

30 kilometres or more

Uprushing gas

Volcano

Figure 1.5.1 A Plinian eruption.

be distributed toward the poles, and volcanoes near the equator can produce clouds that reach both poles.

The important message from the preceding information is that volcanic eruptions can initially change weather patterns by blocking radiation and cooling the surface of the Earth. This effect will be at its greatest for very big eruptions that place material in the stratosphere, where it will remain for a relatively long time. Large eruptions also place more material into the atmosphere, increasing the cooling effect. After the dust and sulfur dioxide are washed out of the atmosphere by rain, the carbon dioxide that is released tends to warm the planet's climate by enhancing the greenhouse effect.

REVIEW ACTIVITIES

EXTENSION ACTIVITY

1

List the products of explosive volcanic eruptions that may change climate.

3

Explain how aerosols can lower temperatures.

4

Sketch a graph to show how aerosols erupted from a volcano may alter temperature in an area. Consider the temperature change and how you will show normal temperatures. Graph a period of eighteen months in which the aerosol causes an effect for six months.

2

Explain how a cloud of ash lowers the temperatures in an area.

IMPACTS OF VOLCANIC ERUPTIONS ON GLOBAL TEMPERATURE AND AGRICULTURE

Perhaps the best example of a volcanic eruption that changed climate is the Tambora eruption of 1815. When Tambora, in Indonesia, exploded it put 50 km³ of material into the air. The explosion killed 92 000 people living near the volcano, either immediately due to the blast or in the months following the eruption due to disease and starvation.

The stratospheric aerosols from the Tambora eruption also travelled around the Earth many times. Volcanic dust and elevated levels of sulfuric acid that rained out from the aerosols are preserved in the 1815 to 1817 levels of ice cores recovered from both the Antarctic and Greenland icesheets. Europe and North America both experienced cold, wet summers following the eruption. People living in northerly locations that receive heavy winter snow described the following summer as more like a green winter than a proper summer. Winter temperatures in the USA were 3 °C cooler than the 200-year average and crops failed in many places, including the north-eastern USA, Ireland and Russia. Many tens of thousands of people died in these countries from the resulting food shortages.

Figure 1.5.2 shows the average summer temperatures for New Haven, USA, for a twelve-year period that includes the year of the Tambora eruption. You should note from the graph that 1814, the year before the eruption, was a cooler year than 1813. This means that the Earth was already experiencing a cooling climate, which was probably due to a slight decrease in the amount of radiation generated by the Sun at that time. We think that the Tambora

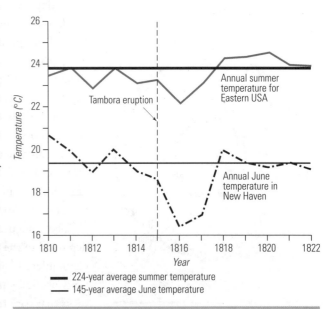

Figure 1.5.2 Summer temperatures in New Haven, USA, 1810–22.

eruption caused further cooling of the planet and increased the effect of a natural cycle of climate change that was already under way.

The Laki eruption in Iceland in 1783 produced terrible changes in local conditions and affected both Europe and North America. Iceland produces fissure eruptions rather than explosive eruptions, but the prolonged eruptions in 1783 put an estimated 100 million tonnes of carbon dioxide into the atmosphere. Locally the crops failed and two-thirds of the livestock died due to carbon dioxide poisoning. The resulting famine killed one-quarter of Iceland's human population. In that year both Europe and North America recorded record low temperatures.

It does not take a great change in atmospheric temperature to cause terrible effects. Tamboura lowered the northern hemisphere's temperature by 0.4–0.7 °C, while the Laki eruption lowered the global temperature by 1 °C. Consider what would happen if Australia's average temperature was to fall by 1 °C this winter. Frost damage to crops may occur in places it normally does not. Colder days would force us to use more energy to stay warm. Now consider the consequences of an event that reduced temperatures for a whole year. Crops may not ripen and the price of some fresh foods would increase due to resulting shortages, energy costs would rise and rainfall patterns would possibly change. The poor would suffer most if this were to happen as the prices of essential items increased beyond their ability to pay for them.

REVIEW ACTIVITIES

EXTENSION ACTIVITY

1
Summarise two examples of eruptions that have altered climates for a period of time.

2
Outline how information similar to that in Figure 1.5.2 (page 55) could be collected for Australia.

3
Tomatoes originated in South America but are grown all over the world. They do not cope well with frosts. Make a list of common vegetables that are frost-sensitive and explain what effect their absence would have on your eating habits.

THE PLATE TECTONIC SUPER-CYCLE AND ICE AGES

The plate tectonic super-cycle was described in Chapter 1.3. The formation of a supercontinent often leads to a long-lived period of extensive glaciation. Such conditions occur when the elevated supercontinent moves over the North Pole or South Pole and a giant icesheet forms on top of it. Figure 1.5.3 shows how the cycle of formation of supercontinents coincides with a long-lived period of cold, or icehouse, conditions while the break-up of the supercontinent coincides with warmer, or greenhouse, conditions. You should also note from the diagram that shorter periods of icehouse conditions can occur. These shorter icehouse periods correspond with times when a continent becomes isolated at one of the poles, just like Antarctica is today. We are actually living through a milder interval of an icehouse period. If you find it hard to believe that the planet is actually in the middle of an ice age, just try surviving a winter on the Antarctic icepack like the emperor penguins do. You'll soon change your mind.

The evidence for the last major icehouse event is found in rocks that were deposited in the Gondwanan section of the supercontinent Pangaea during the Late Carboniferous and Early Permian. These deposits include some of the lower Permian units of the Sydney Basin succession as well as many other Australian sedimentary rocks that date from that time. Gondwanaland had drifted over the South Pole and an enormous icesheet was formed. Such extreme events cause the extinction of some

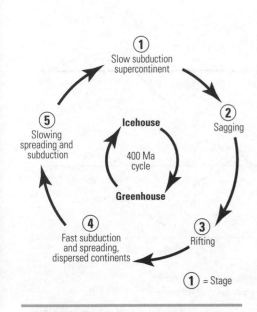

Figure 1.5.3 The plate tectonic super-cycle and the icehouse–greenhouse cycle.

animal groups but also create areas for new forms of life to inhabit. The previous icehouse event occurred around 650 Ma ago when the previously formed supercontinent, called Rodinia, also drifted over one of the poles.

Greenhouse conditions arise in the cycle when sea-floor spreading and subduction are most active during the dispersal of the continental fragments that follows break-up of the supercontinent. At this time in the cycle enough carbon dioxide is being released into the atmosphere that a great deal of greenhouse warming occurs and the climate heats up. According to John Veevers's model, this stage occurred at 160–85 Ma during the Mid-Jurassic to Mid-Cretaceous and 560–480 Ma during the Late Proterozoic to Early Ordovician. The warm conditions and high sea level that followed the melting of the Rodinian icesheet and the break-up of Rodinia coincides with the evolution of complex metazoans and the Cambrian event. (See Section 2.)

REVIEW ACTIVITIES

1
Sketch a diagram to show how an icehouse world is different from a greenhouse world.

2
Outline the relationship between plate tectonics and ice ages in Earth history.

EXTENSION ACTIVITY

3
How would icehouse conditions affect the ecosystems in your local area? Describe how the ecology of your local area would respond to:
a gradual temperature increase
b rapid temperature decrease.

SUMMARY

Volcanic activity can affect small, local areas and global climate.

Volcanic aerosols can reduce temperatures by reflecting incoming solar radiation.

The plate tectonic super-cycle has periods of cold climate (icehouse conditions) when the supercontinent forms and periods of warm climate (greenhouse conditions) when continents are widely distributed.

Ash aerosols and global temperatures

In June 1991, Mt Pinatubo in the Philippines erupted 20 million tonnes of sulfur dioxide into the atmosphere. A satellite called the *Stratospheric Aerosol and Gas Experiment II* identified that a weather system in the South Atlantic caused a band of aerosol to reach Southern Australia in early July.

In this exercise you will identify possible data sources, gather information and analyse it in order to identify any decrease in temperatures that followed the Pinatubo eruption.

Procedure

Begin by deciding where you may find information. The source of the eruption is obvious, but how will you find information on the climate for the period? Consider the sources for weather information, such as temperature and frosts. Also think about scientific journals that may report such events.

If you acquire adequate data consider how they may be analysed. Will a fall in temperature over time indicate the effect of Pinatubo? What will you use as a control?

Prepare a report on the relationship between Pinatubo's eruption and subsequent decreases in global temperature. Make your hypothesis that the eruption did not affect global temperatures. Remember to record the sources of your information.

(*Note:* This is a difficult task. Your teacher may provide information, but you can do this task as a test of your research skills.)

Environments through time

2

In this section we will examine the enormity of geological time by focusing on some of the more critical and interesting events in the evolution of life as well as major changes in Earth's life-supporting environments. We will also investigate interrelationships between Earth's living things and their environments and discover some of the dramatic impacts living things have had on the Earth's environments.

CONTENTS

OUTCOMES

At the end of this chapter you should be able to:

- identify that geological time is divided into eons on the basis of fossil evidence of different life forms

- distinguish between relative and absolute dating.

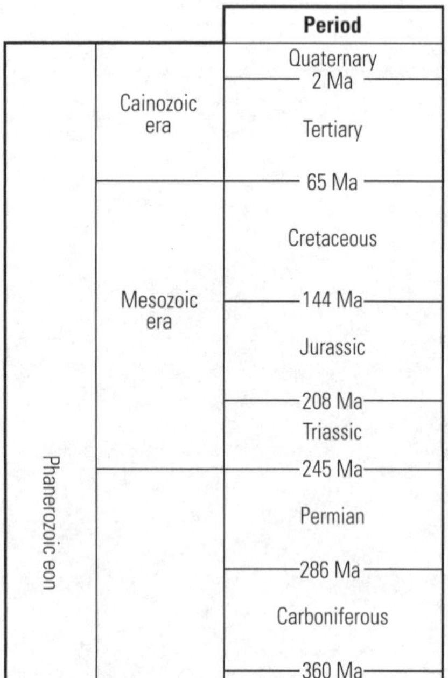

	Period	
Cainozoic era	Quaternary	
		2 Ma
	Tertiary	
		65 Ma
Mesozoic era	Cretaceous	
		144 Ma
	Jurassic	
		208 Ma
	Triassic	
		245 Ma
Palaeozoic era	Permian	
		286 Ma
	Carboniferous	
		360 Ma
	Devonian	
		408 Ma
	Silurian	
		438 Ma
	Ordovician	
		505 Ma
	Cambrian	
		545 Ma
Proterozoic eon		
		2500 Ma
Archaean eon		
		4000 Ma
Hadean eon		
		4550 Ma

(Phanerozoic eon, Cryptozoic eon labels span left column)

Figure 2.1.1 The geological time scale.

THE EONS OF GEOLOGICAL TIME

Initially, the geological time scale was divided into two on the basis of whether or not the rocks contained visible fossils. The Earth's rocks were divided into two groups:
- relatively younger, 'signs-of-life' rocks that contain fossils large enough to be seen without the aid of a microscope or magnifying glass
- older, before 'signs-of-life' rocks that do not seem to contain any fossils.

The younger, **fossiliferous** rocks are called **Phanerozoic** rocks, while the older, non-fossiliferous rocks are called **Cryptozoic** rocks. The terms 'Phanerozoic' and 'Cryptozoic' come from Ancient Greek words that mean 'visible life' and 'hidden life', respectively. Before the term 'Cryptozoic' was coined, these rocks were called Pre-Cambrian rocks. This is because the oldest fossil-containing rocks that were known at that time were called Cambrian rocks. Rocks that are older than the first fossil-bearing rocks obviously predate them. Hence the term 'Pre-Cambrian'.

The Cryptozoic has been divided into three **eons**: the **Hadean**, the **Archaean** and the **Proterozoic**. Adding these three eons to the Phanerozoic eon means that the entire geological time scale has been divided into four distinct parts. (See Figure 2.1.1.) The first eon lasted 500 Ma and represents the time when the Earth initially formed and differentiated. It is called the Hadean, after the Ancient Greek name for 'fires of hell'. The Hadean Earth was extremely active. The rocks that formed the original Hadean crust are thought to have been reworked into the planet's interior, where they were remelted and reprocessed. No rocks from the Hadean have yet been found. However, a few zircon grains dating from that time have been found in the rounded grains that

fossiliferous
fossil bearing; containing obvious fossils

Phanerozoic
a major division of geological time based on the presence of observable fossils, that is, visible life

Cryptozoic
a major division of geological time based on the absence of observable fossils, that is, hidden life

eon
a unit used to divide geological time. There are four eons: the Hadean, Archaean, Proterozoic and Phanerozoic.

Hadean
the oldest eon of the three Cryptozoic eons. No rocks from this time have survived.

Archaean
the middle eon of the Cryptozoic. The most ancient surviving rocks on the Earth formed at this time.

Proterozoic
the youngest of the Cryptozoic eons

definitions

make up 3.6-billion-year-old sedimentary rocks found in Western Australia. These rocks formed during the early stages of the second-oldest eon: the Archaean.

REVIEW
ACTIVITIES

EXTENSION
ACTIVITY

1
State another name for the interval of time commonly called the Pre-Cambrian.

2
Define the word 'eon'.

3
A mnemonic is a device that can be used to aid the memory. Try writing a mnemonic poem that summarises the eons of geological time and their features.

CRYPTOZOIC GEOLOGICAL ENVIRONMENTS

Rocks of the Archaean

The Archaean lasted from about 4 billion to 2.5 billion years ago. The term 'Archaean' is based on the Ancient Greek word for 'ancient' or 'primitive' and was used to name those rocks that were inferred by late-nineteenth-century geologists to be the oldest rocks on the planet. An oldest relative age had been determined for these rocks from their geometrical relationships to other rock bodies using geological principles, such as the **principle of inclusions** and the **principle of cross-cutting relationships**. (See Figure 2.1.2, page 62.) These rocks were also thought to be ancient because many of them had been metamorphosed at very high temperatures and they contained evidence indicating that they had been deformed in many crustal collision events. This evidence was taken to mean that these rocks had undergone several cycles of crustal evolution. Each of these cycles was thought to have lasted many hundreds of millions of years. It followed that the larger the number of crust-forming cycles a group of rocks had endured then the older those rocks must be.

Most areas of Archaean continental crust are composed of cores of granitoid rocks surrounded by belts of distinctive rocks called greenstones. Granitoids generally consist of associations of intrusive igneous rocks (such as granite and granodiorite) and multiply deformed granitoid gneisses that have been metamorphosed at high temperatures (about 750 °C). Extensive blocks of these granitoid rocks are usually separated from one another by belts of greenstones (see Figure 2.1.3, page 63), which are dominantly composed of basaltic and higher-temperature, magnesium-rich, ultramafic volcanic rocks called komatiites. The komatiites often show pillow lava structures, which indicate they erupted into water. Usually the greenstones have been metamorphosed and deformed, but this metamorphism was generally at lower temperatures (about 450 °C) than the metamorphism that affected the granitoid gneisses. This metamorphism is one reason why Archaean fossil evidence is so rare. The existence of Early Archaean sediments and pillow lavas indicates that permanent seas and oceans had formed by this time.

principle of inclusions
a geological rule used to determine the relative age of rocks (See Figure 2.1.2a.)

principle of cross-cutting relationships
a geological rule used to determine the relative age of rocks (See Figure 2.1.2b.)

definitions

Rocks of the Proterozoic

The Proterozoic lasted from 2.5 billion years ago to 545 Ma ago. Strictly speaking, Proterozoic means 'fore-life', but the term is more easily understood as meaning 'proto-life'. Only a very few greenstone belts formed during this eon and there was a gradual change in the quantities and types of rocks that were formed during this eon. Many geologists think this change corresponds with the stabilisation of the plate tectonic cycle into the form it has now.

a Principle of inclusions

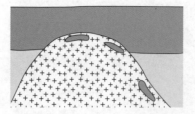

Blocks of sedimentary rock that are included within the granite intrusion existed before the granite did and therefore must be older.

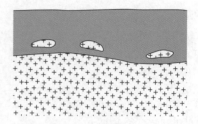

Blocks of granite included within the overlying sedimentary rock layer existed before the sediments did and must be older.

b Principle of cross-cutting relationships

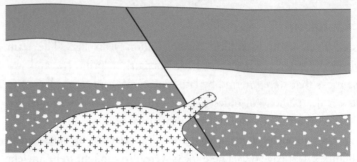

Geological structures or boundaries that cut across other geological structures are younger than the ones they cross. Here the granite boundary cuts across the fault, which in turn cuts the sedimentary bed boundaries. Therefore, the granite is younger than the fault, which is younger than the sedimentary beds.

Figure 2.1.2 The principles of inclusions and cross-cutting relationships.

A wide variety of younger oceanic and continental materials were accreted onto the outside edges of the older Archaean continental cores during the Early Proterozoic. These younger accreted materials are thought to have been deformed and metamorphosed during collisions between the Archaean cratonic cores that amalgamated the Archaean and Proterozoic rocks into large continental shields. Regionally extensive layers of sedimentary rocks were commonly deposited onto the shields, as shown in Figure 2.1.4. These deposits are often called post-cratonisation cover rocks. They are commonly composed of shallow marine limestones and mudstones, as well as mudstones, sandstones and conglomerates deposited by rivers or in lakes. In general, these cover rocks are not obviously metamorphosed or deformed. As the name of this eon would suggest, they commonly contain evidence of microbial life in the form of microscopic fossil bacteria and algae. These microfossils were the precursors of modern complex life forms.

The Phanerozoic begins

The Phanerozoic began 545 Ma ago. The two defining events that mark the beginning of this eon are the first appearance of fossils with hard parts and the worldwide appearance of abundant fossils large enough to be seen easily with the unassisted eye.

Continental material has continued to form throughout the Phanerozoic. It has been added to the Cryptozoic continental cratons at two types of convergent margins: subduction zones and continental convergence zones. A great deal of oceanic lithosphere has also formed during the Phanerozoic. While some of this material has been incorporated into continental lithosphere at convergent margins, most has been subducted back down into the mantle.

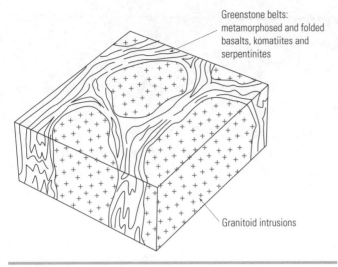

Greenstone belts: metamorphosed and folded basalts, komatiites and serpentinites

Granitoid intrusions

Figure 2.1.3 The structure of greenstone belts.

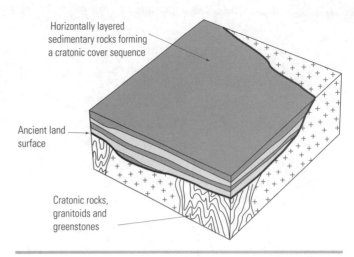

Horizontally layered sedimentary rocks forming a cratonic cover sequence

Ancient land surface

Cratonic rocks, granitoids and greenstones

Figure 2.1.4 The relationship of a cratonic cover sequence to its underlying basement.

REVIEW ACTIVITIES

1
The common assemblages of rocks that make up the Cryptozoic continental cores are:
• *Assemblage 1.* Granitoid gneisses and greenstones.
• *Assemblage 2.* Post-cratonic cover rocks.
a Which of these two assemblages is generally older than the other?

b Draw a sketch that shows the normal relationship between these two rock assemblages.

2
Name the four eons into which the 4.6 billion years of Earth's history is divided.

3
How do we know that seas and oceans were present in the Early Archaean?

EXTENSION ACTIVITY

4
The discovery of radioactive decay revolutionised our scientific understanding of:
a the Earth's history
b the Earth's internal geological processes.
Outline how the discovery of radioactive decay was important in changing scientific understanding of these two phenomena.

SUMMARY

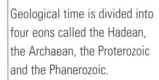

Geological time is divided into four eons called the Hadean, the Archaean, the Proterozoic and the Phanerozoic.

No rocks that formed during the Hadean have survived to the present day.

The cores of the continents formed during the Archaean, and these consist dominantly of granite and greenstone terrains.

The first permanent seas and oceans are thought to have formed at the beginning of the Archaean.

Proterozoic geological sequences are dominated by sedimentary cover assemblages consisting of shallow marine rocks.

PRACTICAL EXERCISE
A geological top ten

This exercise will give you an understanding of the length of the four eons of geological time, an appreciation of the absolute ages of the main divisions of the geological time scale and a feel for the importance of some geological events.

Materials
- A sheet of paper or cardboard at least 55 cm high by 30 cm wide
- A ruler
- Pens

Procedure

1

Draw a vertical time axis, 50 cm high, on the left-hand side of the sheet of paper or cardboard.

2

Divide the vertical axis by centimetres and label the axis so that it shows the present at the top and 5000 Ma ago at the bottom. Each centimetre represents 100 Ma.

3

Draw the outline of a 5 cm wide vertical column just next to your time axis to show when each of the eons (that is, Hadean, Archean, Proterozoic and Phanerozoic) began and ended.

4

Subdivide the Phanerozoic into each of its periods (that is, the Cambrian, Ordovician and so on) and place these boundaries on the column. Use the dates given in Figure 2.1.1 (page 60).

5

Find out about the following events and write them on the column. In the space adjacent to the column, write a brief description of why they are important. (*Note*: Many of these events are described later in this section of the book.)

a oldest known age determined for a meteorite

b oldest known rocks from Earth

c oldest known evidence for life on Earth

d oldest known fossils

e oldest animal fossils

f time when the first permanent seas formed

g appearance of the first land-based vertebrates

h appearance of the first mammals

i appearance of the first hominids (our direct ancestors)

j appearance of the species *Homo sapiens*.

ACTIVITIES

1

Find out about another five geological events that you consider to be important, and place them on the column you have drawn.

2

Determine how long it took for life to first appear on the Earth after the seas first formed.

3

How long after the first evidence of life did the first animals appear?

4

List some of the groups of animals that dominated the different periods of the Phanerozoic.

5

How long did it take for hominids to evolve from the first animals?

6

How much of Earth history have humans experienced? Express your answer as a percentage.

7

Debate the following statement in class: 'The presence and history of the hominids and humans on Earth is a relatively insignificant event in comparison to other prominent events that have occurred during the long history of the planet'.

OUTCOMES

At the end of this chapter you should be able to:

- examine and explain processes involved in fossil formation and the range of fossil types

- interpret the relative age of a fossil from a stratigraphic sequence.

WHAT ARE FOSSILS?

The discovery of fossils within sedimentary rocks and then the discovery of what their existence meant helped to forge geology into a worthwhile natural science. The many propositions that Hutton, Cuvier, Owen, Smith and Darwin made and proved in the late eighteen and nineteenth centuries are today taken for granted. In about 1800 Cuvier demonstrated that fossil mammoth skeletons were the remains of creatures that no longer existed and, therefore, some creatures had not been allowed on to Noah's Ark. By demonstrating this he invented the concept of extinction. This opened the door to a series of investigations that eventually led Charles Darwin to propose his theory of evolution by natural selection, which challenged the orthodox view of biblical creationism and changed the way we perceive our world.

Fossils are the preserved remains and impressions of prehistoric creatures and plants that were buried in soft sediment. An example of a fossil trilobite is shown in Figure 2.2.1. Fossils usually form when biological remains are permanently converted into durable mineralised or carbonaceous films and the soft sediment that encased them is transformed into rock. Fossils of once living things are sometimes called **body fossils** rather than just fossils. In rare instances, body fossils can also form when biological remains are encased in organic deposits and materials (such as tar pits, peatbogs and tree sap, or amber) or, as has happened in relatively recent geological time, within glacial ice and permafrost. A related group of features that are found in sedimentary rocks are called **trace fossils**. (See Figure 2.2.2, page 66.)

HOW FOSSILS FORM

Many animals die by being killed, and then eaten, by other animals. The rest die from natural accidents, disease or old age. Scavengers consume most of the remains of these animals, as well as most of the leftovers that predators don't consume. The soft parts and smaller bones are eaten, larger bones are picked clean by small animals, and then microbial organisms break up and gradually consume the nutrients left in the remaining material. Eventually the last fragmentary remains of the animal mix in with sediment and soil or gradually dissolve away.

Similarly, the soft foliage of plants is continually consumed by a variety of organisms while the plants are alive. The hard, woody parts of plants may rot or they may be consumed while the plant is alive or after it dies.

fossils
the preserved remains and impressions of prehistoric life forms

body fossil
plant or animal fossil

trace fossil
a preserved track, trail or burrow made by a once living animal

definitions

Figure 2.2.1 A fossil trilobite.

Figure 2.2.2 Trace fossils, moulds of dinosaur footprints, at Lake Quarry, Winton, Queensland.

lithification
rock formation

The remains of the vast majority of living things are destroyed rather than preserved. The preservation of an animal or plant after death is therefore an unlikely event. The conditions required for the formation of fossils are quite special and in some ways it is surprising that fossils are a fairly commonplace feature of the geological record.

Most fossils form when living things die naturally and are rapidly buried by sediment soon after death, before scavengers and decomposers can destroy them. A continuing supply of sediment over many millions of years and a place for the sediment to accumulate are required to produce continuous sequences of sedimentary rocks containing enough fossils to record changes in the biota. The continuing sediment supply results in repeated episodes of burial that eventually form a thick pile of sediment. The presence of a thick deposit of sediment is required to bring about the compaction and cementation processes that convert soft sediment into hard rock. These processes are known collectively as **lithification**.

The most likely place for rapid burial to occur is on sea floor that regularly receives sudden influxes of sediment, such as the underwater avalanche shown in Figure 2.2.3. This situation regularly arises in shallow seas and continental shelves where storm waves produce bottom currents that stir up the bottom sediment and suspend it in the water column. After the storm finishes, this material settles out of the water column and is redeposited on the sea floor as a covering blanket of sediment, that is, a newly formed sediment layer.

Such deposits are commonly only a few centimetres, or a few tens of centimetres, thick. Occasionally, however, very violent storms can form sediment layers up to several metres thick in this way. Thicker deposits not only bury recently dead life forms but also bury many bottom-dwelling creatures alive, killing them and preserving them in their life position, almost as if they had been quick frozen. When this happens it provides geologists and palaeontologists with a snapshot of the ecosystem in which these animals lived. Sudden burial events are also common in the deep sea when underwater landslides produce slurries of sand and mud that bury hundreds or thousands of square kilometres of the sea floor. These deposits are known as turbidites and often form sand and mud layers that vary in thickness from several centimetres to several tens of metres, depending on the distance between the source of the flow and the site of burial.

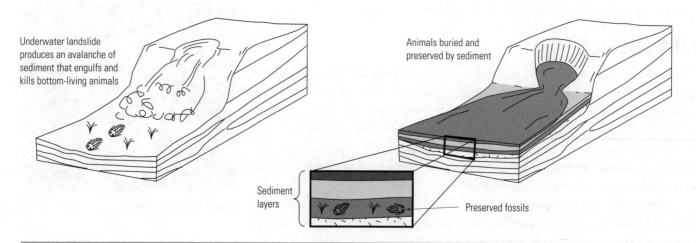

Underwater landslide produces an avalanche of sediment that engulfs and kills bottom-living animals

Animals buried and preserved by sediment

Sediment layers

Preserved fossils

Figure 2.2.3 Burial and preservation of ancient animals in an underwater avalanche.

On land, large floods can quickly erode, transport and then deposit vast amounts of sediment over large areas of floodplain and, in the course of a few days, bury small and large life forms. Large, explosive volcanic eruptions often produce enormous amounts of ash, which rains down out of the sky and can bury large areas of the landscape under several metres of volcanic sediment. These events can also kill and entomb small and large animals.

REVIEW
ACTIVITIES

EXTENSION
ACTIVITY

1

What is a fossil, and how is a trace fossil different from a body fossil?

3

What conditions assist the formation of good sequences of fossil-bearing sediments?

4

Look on the Internet for an example of an ancient volcanic ash fall that has preserved fossils. Outline the important characteristics of this deposit.

2

Name and briefly describe two conditions that are required to form most fossils.

PRESERVING FOSSILS

Sometimes the shelly skeleton of a fossil organism is dissolved soon after burial. This often occurs if the skeletal material is composed of the relatively soluble mineral calcite. Frequently, the surrounding sediment will preserve an impression of the skeleton despite its dissolution. In this way, the original fossil is destroyed but information about it survives. The actual impression that forms can be a direct copy of the original remains (called a cast) or a reversed impression of the preserved parts (a mould of the original parts). (See Figure 2.2.4.)

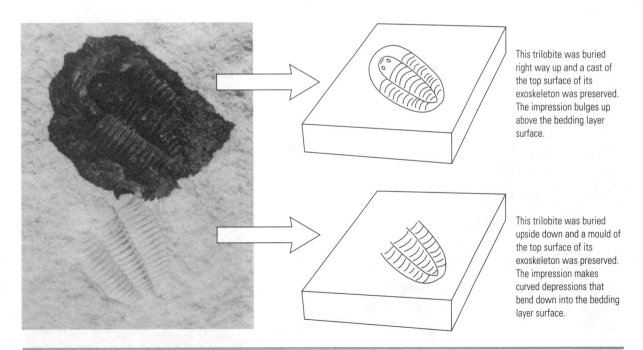

This trilobite was buried right way up and a cast of the top surface of its exoskeleton was preserved. The impression bulges up above the bedding layer surface.

This trilobite was buried upside down and a mould of the top surface of its exoskeleton was preserved. The impression makes curved depressions that bend down into the bedding layer surface.

Figure 2.2.4 The differences between fossil casts and fossil moulds.

petrifaction
fossil preservation where the original skeleton remains unaltered

carbonisation
fossil preservation where the remains are converted into a carbonised film

replacement
fossil preservation where the original skeletal remains are completely converted into some other material

permineralisation
style of fossil preservation in which voids within the skeletal remains are filled by other materials

definitions

Some spectacular but quite rare fossils have formed over relatively recent geological time in which all the soft parts and skeletons of insects, mammoths and people have been preserved. This occurred because these creatures and people were embalmed in tree sap, buried in a peatbog or frozen when covered by ice or permafrost. These styles of preservation retard or prevent decay and the fossilised life forms come to us in almost the same state as when they died. It has even been possible to complete autopsies on these remains, including a 5000-year-old human body found in an Austrian glacier in 1998.

The four main processes that preserve fossils are petrifaction, carbonisation, replacement and permineralisation. **Petrifaction** is the preservation of skeletal material in its original mineralised form (see Figure 2.2.5) within layers of sedimentary rock. Most shelly fossils are made of carbonate minerals, such as calcite or aragonite, and are preserved in this way. It is the most usual form of fossilisation.

Carbonisation is the process by which the soft parts of an organism are converted into a carbon-rich residue of stable coal-like or tar-like organic material. (See Figure 2.2.6.) In the right conditions this process can preserve impressions of the internal organs in exquisite detail. When carbonised fossils are examined using optical and electron microscopes it is often possible to work out a great deal of information about the physiology, biochemistry and anatomy of these ancient living things.

Replacement occurs when the substances that make up the hard parts of a life form gradually dissolve away and are continuously replaced by another mineral. A striking example of this is the replacement of calcite exoskeletons by the mineral pyrite. This happens when acidic waters moving through the sediment gradually dissolve the calcite that formed the original skeleton. The space created when the calcite dissolves is immediately filled by a new mineral, usually pyrite. This occurs because the rotting animal remains release sulfide ions, which are able to react with iron dissolved in the water. This causes pyrite to precipitate into the space that the original calcite skeleton had occupied.

Permineralisation occurs when minerals precipitate in an organism's empty spaces, for example, the space taken up by blood vessels or bone marrow. The minerals added to the fossil harden it and make it more durable. A common example of this type of preservation is petrified wood.

Figure 2.2.5 A preserved and reconstructed Muttaburrasaurus skeleton.

Figure 2.2.6 Fossil fern fronds preserved by carbonisation.

1

Name and describe one of the common ways that fossils are preserved.

2

What is meant by the terms 'cast' and 'mould' in the context of fossil preservation?

3

Outline how fossils can be preserved by pyrite.

4

Outline the similarities and differences between the processes of carbonisation and perimineralisation.

5

Look on the Internet for a picture of a fossil that has been preserved by pyrite replacement. Describe the particularly appealing aspect of the appearance of the fossil.

USING FOSSILS TO DETERMINE THE AGE OF ROCKS

The Danish natural philosopher Nicholas Steno established three of the most important principles of geology in the middle of the seventeenth century. The first is the principle of original horizontality, which states that beds of sedimentary rock are originally deposited in horizontal layers. The second is the principle of superposition, which states that the lowest layer in a stack of undisturbed sedimentary rocks is the oldest while the highest layer is the youngest. The third is the principle of lateral continuity, which states that undisturbed sedimentary rock layers extend laterally in all directions until they thin and pinch out or run into and terminate against the edge of the sedimentary basin in which they are deposited. (See Figure 2.2.7.)

Early in the nineteenth century William Smith and Baron Cuvier expanded on the principle of superposition when they each independently recognised that the fossils appeared in, and then disappeared from, vertical stacks of sedimentary rocks in a distinct order. Smith and Cuvier inferred that rocks containing the same fossils had formed at the same time. They also observed that once a fossil species disappeared from a rock sequence at a particular level then it did not reappear at a higher level. They interpreted the disappearance of fossils to mean that particular species had died out and become extinct.

Smith also determined a distinct sequence in different groups of fossils (called faunas) that he found in progressively higher, and therefore younger, rock layers. Moving up through a sedimentary rock sequence, Smith found that when one fossil fauna became extinct it was succeeded (that is, replaced) by a new and different fossil fauna. This phenomenon is called faunal succession. Smith was able to use the presence of the same fossil fauna to assign the same relative age to different sedimentary rocks. This method is now called correlation. Using the **principle of faunal succession** in this way forms the basis of the method of relative dating called **stratigraphy**. The principles of this method are shown in Figure 2.2.8 (page 70).

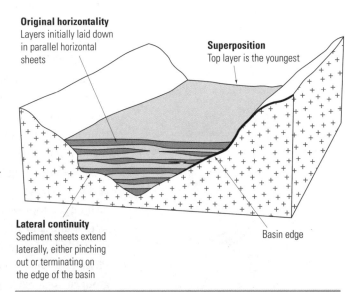

Original horizontality
Layers initially laid down in parallel horizontal sheets

Superposition
Top layer is the youngest

Lateral continuity
Sediment sheets extend laterally, either pinching out or terminating on the edge of the basin

Basin edge

Figure 2.2.7 Principles of original horizontality, superposition and lateral continuity applied to layers of sediment deposited within the confined basin of a river valley.

principle of faunal succession	stratigraphy
a geological rule that forms the basis of stratigraphy	a method used to determine the age of sedimentary rocks based on the fossils they contain

definitions

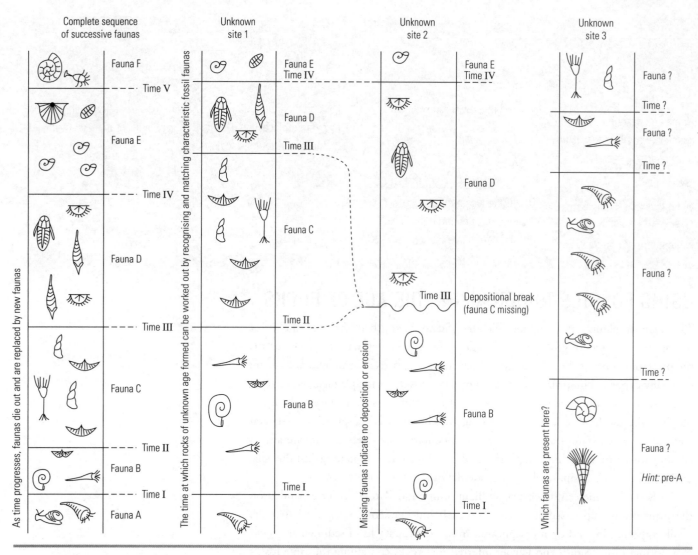

Figure 2.2.8 Faunal succession.

Sedimentary rocks of the Phanerozoic were eventually divided into eleven groups using distinctive, characteristic fossils and fossil faunas. The interval of time each of these faunas lived through was called a **period**. The periods were mostly named after the geographical location where particularly good fossils of that age were found. So we have the Cambrian after the fossils found in the rocks of the county of Cambria, the Devonian after the English county of Devon, the Permian after Perm in Russia, and so on.

The divisions of geological time are often presented in the form of a vertical bar graph called a **stratigraphic column**. It resembles a vertical stack of sedimentary rock layers, with its major divisions being based on the three Cryptozoic eons and the eleven periods of the Phanerozoic. Figure 2.2.9 presents the divisions of geological time and names some of the characteristic fossils and evolutionary events that are used to subdivide the Phanerozoic and Cryptozoic.

The absolute age ranges of the Cryptozoic eons and the Phanerozoic eon's eleven periods were more and more accurately determined as absolute isotopic dating techniques improved during the second half of the twentieth century. As more fossils were found and described the stratigraphic column became increasingly detailed as the periods were divided into shorter time spans called **epochs**. Examples are the Pleistocene and Recent epochs of the Quaternary period. The periods were also grouped together into longer time spans called **eras**, for example, the **Palaeozoic**, **Mesozoic** and **Cainozoic** eras. This grouping reflects the similarities between older, fossil life forms and present-day ones. Relative age dating of rocks using stratigraphy was the main method of dating of rocks until about the 1980s. It continues to be an important and inexpensive technique that is commonly used to determine the age of Phanerozoic rocks.

Eon	Era	Period	Geological events	Biological events		Ma ago
Phanerozoic	Cainozoic	Quarternary	Ice ages	Megafauna extinction		2
		Tertiary	Tibetan Plateau forms Southern Ocean forms–planet cools Tethys Sea closes	First humans	Mobile marine animals dominate	
			Bolide impact	Mammalian radiation		65
	Mesozoic	Cretaceous	Central sea-floor spreading intensifies as sea level rises South America and Africa separate	End-Cretaceous mass extinction Flowering plants evolve		144
		Jurassic	Modern continents begin to separate	First birds	Dinosaurs dominate	208
		Triassic	Pangaea starts to break up	Second reptilian radiation • dinosaurs • ichthyosaurs • crocodilians • pliosaurs • pterosaurs Mammals evolve		245
	Palaeozoic	Permian	Pangaea accumulates	End-Permian extinction: 95% of species; affects all plant and animal groups First reptilian radiation	Eastern Australia: a zone of convergence and mountain building	286
		Carboniferous	Major glaciation Gondwana moves over South Pole Pangaea forming	Amphibians dominate Reptiles evolve		360
		Devonian	Laurasia forms	Extinction of reef builders 'Age of fishes' Plant diversity, first forests	Reefs abundant	408
		Silurian		Earliest land plants and animals First jawed fish		438
		Ordovician	Minor glaciation	Mass-extinction Invertebrate radiation	Attached marine animals dominate	505
		Cambrian	Continents are adrift	Extinction: trilobites particularly affected Earliest vertebrates Cambrian event		545
Cryptozoic	Proterozoic		Glaciation Rodinia breaks up Rodinia forms: • Laurentia • East Gondwana • West Gondwana Terrestrial red beds appear Deposition of banded-iron formations peaks	First marine animals First marine plants Stromatolites become very abundant	Widespread glaciation	2500
	Archaean		Granite and greenstone cratons form Oldest rocks form	First eukaryotes First stromatolites First evidence of organic life		4000
	Hadean		Oldest known zircons form			

Figure 2.2.9 Biological and geological events on Earth.

1

How is the absolute age of a rock determined?

2

Draw labelled sketches that explain the principles of original horizontally, lateral continuity and superposition.

3

How are fossils and stratigraphy used to establish the relative ages of Phanerozoic rocks?

4

Investigate and then explain how William Smith used the principles of original horizontality, lateral continuity and superposition as well as the appearance and disappearance of different fossils to establish the principle of faunal succession.

5

Find out about a group of fossil animals or plants that was prominent during each of the following periods and may be used to identify rocks of that age:

a Cambrian
b Ordovician
c Silurian
d Devonian
e Carboniferous
f Permian
g Triassic
h Jurassic
i Cretaceous
j Tertiary.

---

missing-link fossils

creatures that must have existed that are intermediate in form between younger life forms and their older ancestors

definition

THE INCOMPLETENESS OF THE ROCK AND FOSSIL RECORD

One of the great difficulties that geologists face when trying to work out the detailed history of important geological events is that the geological record of that particular time is often incomplete or, in some cases, entirely absent. A variety of processes constantly rework the Earth's interior and surface. These processes create new rocks by destroying old rocks and this can make the survival of fossils that have formed somewhat haphazard. The evidence we would like to find has been destroyed because the sedimentary rocks containing it have been metamorphosed at high temperatures, melted to form new igneous rocks or, more commonly, eroded away soon after their formation. One consequence of the Earth's continual geological activity is the loss of parts of the geological record.

Understanding this problem is aided by examining a graph such as the one given in Figure 2.2.10, which shows the mass of rocks that currently exist on the planet in relation to the time at which they formed. The graph shows that there is a far greater mass of very young rocks on the planet compared with the mass of very old rocks. The further you go back in time, the smaller the amount of rocks of that age there are. This phenomenon has been summarised neatly by a number of people with the following sentence: 'The older the rock, the greater the chance that it no longer exists'.

The incompleteness of the geological record is the reason why geologists and palaeontologists cannot always provide a simple description or explanation of important events in the history of the Earth's formation and the evolution of life. However, every now and again sets of **missing-link fossils** are found that explain how major changes in the history of life actually took place. Such discoveries allow scientists to hope that, with effort and good fortune, many of the remaining gaps in the geological record and the fossil record will be filled.

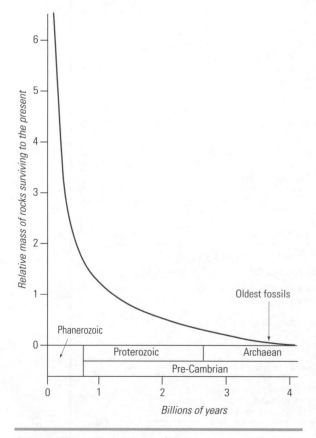

Figure 2.2.10 The masses of rock preserved from different geological times.

1

Explain and discuss the phrase 'the incompleteness of the rock and fossil record'.

2

Explain why missing-link fossils are important.

3

Discuss the reasons why we do not always have particular missing-link fossils.

4

A group of whale-like animals has been discovered recently in Pakistan and Egypt. The animals gradually acquired fluked tails for swimming and their legs changed into flippers during the Middle Tertiary. These are the missing-link fossils between entirely aquatic whales and their four-footed, land-based, wolf-like ancestors. Use the Internet to investigate the evolution of whales. Outline the changes that took place as they evolved from their forebears.

SUMMARY

The recognition that fossils are the remains of animals and plants that are no longer in existence led to the concept of extinction.

Fossils are the remains of prehistoric plants and animals preserved in sedimentary rocks.

Trace fossils are the impressions of tracks, trails and burrows left by animals in soft sediment.

Fossils form when animals or plants are buried in sediment and preserved.

Fossils can be preserved as a cast of the original life form or as a mould. A cast is a copy whereas a mould preserves a reversed impression.

There are four styles of fossil preservation: petrifaction, carbonisation, replacement and permineralisation.

The presence of a fossil that lived at a particular time can be used to date the rock that the fossil occurs in.

A stratigraphic column is a diagram that is used to show the spatial and age relationships of sets of sedimentary rocks.

The appearance and then disappearance of fossils in a stratigraphic sequence led to the concept of evolution.

PRACTICAL EXERCISE
Interpreting the past

In this exercise you will interpret the relative age of fossils found in a stratigraphic sequence with isotopic ages to work out a geological history.

Background

Figure 2.2.11 is an idealised geological cross-section that shows the geometric relationships between a group of rocks that range in age from Proterozoic to Recent. It also shows the isotopic ages of a gneiss and two basalts that occur in the sequence.

The key in Figure 2.2.11 gives the names of eight particular fossils that occur within the sedimentary rocks of the geological section. (*Note:* Fossils B and C are non-specific members of the genus.) Several unconformities are also present in the cross-section. Unconformities are geological surfaces that represent the erosion of a land surface or a period where there was no deposition of sediment.

ACTIVITIES

1

Research the eight fossils given in Figure 2.2.11 using the resources of your library or the Internet. Locate a sketch or photograph of each of these fossils. Find out details about:
a the type of environment in which each lived
b the period of time in which each lived.

2

Use the principal of super-position to rank the fossils from the oldest to the youngest.

3

Fossils D and E are found in rocks that are adjacent to each other, that is, they are spatially close to each other. Does this mean that they are close to each other in age? Briefly explain your answer.

4

The rocks that eventually formed the Proterozoic granite gneisses at the base of the cross-section were originally sandstones. What process or processes caused this transformation?

5

The principle of original horizontality indicates that the mudstone, sandstone, basalt and limestone directly overlying the Proterozoic gneisses were originally deposited as horizontal layers. Describe what has happened to these rocks and explain why this happened.

6

The conglomerate layer is composed of pebbles and boulders of the five rock types found beneath the unconformity. Explain why this might be so.

7

Mosasaurs are a group of organisms that became extinct during a mass extinction event. Find out when this event occurred. Does the age determined for the basalt that overlies the sedimentary layer containing the mosasaurs confirm the age of this particular mass extinction event? What do you think the difference in the age between the basalt and the mass extinction indicates?

8

Write a summary of the geological history of the area that describes major geological events and the environmental changes that the geological cross-section records.

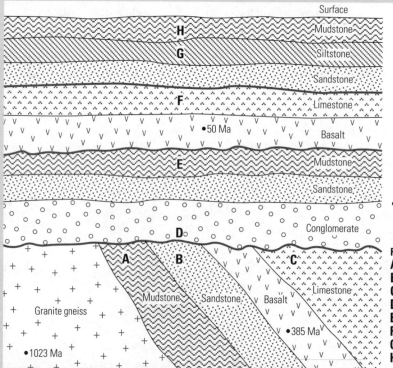

Figure 2.2.11 An idealised geological cross-section showing the fossils found in the area.

OUTCOMES

At the end of this chapter you should be able to:

- outline stable isotope evidence for the first presence of life in 3.8-billion-year-old rocks

- compare and contrast the nature of the first fossil stromatolites and their similarity and relationships to modern examples

- outline the major factors involved in the depositional environment of a banded-iron formation.

- define cyanobacteria as simple photosynthetic organisms and examine the fossil evidence of cyanobacteria in Australia

Figure 2.3.1 The 3.8-billion-year-old Isua chert with its discoverer, Minik Rosing.

stable isotopes
non-radioactive isotopes

graphite
a mineral form of pure carbon that is common in metamorphic rocks

definitions

EVIDENCE OF THE FIRST ORGANIC LIFE ON EARTH

Geologists have found a variety of direct and indirect evidence that simple unicellular creatures were present on the planet fairly early on in the Archaean. Some palaeontologists who deal with the inception of life have argued that life may have originated during the Hadean. Others believe that the heavy bombardment of meteors and comets thought to be occurring at that time would have been too intense for any living thing to survive because they would have been vaporised along with the first seas. The beginning of the Archaean is thought to mark the time that permanent seas and oceans came into existence on the Earth. In any case, there is some evidence that is accepted by most palaeontologists that suggests that life forms were present on the planet by 3.8 billion years ago. This evidence consists of the ratio of the **stable isotopes** of carbon 12 and carbon 13 in **graphite** extracted from metamorphosed sedimentary rocks collected at Isua in Western Greenland. (See Figure 2.3.1.)

When organisms metabolise carbon compounds they take up relatively more carbon 12 into their cells than is present in their environment. This means that the ratio of the stable carbon isotopes carbon 12 and carbon 13 in organic carbon compounds is different from the ratio of carbon 12 and carbon 13 present in inorganic carbon compounds. The carbon compounds found in the Isua rocks are so enriched in carbon 12 that they have a distinctive biological carbon ratio. The inference that follows from this information is that primitive unicellular organisms were alive at this time and metabolised and concentrated carbon 12 within their cells, and that the remains of these cells is the dominant form of carbon in the Isua sediments. The Isua sediments have been metamorphosed since their initial deposition. Despite this, the original carbon compounds, now present in the form of graphite, have retained their original biological, isotopic signature.

Palaeontologists and biologists have worked out the evolutionary relationships between all the main groups of life. These relationships are usually presented as a 'universal tree of life'. Although it is hard to know what the Isua life forms were, we are fairly sure that because they are so very old they must occur near or at the bottom of the tree. They may have even been the common ancestor of all life forms.

Examine the universal tree of life shown in Figure 2.3.2. It divides all the Earth's organisms currently recognised as being distinct from each other into two groups of **prokaryotes** (**bacteria** and **archaea**), which are simpler unicellular organisms without nuclei, and the more complex and somewhat younger **eucarya** (eukaryote), whose cells have a distinct nucleus that contains their genetic material.

Recently it has become possible to date particular branchings of the tree of life because recognisable members of it have been reliably identified in well-dated Archaean and Proterozoic rocks. This has been an important development because it greatly improves our understanding of many geological changes that occurred during the Archaean and Proterozoic.

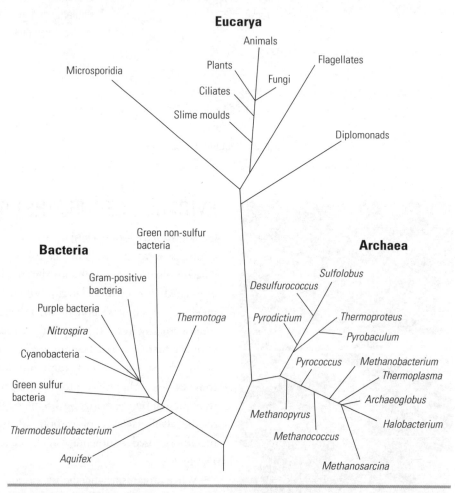

Figure 2.3.2 The universal tree of life.

1
When did the first living things appear on planet Earth and what do we think they might have been?

2
Outline the evidence that is used to show that there were living things on Earth before the time that the world's first fossils formed.

3
Outline why the formation of permanent bodies of water on the surface of the Earth was an important event in our planet's geological and biological history.

Figure 2.3.3 A top view (left) and a cross-section (right) through the North Pole stromatolites.

STROMATOLITES: THE OLDEST FOSSILS

The oldest reliably dated fossil remains are Australian. They are 3.46 billion years old and are found at a remote place called North Pole near Marble Bar in the Pilbara region of Western Australia. (See Figure 2.3.3.) They are found in cherts of the Warrawoona Group and consist of quite large, layered organic structures called **stromatolites**, which were formed by ancient bacteria of some kind. The North Pole stromatolites have been dated very accurately because the cherts they are found in are sandwiched between volcanic rocks containing zircons dated to within an error of plus or minus 2 Ma.

An interesting point to remember when considering the first stromatolites is that there was probably no free oxygen in the Earth's atmosphere 3.5 billion years ago. Consequently, unlike today, there was no ozone to filter out the Sun's ultraviolet light, which kills cells very effectively. The atmosphere may have been thicker and somewhat more absorbent of ultraviolet light than it is today. However, it seems likely that the bacteria that built the first stromatolites would have required the protection of a few metres of sea water above them in order to survive. So, these first stromatolites probably formed under water.

Modern-day stromatolites do exist, but they are relatively rare and occupy very harsh environments in which few other life forms can survive. Indeed, modern-day stromatolites are so rare that they were only discovered in the early 1960s, at a site in Shark Bay. (See Figure 2.3.4.) By coincidence, this is in Western Australia, at a site not far from the world's oldest known stromatolites. Modern-day stromatolites have subsequently been identified at other sites throughout the world, for example, in the Bahamas and in parts of the Mexican Baja California. The Shark Bay stromatolites are formed by a simple ecosystem of several types of bacteria that build domical and columnar rocky masses in highly salty, partly isolated embayments adjacent to the open waters of the bay. The saltiness is due to the high rates of evaporation, the tiny volume of freshwater that flows into the bay (it is a desert environment) and the relatively low amounts of tidal flushing. The rocky material of these stromatolites is formed from the mineral calcite, which commonly precipitates directly from evaporating sea water in these conditions. The calcite accumulates on sticky slime secreted by the uppermost bacterial layer of the stromatolite.

Typically, modern-day stromatolites consist of three distinct bacterial layers that form an ecosystem of interdependent producers and consumers. The layered structure of stromatolites is shown in Figure 2.3.5 (page 78). Within the two upper layers individual bacteria form filaments of joined cells that are usually woven into horizontal, cloth-like mats. The uppermost layer of a modern stromatolite colony is dominantly composed of photosynthetic **cyanobacteria**, which use chlorophyll to power the conversion of water and carbon dioxide into carbohydrates and oxygen. The cyanobacterial layer is underlain by a mat of other photosynthetic bacteria,

Figure 2.3.4 The Shark Bay stromatolites.

a Stromatolite layers

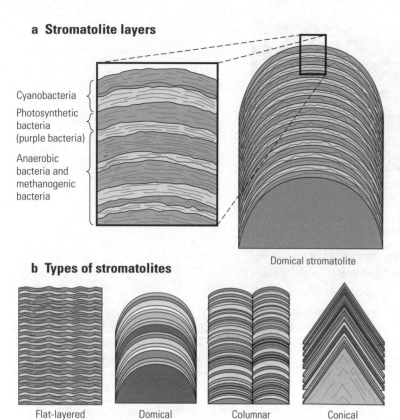

Cyanobacteria

Photosynthetic
bacteria
(purple bacteria)

Anaerobic
bacteria and
methanogenic
bacteria

Domical stromatolite

b Types of stromatolites

Flat-layered   Domical   Columnar   Conical

Figure 2.3.5 Layer formation and structural types of modern-day stromatolites.

Figure 2.3.6 Currents and sands breaking up stromatolitic mats.

definition

methanogenic archaea
a group of prokaryotes that use methane for respiration

known as purple bacteria. These bacteria use photosynthesis and chlorophyll-like materials to power reactions that convert hydrogen and carbon dioxide directly into carbohydrates, or transform hydrogen sulfide and carbon dioxide into carbohydrates by releasing sulfur instead of oxygen. The lowest layer of the stromatolitic ecosystem consists of anaerobic bacteria, sulfate-reducing bacteria and **methanogenic archaea**. These bacteria and archaea use the energy present in the waste products excreted by the upper two photosynthetic bacterial layers.

Because the upper two photosynthetic layers need light to survive and produce slime that can trap calcite-rich mud, the upper surface of the stromatolite grows upwards towards the light. Over time the calcite layer becomes thick enough to block out the light; sometimes the mud that builds up in the slime can also block out the light. When this happens the photosynthetic layer re-establishes above the opaque material and the whole cycle repeats over and over again, producing layer after thin layer of carbonated, cemented rock.

Stromatolites tend to grow in one of four ways: flat layers, domes, columns or cones. Flat-layered stromatolites grow in very quiet, isolated environments (such as saline ponds or salt marshes) where extensive bacterial layers or sheets can grow undisturbed. Domical and columnar stromatolites form in shallow, open water where the daily tides or large monthly spring tides produce currents strong enough to break up the stromatolitic mats. (See Figure 2.3.6.) This isolates individual stromatolites and produces domes, or disturbs the growth of adjacent colonies so that columns form. Conical stromatolites form in deeper water, well below the level of low tide where there is less light. The cone shape probably results from the photosynthetic bacteria congregating at the peak of the cone, which is nearest the light.

Were cyanobacteria present in the oldest stromatolites?

Nobody is certain of the answer to this question because the fossil record is just too incomplete. The answer is very important because the presence of cyanobacteria in the first stromatolites means that they would have been using chlorophyll to produce oxygen, which was released into the oceans and eventually the atmosphere. The production of oxygen by this process is often called oxygenic photosynthesis. We can confidently infer that bacteria did build the first stromatolites and that these bacteria were photosynthetic because we know that stromatolites are photosynthetic structures built by bacteria. Yet we do not know if they used chlorophyll to produce oxygen. They may have used sunlight, chlorophyll-like materials and hydrogen or hydrogen sulfide, rather than water, to produce carbohydrates. Consequently, there are two competing points of view on the presence of cyanobacteria in the first stromatolites.

Some palaeontologists believe that the existence of these structures and the presence of microscopic objects in nearby rocks that resemble bacteria is sufficient evidence that cyanobacteria had evolved by 3.5 billion years ago. Other, more conservative, palaeontologists feel that 1000 Ma is just too short a period of time for bacterial evolution to arrive at chlorophyll and cyanobacteria.

1
What is the name of the place where the world's oldest fossil remains were found and what sort of fossils are they?

2
How old are the world's oldest fossil remains and what sort of environment do we think that these fossils formed in?

3
Name four different stromatolite growth habits and the mineral that stromatolites are mostly made from.

4
What are cyanobacteria? Describe their role in the formation and growth of modern-day stromatolites.

5
Name the biologically and chemically important process that cyanobacteria perform. State the two possible times palaeontologists have suggested that cyanobacteria may have evolved and explain why their evolution was so important.

6
Branching stromatolites are reported from some Proterozoic stratigraphic successions. Research this growth form and use the information you find, or your knowledge of present-day stromatolite growth habits, to explain how branching stromatolites may have formed.

7
Research two places where living stromatolites can be found. Describe the sort of environments that modern-day stromatolites form in and find some representative images of them.

BANDED-IRON FORMATIONS AND THE OXYGENATION OF THE ATMOSPHERE

Banded-iron formations, which are commonly called BIFs, are sedimentary deposits of banded ironstones that are composed of alternating layers of chemically precipitated iron-oxide-rich chert and iron-oxide-poor chert. Chert is rock composed of very fine grained quartz. Most BIFs were deposited in the Early Proterozoic between 2.5 and 2.0 billion years ago, while virtually none of these types of rocks formed after 1.7 billion years ago.

Ironstone bands vary in thickness but are typically 2–20 mm thick. They alternate between iron-rich bands that are bright-red, dark-red or black in colour and silica-rich bands that are light or dark in colour. The bands are usually very continuous, and individual bands can sometimes be traced for several kilometres. Large parts of some of the Pre-Cambrian continental shields are buried under these deposits.

Most of the Early Proterozoic BIFs were deposited in shallow marine environments on stable continental crust that was submerged 200–300 m beneath the sea surface. (See Figure 2.3.7, page 80.) The alternate banding is due to fluctuations in the supply or concentration of both oxygen and iron dissolved in sea water. The reduced form of iron (Fe^{2+}) is quite soluble in deoxygenated water, but if oxygen is introduced into the water it quickly reacts with the Fe^{2+} ions and converts them to Fe^{3+} ions. This leads to a precipitation reaction that forms either the cherry red coloured mineral hematite (Fe_2O_3) in relatively more oxygenated water or the black coloured mineral magnetite (Fe_3O_4) in relatively less oxygenated water.

The concentrations of oxygen and iron in these shallow Early Proterozoic seas varied over time in cyclic patterns. We know this because the thickness of the bands varies in such a way as to produce recognisable patterns that repeat at different scales, that is, in cycles. It has not been possible to determine the absolute length of the cycles evident in the ironstone layers with any certainty because it has not been possible to date a thick set of undisturbed BIF layers with sufficient precision. While it is likely that some of the finer banding cycles may have been seasonal or yearly

banded-iron formations
sedimentary deposits of banded ironstones, the majority of which formed in the Early Proterozoic
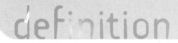

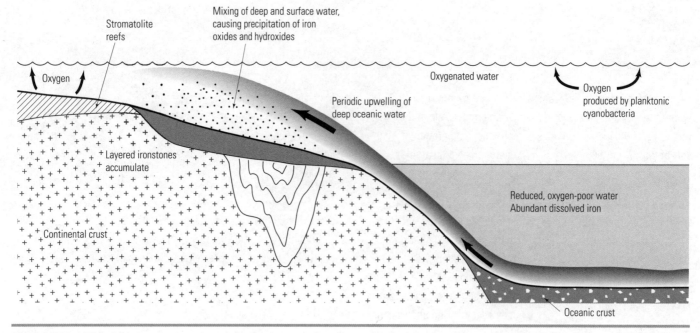

Figure 2.3.7 How BIFs probably formed.

events it seems more likely that most of the cycles that formed the banding lasted for decades or even centuries.

Oxygenating the atmosphere

Several models for BIF formation have been proposed but they all involve two main features. First, oxygen produced by bacterial photosynthesis of carbon dioxide oxygenated the upper layers of the oceans and probably began oxygenating the Earth's atmosphere. This oxygen was not just produced by the vast numbers of newly arrived, Early Proterozoic stromatolitic cyanobacteria. It is highly probable that oxygen was also produced by many related species of free-floating and stromatolitic cyano-bacteria. The free oxygen that the cyanobacteria released caused any dissolved Fe^{2+} to precipitate when the two materials mixed with each other.

Secondly, the deep Early Proterozoic oceans were not yet oxygenated and they still contained vast amounts of dissolved Fe^{2+}. At the same time as the near surface cyanobacteria were oxygenating the upper ocean, large-scale water-mass circulation caused deep oceanic water to rise towards the surface at the continental margins. We call this process upwelling. It transported deep water rich in Fe^{2+} from the ocean depths and mixed it with oxygenated surface water in the shallow seas at the edges of the continents. These mixing events caused enormous amounts of iron oxides to precipitate and then, because these materials are quite dense, settle quickly to the bottom of the shallow continental seas. This formed an extensive, laterally continuous layer of iron-oxide-rich chert. The iron-oxide deposition ceased when all the dissolved iron reacted with the oxygen present in the upper ocean layers. Sedimentation in the shallow seas returned to being dominated by the deposition of iron-oxide-poor chert.

It took about 500 Ma for cyanobacteria to produce enough photosynthetic oxygen to precipitate all the iron that had been dissolved in the world's oceans. When large-scale BIF deposition ceased about 2 billion years ago, it seems likely that the world's oceans had become relatively oxygenated and certainly too rich in oxygen for Fe^{2+} to be stable. The absence of dissolved iron in sea water would not have concerned the cyanobacteria at all. They would have kept on producing oxygen as they had all through the Early Proterozoic. However, instead of this oxygen being used up in precipitating dissolved oceanic iron, the oxygen produced by bacterial photosynthesis would have begun to build up in the oceans and in the atmosphere.

Around the same time that widespread BIF formation ceased, the sediment deposits formed on the land began to contain large enough amounts of the oxidised iron mineral hematite to make them red. The abundance of these terrestrial red beds indicates that oxygen production continued and that there was oxygenation of the atmosphere. At about the same time, pebbles and sand grains of the minerals pyrite (FeS_2) and siderite ($FeCO_3$) ceased to be deposited as particles of sediment. This happened because pyrite and siderite weather rapidly on exposure to oxygen; they both quickly rust just like steel does. They were no longer stable on the Earth's surface when free oxygen became a significant component (at least 1%) of the Earth's atmosphere.

So, vital steps in producing an ocean that was rich enough in oxygen for large multicellular organisms were:

- the change in composition of the Earth's sea water due to the precipitation of dissolved oceanic iron during the Early Proterozoic
- the continuing oxygenation of the oceans during the rest of the Proterozoic, both by cyanobacteria and by eukaryotic algae during the Middle and Late Proterozoic.

Oceanic oxygenation was therefore one of the major environmental changes, if not the major environmental change, to occur in the oceans after they first formed in the Hadean; and this oxygenation occurred because of biological activity. In other words, bacterial life made it possible for oxygen breathers like us to evolve.

REVIEW
ACTIVITIES

EXTENSION
ACTIVITY

1
What are BIFs and what are the minerals they are composed of?

2
How did BIFs form and what type of environment did they form in?

3
When did the majority of BIFs form and why might the formation of these deposits be related to an increase in the worldwide abundance of stromatolites?

4
When did terrestrial red beds first form and when did pebbles composed of pyrite disappear from sedimentary rocks? What is the significance of these two events?

5
Search the Internet for photos of samples of banded ironstones as well as photos of the impressive and colourful outcrops that these rocks can form. Describe, in your own words, the appearance of these rocks and the outcrops. See if you can identify banding at different scales in both the photos of samples and the photos of outcrops.

SUMMARY

The enrichment of the stable isotope carbon 12 in the 3.8-billion-year-old graphitic cherts at Isua is interpreted to indicate that there was microbial life on Earth at that time.

Stromatolites are rocky, layered structures built by an ecosystem dominated by different types of photo-synthetic and anaerobic bacteria.

Fossil evidence for the earliest stromatolites (3.46 billion years old) is found at North Pole in Western Australia.

Cyanobacteria are a group of photosynthetic bacteria that use chlorophyll and sunlight to convert carbon dioxide and water into carbohydrates and oxygen.

The earliest stromatolites were not particularly common and may not have contained cyanobacteria.

BIFs are thick, cyclic sequences of alternating iron-oxide-rich chert and iron-oxide-poor chert that were deposited in shallow seas and continental shelves during the Early Proterozoic.

BIF formation occurred when the global ocean's dissolved Fe^{2+} was converted to Fe^{3+} and precipitated as oxides and hydroxides and coincided with the time that stromatolites became widespread.

PRACTICAL EXERCISE
The importance of stromatolites

This exercise will improve your understanding of stromatolites.

Procedure

1

Use the material in this chapter, along with resources on the Internet or in your library, to research modern and ancient stromatolites.

2

Locate photos and diagrams of modern and ancient stromatolites that show details of their internal layering and give you an idea of their dimensions and growth habits.

3

Find information on:
a the types of geographical locations in which stromatolites formed
b the abundance of stromatolites during the Archaean and the Proterozoic
c where stromatolites can be found today.

4

Organise your findings into a table under the following headings:
• age
• growth habit (for example, domical and branching)
• geographical distribution (for example, restricted)
• abundance.

ACTIVITIES

1

Compare the fossil stromatolites of the North Pole site in Western Australia with modern stromatolites and list any obvious similarities or differences in their appearance.

2

What environments favour the growth of present-day stromatolites?

3

What sorts of environments do we think that Proterozoic and Archaean stromatolites grew in?

4

During what times do we think that stromatolites were more abundant and more widely distributed than they are today?

5

Suggest some reasons that could explain why the distribution of modern-day stromatolites differs from the distribution of stromatolites in the past.

OUTCOMES

At the end of this chapter you should be able to:

discuss the relevance of hard shells, preservable armour and skeletons in explaining the apparent possible explosion of life in the Cambrian period

deduce possible advantages that hard shells and armouring would have given these life-forms in comparison with the soft-bodied Ediacaran metazoans of the Late Proterozoic, in terms of predation, protection and defence

explain the relationship between changes in oxygen concentrations and the development of the ozone layer

outline the chemical relationship between oxygen and ozone

recall the theory of evolution by natural selection.

LIFE DIVERSIFIES

Life made several important advances during the Late Archaean and Proterozoic, including:

- the evolution of eukaryotes from Archaean forebears, which required the development of cell nuclei and the symbiotic inclusion of species of bacteria inside eukaryotic cells
- the evolution of multicellular organisms
- the development of multicellular plants with discrete organs dedicated to performing specific tasks
- the development, late in the Proterozoic, of animals with discrete organs.

The development of multicellular, multi-organ life forms during the latter half of the Proterozoic paved the way for the dramatic appearance of visible body fossils in the rock record at the beginning of the Cambrian. To understand this event we will examine the oldest known animal fossils: the Ediacaran fauna.

THE FIRST ANIMALS: EDIACARAN METAZOANS

Metazoans are many-celled animals with separate and distinct organs. The first metazoan fossils appear in the geological record at the very end of the Proterozoic eon, in the Vendian period. This period lasted from 620 to 545 Ma ago.

In 1946 the Assistant Government Geologist of South Australia, Reg Sprigg, discovered fossils that he thought were Early Cambrian jellyfish and algae. They were found while he was investigating some old silver and lead mines in the Ediacara Hills in South Australia's Flinders Ranges. Investigations by South Australian palaeontologists in the late 1950s led to the collection of 1500 specimens of what palaeontologist Martin Glaessner called 'entirely new animals that had never been seen before'. Other geological investigations indicated that these unusual impressions were definitely Proterozoic in age. The reason is that the upper parts of the rock unit they were found in (called the Pound Quartzite) had been eroded away during the

definition

metazoans
multicellular animals with separate and distinct organs

Figure 2.4.1 *Mawsonites.*

Figure 2.4.2 *Tribrachidium.*

Figure 2.4.3 *Dickinsonia costata.*

Figure 2.4.4 *Spriggina.*

formation of an ancient land surface, which was then buried by sediments containing fossils of the very earliest Cambrian age.

The first discovery of these Vendian metazoans, which are known as Ediacaran organisms or the **Ediacaran fauna**, has been followed by their recognition in rocks of similar age all around the world. Australia's Ediacaran fossils are about 570 Ma old, while the oldest Ediacaran fossils are dated at about 590 Ma and appear in rocks that were deposited after the end of a worldwide glaciation (the Varangian Glaciation) that finished at this time.

You can see from the photos of them (Figures 2.4.1 to 2.4.4) that Ediacaran organisms are strange beasts indeed. They were soft-bodied **invertebrates** that were really quite large, varying in size from about 1 cm (*Tribrachidium*) to about 1 m (*Dickinsonia*). Many of these fossils don't look much like modern creatures, or other younger fossil creatures for that matter. However, they do show the sorts of shapes and symmetrical organisation of furrows and lobes that strongly suggest that they were formed by animals of some sort.

Ediacarans are usually preserved as moulds of the upper surfaces of the animals in impressions they left in the base of fine-grained sands that buried shallow marine or intertidal muds. Preservation of soft-bodied fossils by sand is very rare and indicates that the Ediacarans probably had a relatively tough or leathery skin. We also think they may have been preserved because the burrowing sediment feeders that would have eaten their remains had not evolved yet.

If Ediacarans were the ancestors of modern creatures then we must ask an important question: what can they possibly be? The conventional view of the thirty or so genera of Ediacarans discovered so far has divided them into four groups: jellyfish; soft corals; worms; and problematica, which is a wonderful name given to animals of unknown origins and relationships. If the conventional view is correct, then most Ediacaran fossils are of jellyfish, soft corals and worms. Figure 2.4.5 shows what their community might have looked like.

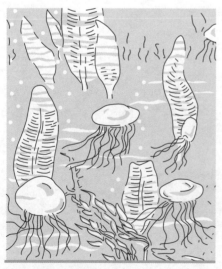

Figure 2.4.5 The Ediacaran community.

REVIEW
ACTIVITIES

EXTENSION
ACTIVITIES

1
How old is the oldest evidence for animals and what is this evidence?

2
Name and sketch two members of the Ediacaran fauna.

3
What evidence do we use to infer that the Edicaran fossils were actually animals of some sort?

4
When did most of the Ediacarans die out?

5
Research the evidence that indicates there were non-Ediacaran animals alive at the same time as the Ediacarans. Outline what we think most of these non-Ediacaran animals were and what they looked like.

THE CAMBRIAN EVENT

The boundary between the end of the Proterozoic and the beginning of the Phanerozoic is, for many reasons, the most important of all the geological boundaries that we recognise. The change that occurred at this boundary, which is more properly called the Cryptozoic–Phanerozoic boundary, is the apparently instantaneous appearance of abundant, large and small, shelly animal fossils that we know to be the ancestors of present-day life forms. These fossils include one of a small, swimming, chordate worm called *Pikaia graciens* (see Figure 2.4.6), whose relatives or descendants eventually evolved into **tetrapod vertebrates**, then into mammals and then, finally, into us.

The seemingly sudden appearance of visible shelly fossils in sedimentary rocks of exactly the same age from sites located all around the world has puzzled geologists and biologists since the middle of the nineteenth century. The vast abundance, variety and size of creatures that are found in Cambrian rocks provide a striking contrast to the apparent absence of skeletonised life in the rocks immediately underneath them. If you examine a cliff face containing a set of rocks that includes the Pre-Cambrian–Cambrian boundary it seems as if life has almost exploded into being. This characteristic gave rise to a popular name for this event: the **Cambrian explosion**.

Many nineteenth-century creationists used the sudden appearance of life in the rock record to directly challenge Darwin's theories about the origin of species. They argued quite effectively that the suddenness of the appearance of life was direct proof for the accuracy of the biblical description of Creation; for the Bible does describe the beginning of life as an instantaneous event. Briefly summarised, Darwin's **theory of evolution by natural selection** states that individuals from particular animal and plant species that are well adapted to the existing environmental conditions tend to do better. As a result, they will come to dominate the population at the expense of the individuals that are not as well adapted. This happens because the well-adapted individuals are more likely to produce offspring than those individuals that are less well adapted. Eventually the population changes so that the characteristic that gave the environmental advantage to the better-adapted individuals is present in all the surviving individuals. Those without that characteristic may even die out.

Darwin recognised that his theory of the descent of species by natural selection required that life had diversified over very long periods of time from common ancestors and that all life had probably come from a single, common ancestor. Darwin thought, quite correctly as it turns out, that a great deal of evolution and natural selection must have occurred to create the oldest hard-shelled and large fossil

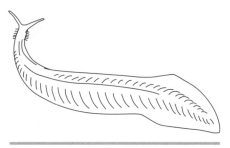

Figure 2.4.6 *Pikaia graciens*, probably our most important ancestor.

life forms he was familiar with. This was one of the reasons why geologists argued so hard that the figure of 20 Ma that Lord Kelvin determined for the age of the Earth was wrong. They could not conceive that this was anywhere near enough time to create the diversity apparent at the beginning of the fossil record or to allow the evolution of present-day life.

1

Outline how the process of 'evolution by natural selection' works.

2

What is meant by the term 'Cambrian explosion' and why was it so important for Darwin and other palaeontologists to explain it?

3

Discuss the following statement: 'Given the fact that large metazoans exist today but did not exist a billion years ago, the Cambrian explosion was an inevitable consequence of evolution'.

Figure 2.4.7 The Cambrian begins, *Trichophycus pedum*.

brachiopod
a diverse group of shellfish that were extremely abundant early in the Phanerozoic. Only a few species are still living today.

extinct
the event of a species dying out

definitions

THE PROTEROZOIC–CAMBRIAN BOUNDARY

The set of body fossils that appeared in the rocks laid down at progressively higher levels above the Proterozoic–Cambrian boundary follows a pattern that is repeated quite consistently in all the sequences of these rocks found around the world. The specific fossil that geologists use to mark the beginning of the Cambrian is actually a trace fossil with a characteristic shape, called *Trichophycus pedum*. (See Figure 2.4.7.) It is a distinctive horizontal burrow that is quite different from the earlier tracks of Late Proterozoic, soft-bodied metazoans and this trace fossil was made by some sort of small worm. A few metres above this level, in the lowermost Cambrian rocks, we find some tiny cones of calcite a few millimetres long. These cones are called sclerites and they acted like chain-mail armour to protect the worm-like animal that secreted them. Sclerites become more common, a little larger and more varied in shape in the rocks a short distance above the Cambrian boundary. They are known collectively as the 'small shelly fauna'. (See Figure 2.4.8.)

The rocks that occur higher and higher above the Cambrian boundary not only contain a greater range of different fossils but increasing numbers of each individual type of fossil. There is a similar increase in the complexity and number of trace fossils at higher and higher levels above the boundary. Fifteen million years after the Cambrian began, the Earth's seas were inhabited by about eighty different types of animals that had hard parts. These include spiral shells made by sea snails, the exoskeletons of primitive sea urchins, trilobites and **brachiopod** shells. Figure 2.4.9 shows the variety of sea animals from this time, based on the Burgess Shale fauna of Western Canada.

Amazingly, ancestral forms of almost all the major groups of animals that are present on the Earth today had evolved only 30 Ma after the beginning of the Cambrian. In addition to these creatures, we have found fossils of other groups of animals that no longer exist because they became **extinct**, like the dinosaurs did at the end of the Cretaceous.

The sudden appearance of complex life indicated by the fossil record therefore presented an enormous problem for Darwin's theory of evolution by natural selection. Darwin was sure that the first fossils that he was familiar with had descended from a single creature. Yet the oldest fossils Darwin was familiar with were the many and varied Cambrian trilobites of Wales. Trilobites were arthropods, a group of segmented animals that includes spiders, insects, crabs and centipedes. Darwin was sure that trilobites had arisen from 'a single crustacean ancestor' that must have lived long before them. He felt that the process of natural selection had led to a variety of divergent changes in the original prototype trilobite and that many successive modifications had produced the great variety of the known trilobites. But there are no positively identified body fossils of the original ancestral trilobite. Neither are there body fossils of the series of animals that indicate the evolutionary changes between the original ancestral form and the first trilobites that Darwin was familiar with.

The absence of this record is partly due to the incompleteness of the rock record (remember that fewer than one in a million animals ends up being fossilised), but mostly to do with the fact that these ancestral forms did not have preservable hard parts. About 60% of modern marine animal species have no hard parts and they just do not get fossilised. Their remains are eaten by sea-floor scavengers and other burrowing scavengers that mine the sea-floor sediments for food. Study of the extraordinary Burgess Shale fossil deposit in Canada indicates that about 85% of Cambrian animals were soft-bodied. Except for this deposit, there is very little evidence that these soft-bodied animals ever existed. The Burgess Shale is one of the world's few unusually good deposits of perfectly preserved fossils that give us an insight into ancient life. Burrowing scavengers got better at burrowing by the end of the Cambrian and, as a consequence, it seems that only a very few fossils of soft-bodied animals formed after this time.

To deal with this absence of original, ancestral animals and the missing links between them and their descendants, Darwin suggested that there was simply no geological record of the ancestor or of the long period of time during which its evolution had taken place. He thought that either there had been no deposition during this time or that any deposited rocks had been eroded away. We know now that this is not the case and that many sequences of sedimentary rocks were deposited continuously throughout this period of Earth history. The interesting feature about these sequences is the gradual but consistent expansion in the variety and abundance of trace fossils through the Late Proterozoic and into the Early Cambrian.

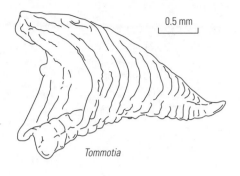

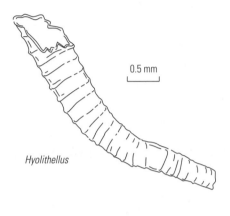

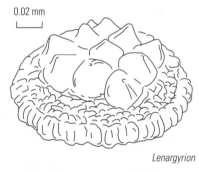

Figure 2.4.8 Examples of the small shelly fauna.

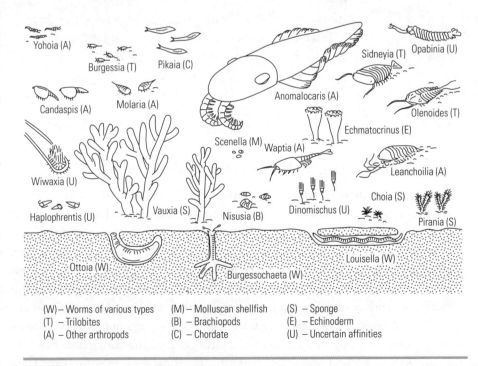

Figure 2.4.9 Some Cambrian animals of the Burgess Shale.

Palaeontologists, biologists and geologists are gradually finding more and more evidence that indicates that the Cambrian explosion was a very real and special event. The increasing complexity and abundance of trace fossils is interpreted by many palaeontologists to mean that there was a sudden and large-scale diversification of soft-bodied animals in the Late Proterozoic and Early Cambrian. This led to an increase in the number of different species of animals present at that time, as well as the appearance of hard parts in some of them.

So while we cannot point to a body fossil and convincingly state that it is the ancestor of all metazoan animal life, we can point to many Pre-Cambrian trace fossils that are likely candidates for this honour. Biologists have recently shown that essentially the same genes control the growth of all animal embryos. It may come as a surprise to you, but the embryonic development of the humble house mouse and fruit fly (as well as the genetic coding that drives the growth of these embryos) shows many fundamental similarities despite the great differences in appearance between the adult forms. The embryology of these two apparently different animals is actually so similar that the famous palaeontologist Simon Conway Morris has suggested that 'in a certain sense their differences can only be superficial'. This embryonic similarity provides a clue to identifying Darwin's ultimate ancestor animal and goes a long way towards explaining the puzzling explosion in the variety of large life forms that became established in the astoundingly brief 40 Ma in which the Cambrian took place. If the embryology of insects and mammals is so similar, then the diversity we perceive is better thought of as minor modifications of a basic design.

Many palaeontologists now think that the ancient ancestors Darwin was hoping to find are probably the tracks and trails of a few worm-like Proterozoic creatures. We will investigate the possible reasons why these soft-bodied creatures diversified so explosively at the end of the Proterozoic and in the Early Cambrian a little later.

REVIEW ACTIVITIES

1

What fossil marks the base of the Cambrian?

2

What is the small shelly fauna and why is it so important?

3

Why is the presence of the Burgess Shale important and what does it indicate about the incompleteness of the entire fossil record?

EXTENSION ACTIVITY

4

Investigate and research the Burgess Shale fauna using the Internet and resources in your school or local library. Outline the features that make the Burgess Shale fauna so very special.

THE ADVANTAGES OF HAVING HARD PARTS

Two interesting characteristics of Ediacarans is that their fossils never contain any bite marks and that none seem to have left behind any trace fossils. There are no tracks and trails of the right size or shape that could have been formed by any of the larger Ediacarans. These facts indicate that is quite likely that the Ediacarans did not move about and were not eaten by predatory animals. Perhaps aggressive predation as we know it had not yet evolved when these creatures lived. In contrast, we know that there were both small and large predators present in the Early Cambrian seas and oceans. It is obvious that a hard exoskeleton gives some protection from being eaten by predators. Not only are there fossils of a variety of predators dating from this time, but we have also found healed bite marks that predators made in arthropods. These bite marks provide direct evidence of predatory hunting. (See Figure 2.4.10.)

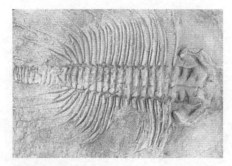

Figure 2.4.10 A healed bite mark in a Cambrian trilobite.

Skeletal hard parts provide animals with other advantages. For example, skeletons provide a framework, which supports and spreads out the feeding organs of suspension feeders and filter feeders, such as brachiopods, corals and bivalves. In general, skeletons allow animals to grow larger than they would otherwise be, and large size is also a form of protection from predators. Similarly, skeletons provide firm points for muscle attachment, which may improve animals' ability to move, as well as a framework that makes it easier for animals to organise and support their internal organs. Exoskeletons also provide protection from damaging ultraviolet light and can help seal off animals from their environment. These are advantages that help animals living in the intertidal zone to survive and not dry out during low tides.

REVIEW ACTIVITIES

EXTENSION ACTIVITY

1
What is the significance of healed bites marks that can be found in Cambrian trilobites?

2
What advantages did hard parts give to those animals that evolved them?

3
Research a Cambrian animal, such as the brachiopod *Lingula*, that is still alive today. Suggest why it has endured for so long.

WHY DID THE CAMBRIAN EXPLOSION OCCUR?

Most palaeontologists agree that the Cambrian explosion occurred, and occurred when it did, because of oxygen supply and the invention of predation. Some palaeontologists would add to these reasons: a chance genetic mutation that increased the rate of evolution; and the evolution of gills or similar organs that are specialised for the task of extracting oxygen from water.

There was an increase in the world's oxygen supply

All animals require oxygen, and the larger the animal is the better it has to be at extracting oxygen from its environment. Large animals that move require lots of oxygen. The earliest, simpler metazoans did not have lungs or gills. Instead they acquired their oxygen by diffusion, which involves the oxygen present in the water around them passing through their outer membranes and directly into their body tissues. Jellyfish, corals and sponges still use diffusion to acquire oxygen and, while some of them grow quite large, they are not very active or complex in comparison with the metazoans that do have gills or lungs. The internal organisation of these metazoans (two layers: gut and skin) is simpler than ours (three layers: gut, muscles and organs, and skin). Animals that use diffusion are unable to survive in environments where the oxygen content of the water is low because they suffocate. The development of large metazoans means that the oxygen content of the oceans must have reached a level that was high enough to sustain them. Some palaeontologists have suggested that the first time these high oxygen concentration levels were established on the Earth was just before metazoans arose.

Some animals invented ways to eat other animals

We call the behaviour of animals that live by killing and eating other animals predation and we think that the invention of predation may be one of the factors that drove the Cambrian explosion. Palaeontologists recognised early in the twentieth century that hunting and predation was quite probably a major cause of the skeletonisation of animals during the Cambrian. Darwinian natural selection would have been the driving force responsible for this adaptation. Animals that were able to catch and consume the scavengers and filter feeders were able to exploit a much more nutritious food supply. This is, of course, the great advantage that predators have over browsers, filter feeders and scavengers.

After the first predators appeared and began eating their way through the available prey species, the prey evolved by the process of natural selection and invented a number of different ways to survive. These included better ways of getting away (escape), better ways of hiding (burrowing), and protecting themselves with body armour. There must have been other ways of not being eaten (such as tasting bad, camouflage and secreting poisons) but we cannot detect this from the fossil record. The animals that were better adapted (that is, naturally favoured) along these lines were more likely to survive the efforts of the predator to eat them and they could therefore pass these advantages on to their offspring.

Two of these protective strategies form the evidence that is used to identify the Cambrian explosion. These are more complicated burrows and, secondly, armoured hard parts, such as shells and other exoskeletons. Hard parts are more difficult to bite through than soft tissue, and the development of an exoskeleton or enclosing shell is an obvious response to being eaten. Predators quickly responded to armour with better and stronger jaws. More and more complex anti-predator defence methods evolved in response to more complex and better-adapted predators in an arms race of ever-increasing intensity. Larger predators that hunted the small predators evolved and, with time, the animal world became increasingly complex.

Palaeontologists regard predation as the most plausible explanation for the Cambrian explosion. Whatever the cause of the Cambrian explosion it could not have happened without enough oxygen, because without the oxygen it would not have been possible for animals to develop at all.

REVIEW ACTIVITIES

1
Name and briefly describe the four possible events that caused the Cambrian explosion to occur.

2
Outline the two most likely explanations for the Cambrian explosion.

3
What does the word 'predation' mean and how do predators survive?

4
Explain why an increase in oxygen content in the seas and oceans was a necessary requirement for the evolution of animals.

EXTENSION ACTIVITIES

5
Outline other strategies that animals without hard parts may have used to survive in the Cambrian seas.

6
Research a group of fossil animals with hard parts that appeared in the Cambrian explosion and has modern descendants. Find out how the Cambrian species lived and how the group evolved to produce the present-day species. Present the results of your research in an illustrated essay. It should include a description of the group and their life habits and outline the evolution of the modern representatives.

THE EARTH'S OXYGEN LEVELS THROUGH TIME

It is quite difficult to establish what the levels of oxygen were in ancient times. Until recently, any attempt to do this had been based on just a few points in time when oxygen levels in the oceans and atmosphere can be determined from specific events in the geological record. These events are the appearance of red beds, the appearance of living things in the oceans and the current oxygen levels, which we can measure directly.

The disappearance of pyrite and the appearance of widespread terrestrial red beds in the Middle Proterozoic about 2 billion years ago means that the oxygen in the air and oceans had reached at least 1% of the modern-day levels. These levels were reached after enough cyanobacterial oxygen production had occurred to oxidise the enormous quantities of reduced iron that had been dissolved in the Earth's oceans.

The next determinable point for such estimates is from the Cambrian explosion itself. Some modern invertebrates can survive in water that contains about 10% of the present-day oxygen level. Most invertebrates require more oxygen than this, but the figure of 10% of the current levels gives a minimum estimate for oxygen levels 550 Ma ago, which is when these creatures first appeared. We can push the timing of this minimum estimate back to about 600 Ma ago if we assume that their Proterozoic forebears developed in the same conditions. Early estimates of the Earth's changing oxygen content assumed a starting level of no oxygen and used the two points at 2 billion and 600 Ma ago as well as modern levels to construct a simple, continuously increasing curve to provide an estimate of past oxygen levels. This curve is given as the traditional oxygen estimate curve on the graph in Figure 2.4.11.

More recently, three different studies have greatly improved upon the traditional estimate of the Earth's oxygen contents as a simple, gradually increasing rise and have also shown that the atmosphere was much more oxygen-rich than was traditionally thought. The first study, by Dick Holland of Harvard University, has shown that there is a marked change in the composition of the soils that developed on iron-rich rocks about 2 billion years ago. Soils that developed from iron-rich rocks before this time lost much of the rock's original iron content, while soils that developed on the same rock types after 2 billion years ago retained the original iron. The smallest amount of atmospheric oxygen that can cause iron to be retained in the soil is about 10% of present-day levels. The simplest explanation for the presence of **detrital pyrite** until about 2 billion years ago, its disappearance from then on, and the subsequent development of soils rich in iron oxide is that atmospheric oxygen increased. It increased from less than 1% of the present-day level to more than 10% of the present-day level over a period of 100 Ma in the Early Proterozoic.

A second study, by geochemist Don Canefield, uses sulfur and oxygen isotope data to estimate the biological productivity of oxygen production and the quantity of organic carbon burial during the Proterozoic. Two consequences of the burial of organic carbon in limestones and as petroleum or hydrocarbons in black shales are:
• the removal of carbon dioxide from the oceans and atmosphere
• an increase in oxygen levels in the oceans and atmosphere.

detrital pyrite
loose, eroded fragments of pyrite transported by wind or water

definition

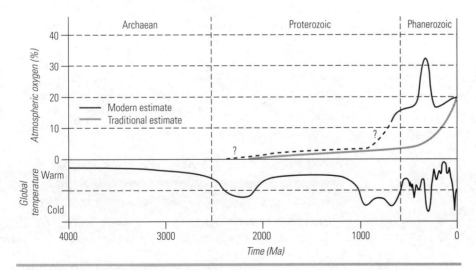

Figure 2.4.11 Oxygen content and global temperature of the Earth's atmosphere through time.

This geochemical study shows that a great deal of organic carbon was buried during the Late Proterozoic and, as a consequence, there should have been a corresponding sharp rise in the oxygen contents of the sea and air.

The third study uses mathematical models, again checked against isotope data, to determine the concentration of oxygen in the atmosphere during the Phanerozoic. These models were developed by the geochemist Bob Berner of Yale University. They show that oxygen contents were probably similar to today's levels at the beginning of the Cambrian. They also show that oxygen levels actually peaked at about 1.5 times present-day levels during the Carboniferous as large quantities of land plants were buried in the first large coal deposits. The more complex picture of atmospheric oxygen levels during the past 4 billion years resulting from these studies is shown as the modern estimate curve in Figure 2.4.11.

The implications of oxygenation of the atmosphere

There are several other important aspects of the oxygenation of the atmosphere during the Proterozoic. First, the past increases in atmospheric oxygen content should have led to increased ozone production and, therefore, decreased amounts of the Sun's ultraviolet radiation reaching the surface of our planet.

Secondly, because much of the atmospheric oxygen was made from converting carbon dioxide to carbohydrates by photosynthesis, there was a consequent and significant decrease in the amount of carbon dioxide present in the atmosphere. Twenty per cent of the Earth's original atmosphere was probably carbon dioxide. Today carbon dioxide makes up 0.035% of the atmosphere's gases.

Thirdly, as you are probably already aware, carbon dioxide traps heat in the atmosphere. It follows that the higher level of carbon dioxide present in the atmosphere during the Archaean acted to keep the planet warmer than it is today, and that the jumps in atmospheric oxygen were probably accompanied by drops in the carbon dioxide level, which should have caused global cooling events. Indeed, the first large-scale ice ages accompanied or followed the atmospheric oxygenation events of the Early Proterozoic and the Late Proterozoic. Figure 2.4.11 also shows a generalised estimate of global temperature based mainly on the record of widespread glaciation. (Compare and relate it to the oxygenation curve.) Harvard University's Paul Hoffman and his coworkers have even suggested that the Late Proterozoic glaciation was so intense that permanent sea ice was present all the way to the equator and created a 'snowball Earth'.

REVIEW ACTIVITIES

EXTENSION ACTIVITIES

1
Outline the significance of the 2-billion-year-old iron-rich soils that are found around the world.

2
Outline the changes in the amount of oxygen present in the Earth's atmosphere over the last 4 billion years.

3
Research the 'snowball Earth' concept and explain how increases in atmospheric oxygen content may be related to these worldwide glacial periods.

4
Write a general outline of the changes in surface temperature conditions on planet Earth between the earliest Archean and the present day. Then discuss the role that the Earth's biota has played in bringing about those changes.

SUMMARY

The oldest known metazoan or animal fossils are Vendian in age, that is, between 620 and 550 Ma old.

The first Vendian fossils that were ever recognised were found in the Ediacara Hills and are called the Ediacaran fauna. These fossils are about 570 Ma old.

It is unclear what the Ediacarans actually were, but jellyfish, soft corals and worms are likely candidates.

The Cambrian explosion is the name given to the seemingly instantaneous appearance of a diverse group of fossil animals in the fossil record.

Hard parts, that is, the armouring of animals, probably evolved in response to predation.

Oxygen contents in the oceans and atmosphere have varied over geological time but have increased, in a general sense, since the Early Proterozoic as a result of the consumption of atmospheric carbon dioxide by photosynthesising cyanobacteria and plants.

Abundant oxygen was a necessary condition for the development of animals.

PRACTICAL EXERCISE
The evolution of animals

This exercise will improve your understanding of the Cambrian explosion and the appearance of fossils.

Procedure

1
Use the material in this chapter along with resources on the Internet or in your library to research how Ediacaran fossil animals and Cambrian fossils are thought to have lived.

2
Locate photos and diagrams of Ediacaran fossil animals and Cambrian fossil animals that show their internal and external anatomy.

3
Compare the lifestyle, appearance and organisation of the Ediacaran fossil animals and Cambrian fossil animals.

4
Organise your findings into a table under headings that indicate the presence or absence of features (such as hard parts, a mouth, legs or the ability to move) and complex organs (such as eyes and gills). Also include:
- the name of the animal
- the time in which it lived
- a rough indication of its relative size, that is, small (less than 10 cm long), moderate (10–50 cm long) or large (more than 50 cm long).

ACTIVITIES

1
Do any of the animals of Ediacaran age have hard parts?

2
Are mouths or mouth-like structures present in animals of Ediacaran age?

3
How often are bite marks recorded in Ediacaran fossils?

4
Describe an animal that you think is typical of Ediacaran times.

5
List the different sorts of hard parts found in Cambrian animals.

6
What body part or organ makes bite marks and when did it probably evolve?

7
Discuss the role of evolution by natural selection in the 'invention' of the exoskeleton.

8
How would eyes improve the chances of survival of:
a predators
b prey animals?

9
Outline the advantages that Cambrian animals apparently had over Ediacaran animals.

10
Debate the following topics in class:
a 'Which came first: teeth or armour?'
b 'Better eyes make better hunters.'
c 'Hunger drove the evolution of animals.'

OUTCOMES

At the end of this chapter you should be able to:

- recall evidence that present-day organisms have developed from different organisms in the distant past

- summarise environmental pressures faced when living things evolved for terrestrial environments

- describe the role of ozone in filtering ultraviolet radiation and the importance of this for life that developed during the Phanerozoic eon

- outline the major steps in the expansion to the terrestrial environments by land plants, the first land insects, lungfish and amphibians

- identify advantages enjoyed by the first land-dwellers.

aquatic
water-based

terrestrial
land-based

definitions

LIFE INVADES THE LAND

As far as it is possible to tell, it seems that life was restricted to **aquatic** settings for all of the Cryptozoic and the first 100 Ma or so of the Early Phanerozoic. The first **terrestrial** fossils are spores of land plants, probably liverworts, that are found in Ordovician rocks. The first upright plants appeared in the Late Silurian. A moderately diverse land flora of plants with trunks around 10 cm wide that stood around 1 m high were present on land by the Early Devonian. Land-living invertebrates, such as millipedes and primitive scorpions and spiders, first appear in Late Silurian rocks. These evolved from similar marine animals that must have lived in near shore environments and were encouraged onto the land to take advantage of the food supply provided by the early land plants. The earliest known land-living vertebrates appeared in the Late Devonian after they had evolved from members of the lobe-fin fish group.

Advantages and challenges for land-based life

Plants and animals that made the move from the seas to the land all evolved from aquatic forebears that probably lived in shallow water or tidal environments. There were a number of advantages for those life forms that could move onto the land. Oxygen is much more abundant in air than it is in water. Also, much more light is available to drive photosynthesis on the surface of the land than under water because water is an effective filter of light. The greater availability of light to land plants in comparison with submerged plants is especially evident during the early morning and the late afternoon. So, the food-production advantage available to plants that can survive above the surface of the water in comparison to those that had to remain submerged is obvious. When land plants evolved to exploit this food-production advantage they created a land-based food supply for animals to eat. It seems likely that the ancestors of those animals that made the shift onto the land were probably aquatic species that had begun to make brief excursions onto the land surface in search of safety from predators or to forage.

All those species that made the shift from the water to the land faced and overcame a number of challenges in order to survive. The most difficult of these challenges was direct exposure to air. This generally leads to dehydration and tissue degradation as the cells dry out and the fat-like lipid molecules of cell membranes react with the atmosphere's much more abundant oxygen and break down. The outermost layer of our skin consists of dead cells that form a protective, enclosing barrier that prevents the air's oxygen from coming into contact with our active and alive inner cells. If our skin did not do this the air's oxygen would attack our living cells and we would decompose in a sort of slow burn.

To survive on the land, plants and animals also had to develop different mechanisms to move oxygen and carbon dioxide back and forth between themselves and their environment. This is because atmospheric oxygen and carbon dioxide are gases, whereas oxygen and carbon dioxide occur in water as dissolved ions.

The other major difficulty land plants and animals had to overcome was supporting themselves against gravitational force in their new out-of-water environment. Air is of much lower density than water and does not provide the support that an enclosing body of water does. Consequently, land plants and animals had to develop ways of holding themselves up.

The importance of ozone in the atmosphere

Intense **ultraviolet radiation** was probably not one of the challenges that the first pioneers onto the land had to face. By the time the Ordovician began, the atmosphere had probably contained enough oxygen to produce **ozone** for hundreds of millions of years. Photochemical reactions in the upper atmosphere form ozone from oxygen. The small amount of ozone formed in this process is enough to filter out most of the dangerous ultraviolet radiation from light that reaches the Earth. Ultraviolet light is commonly used as a sterilising agent and the quantity of ultraviolet light in unfiltered sunlight is powerful enough to attack and kill cells. At lower, non-lethal levels, ultraviolet light can degrade DNA in cells. If ozone disappeared from our atmosphere today, humans would suffer horrible sunburn that would probably prove fatal because our cells would simply break down. The rate of mutation and cancer in other eukaryotic animals and plants would be too great for many of them to survive in the open.

ultraviolet radiation
non-visible electromagnetic radiation with a shorter wavelength than visible violet light

ozone
a form of oxygen that forms a molecule consisting of three oxygen atoms

definitions

REVIEW
ACTIVITIES

EXTENSION
ACTIVITIES

1
Name the sorts of organisms that were the first to inhabit the land.

2
List the advantages that land-based plants had in comparison to water-based plants.

3
Name and describe three challenges that organisms had to overcome to move from aquatic environments to terrestrial environments.

4
What protection does atmospheric ozone provide for terrestrial organisms?

5
Research and outline some of the strategies that terrestrial plants and animals used to prevent dehydration.

6
Research the effects that ultraviolet light has on unprotected cells.

THE FIRST LAND PLANTS

Two major groups of plants live on the land. The first group consists of the small non-vascular plants, such as moss and fungi, which generally hug the ground in moist environments. The second group consists of the much more abundant and larger vascular plants, which have developed a specialised system of tubes to transport water and nutrients around the plant.

Although there are no fossils of the actual plants, fossil spores indicate that simple, ground-covering, non-vascular plants probably colonised the land surface during the Late Ordovician and Early Silurian. The first plants that made the shift from water onto the land may have migrated through the intertidal zones of deltas, estuaries and salt marshes into positions increasingly closer to the high-tide mark, and then beyond the high-tide line onto the nearby shore. To complete this migration they would have needed to gradually evolve resistance to drying out. However, it seems more likely that the seasonal lowering of freshwater lakes encouraged a species of multicellular green algae to adapt to life out of water. In support of this argument, we know that green algae were the only group of marine plants to adapt to freshwater environments and that all land plants have evolved from green algal ancestors.

The first land plants may have survived by forming marsh-like colonies in swamps or in areas of low-lying wet and boggy ground in regions of high rainfall. It is almost certain that constantly wet conditions would have been necessary for the survival of the first land plants because these conditions are the closest imitation of complete immersion in water that land environments can provide. Tropical, swampy lowlands are the more likely setting for the invasion of the land by plants because all the early groups of land plants lived in just such an environment. We know this because their fossils are found preserved in sediments that were deposited in the floodplains and bogs of tropical lowland rivers.

The First vascular plants

The earliest known vascular plants are called Rhyniophytes and they first appear in Late Silurian rocks. They were quite small plants. An example is *Cooksonia* (see Figure 2.5.1), which was little more than a simple branched stick. *Cooksonia* fossils are found in many parts of the world, including Wales and Eastern Australia, which were both located near the equator at that point in time. *Cooksonia* did not have a true root system or leaves and was only a few centimetres high. It was anchored to moist earth by a short underground extension of its stem, but used the innovation of a hollow tube to transfer water from the ground to its stem and branches. The tube is called xylem and is one of the major adaptations that characterises the terrestrial vascular plants.

The other adaptations that *Cooksonia* used to solve the problems of drying out and gas exchange were:
• a cuticle—a waxy, waterproof outer surface layer
• stomates—pores in the surface of plants that open and close, allowing gases to move in and out of the plant tissue.

Water evaporating from the top of the plant helped to suck water up through the xylem from the stubby root into the branching stems. The plant's reproductive nodes are called sporangia and were located at the tips of the branches, from where they could release spores into the wind. The spores would reproduce the species when they fell onto damp earth. Later plants improved upon the early form of xylem by strengthening it with lignin, which provided better support and allowed plants to grow taller and larger.

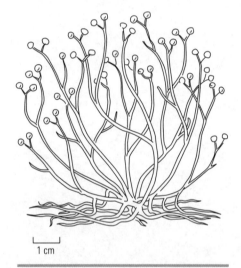

1 cm

Figure 2.5.1 *Cooksonia.*

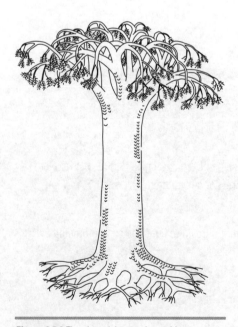

Figure 2.5.2 The giant club moss *Leptophloem.*

The diversity of vascular plants

Several groups of more complex and larger vascular plants evolved from plants like *Cooksonia* during the Early and Middle Devonian, and these all improved upon the adaptations the Rhyniophytes had used to move onto the land. They developed leaves, more substantial and complex roots, as well as stronger xylem. These adaptations allowed plants to gradually get larger and larger as well as survive in locations that were not as moist. Most of the early plants were seedless and reproduced like ferns do, that is, by releasing male (sperm) and female (egg) spores, which need to fall upon moist ground in order to survive and produce new plants. The major groups of vascular plants to evolve during the Devonian were the sphenopsids (horsetails), the lycopods (club mosses, see Figure 2.5.2) and the filicopsids (ferns). The filicopsids developed wood to strengthen the xylem, making them the first true trees.

Also evolving at about this time were the progymnosperms (seed ferns), which were the first seed-bearing plants. The innovation of seeds made it possible for plants to reproduce without needing male sperm to fertilise a female spore in water. This innovation helped plants to spread out from moist lowland areas onto drier parts of the land. During the Permian, gymnosperms (true seed plants, see Figure 2.5.3) became abundant after evolving from progymnosperms. The gymnosperms include conifers and cycads.

Angiosperms appeared early in the Cretaceous. These are true flowering plants, and they attract insects and birds with their flowers and use them as carriers who spread the pollen produced in the flowers to other plants. This system of dispersing pollen (male reproductive cells) is an adaptation that improves upon the wind dispersal of pollen used by non-flowering gymnosperms.

Figure 2.5.3 The early gymnosperm tree, *Glossopteris*, which lived in the Permian.

REVIEW
ACTIVITIES

EXTENSION
ACTIVITY

1
When do the first land plants appear in the geological record?

2
Explain the reasons why palaeontologists and palaeobotanists think that the first terrestrial plants occupied low-lying moist ground or swamps.

3
Name the first group of terrestrial vascular plants.

4
Outline the major structural difference between vascular plants and non-vascular plants.

5
List two groups of plants that evolved from plants like *Cooksonia*.

6
Research the evolution of angiosperms or grasses. Outline the differences between this group of plants and the plants they evolved from, and indicate what advantages these differences gave them in comparison with their ancestors.

THE FIRST LAND ANIMALS

The first animals to follow plants onto the land did so relatively quickly after the first land plants appeared. Arthropods (jointed-leg animals) made the switch to land-based existence relatively easily. This group of animals is the most diverse and abundant group of animals; its members make up about 80% of animal species.

Figure 2.5.4 *Alkenia*, an Early Devonian spider.

Some of its representatives are the scorpions, spiders, trilobites, myriapods (millipedes and centipedes), insects and crustaceans. The first land arthropods are found in Late Silurian rocks along with fossils of early land plants. (See Figure 2.5.4.) These animals were deposit feeders that fed upon the dead plant litter and were mostly small mites and springtails. But by the Early Devonian at least one large vegetarian myriapod, called *Eoarthropleura*, had evolved from aquatic relatives and taken to land-based life.

Of course, these deposit feeders and grazers provided a food supply for intrepid predators, which could follow them onto the land. The fossils of the first land-based vegetarians are accompanied by the fossils of the primitive spiders and scorpions that ate them and were also evolved from close aquatic relatives.

The first land animals fed upon the first land plants, and so it will come as no surprise that the first land animals lived in the places where the first land plants lived. As we have just seen, these places were tropical lowland floodplains that probably contained quite a lot of moist earth and boggy ground and were not as different from the aquatic realm as many of the dry-land places we live in. They were swampy lowlands like the Everglades of Florida, but with low trees only about 1 m high.

Aquatic arthropods are enclosed in exoskeletons made from material that is already reasonably watertight. This material is called chitin and it is also quite resistant to attack from atmospheric oxygen. Modern-day terrestrial arthropods also coat their chitinous coats with a thin layer of wax to further enhance water retention. Arthropods were also already well adapted for living on the land in that their exoskeletons provide support for their internal organs. Arthropods already had limbs, which they had used to swim and crawl on the sea floor for over 100 Ma. These appendages were easily adapted into legs, which could be used to move around on the land.

So, as you can see, arthropods were already reasonably well adapted to terrestrial environments before they tried to move on to the land. In fact, some palaeontologists have suggested that arthropods were virtually pre-adapted for life on land. The arthropods' body design and skeletal characteristics gave them something of an advantage over other groups of animals when it came to exploiting the land as a new place to live. This is probably why they were the first group of animals to move onto the land. They had already pretty much solved the drying-out challenge and the internal-support challenge that all terrestrial life forms have to solve in order to make the move from the sea to the land.

The only major change that the first arthropods had to make to the design of their bodies in order to survive on land was to evolve a way of breathing out of water. They apparently made this change fairly easily as well because we know that the arthropods appeared on land almost as soon as there was land plant detritus to eat.

The first land-based spiders and scorpions found that developing ways of breathing air instead of water only required a relatively minor modification to their existing anatomy. These animals converted their gills into lungs by sealing together covering plates that had been previously used to protect the gills. A small hole was left in this 'new' structure, which allowed air to enter and pass over thin sheets of oxygen-exchanging tissue. Later on, some spiders and scorpions developed a different way of breathing air by using special tubes that pierce the exoskeleton all over their bodies. These tubes are called tracheae and lead directly to the internal tissues, where oxygen and carbon dioxide are exchanged directly with the cells by diffusion. Myriapods and insects also use tracheae to breathe in air.

The **insects** appeared later in the Devonian after evolving from a myriapod or a myriapod-like arthropod. The first insects were flightless, but members of this group developed flight by the Carboniferous. During this period some enormous dragonflies appeared that had wingspans up to 1 m across. These giants evolved during a period of time when the amount of oxygen in the atmosphere was about 1.5 times more abundant than it is today.

The first terrestrial vertebrates

The first vertebrates to leave the water and take up permanent residence on the land were labyrinthodont amphibians, which had evolved from members of the group of lobe-finned fish, such as the Australian lungfish. The group of lobe-fin fish that gave rise to the amphibians is called the rhipidistians. Study of their anatomy indicates that these fish hunted other fish in shallow water along coasts, lakes and rivers.

Rhipidistians have long been extinct, but some idea of their habits can be gained from their closest living relative: the coelacanth. Interestingly, the coelacanth was also thought to have died out and become extinct in the Cretaceous until a living one was pulled up in a fishing net off Madagascar in the 1930s. Studies of coelacanth behaviour have shown that the fins located on the underside of their body move in the same pattern that modern amphibians use for walking. It is thought that some rhipidistian species may have chased their prey into very shallow water, as some species of sharks and dolphins do today. The rhipidistians may have even forced their prey out of the water, where the poor fish would have wriggled across the shore in an attempt to get away. Once a rhipidistian trapped its prey in the shallows or had it flopping about on the shore, the rhipidistian is thought to have used stout underside fins to walk over to the beached fish to feed on it.

We know from their fossils that a group of rhipidistians had nostrils as well as gills, which means that they breathed air into lungs in a similar way to their relatives, the modern lungfish. Lungs and nostrils probably evolved as a response to loss of oxygen from lake or river water when it warms up in the summer. Some modern fish that find themselves in this situation take bites of air, which they hold in their mouths so that the oxygen can diffuse directly into their blood through the tissue lining the mouth. A group of rhipidistians greatly improved upon this strategy by developing a special sac at the back of the mouth to hold the bite of air and extract its oxygen. This sac eventually evolved into lungs.

You can see from Figure 2.5.5 that the anatomy of these fish is surprisingly similar to the early amphibians. When you compare their skeletons it is not difficult to see how the rhipidistians evolved into the first amphibians. As with the other forms of life that moved from the water to the land, the earliest amphibians solved the major problems of living on the land. Support was provided by strengthening their skeletons, lobe-fins evolved into feet to enable movement and scales modified to become more watertight. Amphibians did not completely leave the water though. You probably already know two important facts about this group of vertebrates. First, they generally spend quite a lot of time in the water to maintain their body moisture. Secondly, they reproduce in the water in much the same way as

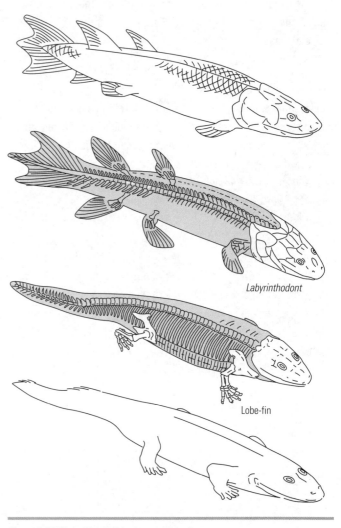

Labyrinthodont

Lobe-fin

Figure 2.5.5 Skeletal similarities between lobe-fins and the early amphibians.

many fish do, that is, by producing entirely aquatic young, such as frog tadpoles, which swim with a finned tail and use feathery gills for respiration before they develop feet and lungs.

Vertebrates became properly terrestrial during the Carboniferous when the first reptiles appeared. Like their amphibian forebears, they were four-footed (tetrapods), which is the characteristic that dominates the basic design of terrestrial vertebrate bodies. The reptiles evolved from amphibians but were completely independent of water. They did not need to return to rivers or lakes because they had developed a properly watertight skin and because they laid shelled, watertight eggs. The early reptiles were cold-blooded, like their modern lizard descendants. However, some groups eventually gave rise to more evolved creatures, such as the dinosaurs and their warm-blooded descendants the birds; the crocodilians; the icthyosaurs and plesiosaurs (swimming reptiles), pterosaurs (flying reptiles); and the mammals.

REVIEW
ACTIVITIES

EXTENSION
ACTIVITIES

1
When did animals first move on to the land? Name some animals that were in this first group of invaders.

2
Name the major group of animals that were the first to occupy terrestrial habitats and indicate why some palaeontologists suggest these animals were virtually pre-adapted for life on land.

3
Name the group of vertebrates that were the first terrestrial vertebrates and the group of vertebrates that these early walkers evolved from.

4
What does the word 'tetrapod' mean?

5
While it is true that adult amphibians breathe air and walk on the land very effectively, all members of this group of animals rely on access to an aquatic environment of some sort at various stages of their life. Explain why this is so.

6
What was the next major group of vertebrates to evolve from amphibians and why can it be said that these animals are the first truly terrestrial vertebrates?

7
Find some pictures and descriptions of the early terrestrial plants and animals in books from a library or on the Internet.

8
Explain why nearly all the terrestrial vertebrates are four-footed. Select three species of modern terrestrial vertebrates that move around on two feet and indicate what has happened to their two missing feet.

9
List and briefly describe the set of evolutionary changes that occurred to produce amphibians from their fish ancestors. Suggest reasons why these adaptations evolved.

SUMMARY

The animals and plants present on Earth today evolved from other forms that lived in the near past, which, in turn, descended from forms that lived in the distant past.

The aquatic animals that evolved into forms that could survive on land had to develop ways to keep moist, support themselves, deal with atmospheric oxygen, and breathe.

The ultraviolet component of unfiltered sunlight is sufficient to quickly kill unprotected cells in both microbes and large organisms.

Atmospheric ozone filters out most of the ultraviolet radiation in sunlight.

There is much more oxygen available for animal respiration in air than there is in water.

Because water absorbs light there is much more sunlight energy available to land plants than there is to aquatic plants.

The first land animals probably enjoyed an abundant supply of plants to eat and comparative safety due to a smaller variety of predators.

The aquatic plants that evolved into forms that could survive on land had to develop ways to keep moist, support themselves, deal with atmospheric oxygen, and transfer oxygen and carbon dioxide between themselves and their environment.

The first terrestrial environments inhabited by plants and animals are thought to have been areas of flat, low-lying and permanently wet ground.

The very first land plants probably evolved in the absence of animals to eat them.

The first forests of vascular plants were only about 1 m high and they were inhabited by a variety of arthropods, such as spiders, scorpions, centipedes and millipedes.

All terrestrial vertebrates probably evolved from a single amphibian tetrapod ancestor.

PRACTICAL EXERCISE
Floodplain forests through time

This exercise will help you to understand how landscapes have changed during geological time.

Procedure

1
Use the material in this chapter, along with resources on the Internet or in your library, to research how a coastal river floodplain or swamp might have appeared in the Late Silurian (before the evolution of terrestrial vertebrates) and in the Carboniferous. Also research what it looks like today.

2
Locate diorama images or other pictorial reconstructions of terrestrial environments of these times. Compare them with one another as well as with pictures of similar, present-day environments.

ACTIVITIES

1
Summarise the differences in the size and general appearance of Late Silurian flora and fauna and Carboniferous flora and fauna.

2
Summarise the differences between the size and appearance of the flora and fauna of the Carboniferous and those of the present.

3
Research the Cambrian terrestrial landscape and write a brief description of what a Cambrian river floodplain would have probably looked like.

4
Briefly outline how river floodplains and swamps have changed during the Phanerozoic.

OUTCOMES

At the end of this chapter you should be able to:

distinguish between mass extinctions and smaller extinctions

compare models of explosive and gradual adaptations and radiations of new genera and species following mass extinction events

analyse smaller extinction events involving several larger species, such as the recent extinctions of the marsupial, bird and reptile megafauna in Australia, and compare such events with widespread and 'catastrophic' events in which entire ecosystems collapse with the extinction of many entire classes and orders

assess the variety of hypotheses proposed for the end-Permian mass extinction with the popular bolide impact theory for the end-Cretaceous extinction event

compare and contrast the scale and timeframe of current rates of extinction of species within Australia with those that have occurred in geological time.

EXTINCTION AND RADIATION

When fossils of a plant or animal cease to appear in the sedimentary rock record we infer that the species has died out, that there are no more of them left alive, and we say that the plant or animal has become extinct. Individual species become extinct all the time and for a variety of reasons. The geological record indicates that there is a constant turnover of plants and animal species, where species die out and are then replaced by others. The activities undertaken by a species, its style of life, the place it occupies in a habitat, and where and how it lives and survives are said to be its **ecological niche.**

In the normal course of events, the ecological niches occupied by species that become extinct are often exploited by other, often new, species. It is thought that many **extinctions** occur because of direct competition for an ecological niche by two or more species that can live in the same way and do the same sorts of things. The better-adapted species survives while the less well-adapted species dies out. Competition between rival species for an ecological niche is the basis of Darwin's theory of evolution by natural selection and can be summarised in the phrase 'survival of the fittest'.

Natural environmental change can modify or even destroy habitats and ecosystems and cause a more widespread extinction that affects a large number of

ecological niche
the place of a plant or animal within its community or ecosystem

extinction
the act of a species dying out

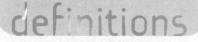

different species at about the same time. Such natural environmental changes can occur quickly or slowly and include changes in climate and sea level, which can be caused by a number of geological processes. These processes reduce the areas occupied by a habitat or change the nature of the habitat. This results in more intense competition between species, which, in turn, causes extinction of individual species or groups of species. Highly specialised animals and larger animals are often more greatly affected by a shrinking habitat.

Regional extinctions

When a large number of species that are unique to a particular geographical area become extinct at about the same time, then these multiple, related extinctions are known as a regional extinction event. This can happen when an ecosystem or habitat disappears due to gradual or sudden environmental change.

Sea-level change can cause regional extinctions. It is easy for us to understand such extinctions if we think about what happens when the sea level rises and drowns an island or a large continental area. As the lowland habitats disappear under water, many of the species that live in them will become extinct if they do not move and adapt to new habitats. If they do move to higher ground they will face increased competition from the species already living there: partly because there is a reduced area for everything to live in, and partly because the displaced animals are competing to take over niches that are already occupied.

Geologists call sea-level rises transgressions and drops in sea level regressions. Regressions can also cause regional extinctions in a similar way to transgressions by reducing the amount of space occupied by shallow continental seas (such as the Gulf of Carpentaria) and the continental shelves. Such events dramatically increase competition for space by the marine plants and animals that live in these productive shallow-water environments and this intensely increased competition often drives large-scale extinctions as the ecosystem adjusts to the reduced space available.

Radiation

Environmental change not only drives large-scale extinction but can also drive the opposite of extinction, that is, the evolution of large numbers of new species. These events are called radiations and happen when plants and animals evolve as they move into and exploit a newly formed or modified habitat. So, when new environments and habitats result from environmental change (for example, if new land is exposed by a fall in sea level) new species of plants and animals will appear as they adapt to the specific environmental characteristics of the newly created habitat. The plants and animals that are the first to move in are usually unspecialised ones that are capable of living pretty much anywhere and in a range of conditions, for example, the hardy cockroaches and small omnivorous mammals, such as rats.

After the first species moves in, their descendants can evolve into new species by enhancing those characteristics that improve their chances of survival. If they are lucky, chance mutations will speed up this process by developing new characteristics. For instance, one species of omnivores may diversify into many others when groups of individuals that are better at exploiting one niche rather than another specialise and adapt to maximise their chances of survival. Some become grazers rather than omnivores and develop mouths and teeth that chew plants well. Some grazers specialise to eat particular plants, others groups of individuals remain as omnivores, while others become predators and develop meat-slicing teeth, and so on. This process usually leads to a fairly rapid increase in species as they radiate out to fill the many different niches provided by a newly formed environment.

1
What is an ecological niche?

2
Name the theory that the phrase 'survival of the fittest' refers to and name the person who proposed it. Outline how the phrase summarises this theory.

3
What do the terms 'extinction' and 'radiation' mean?

4
Draw a sketch that shows how sea level changes during a regression and a transgression.

5
Explain how a regression can directly cause widespread extinction of shallow marine organisms and indirectly cause the extinction of many organisms that live in terrestrial habitats.

MASS EXTINCTIONS: ANCIENT AND RECENT

Mass extinctions occur when a relatively large number of species (30% or more) from many **phyla** disappear from a wide variety of environments located all over the world during a geologically brief period of time (less than 2 Ma). Close examination of the fossil record indicates that, apart from the many regional extinctions, there have been five separate extinction events that definitely qualify as mass extinctions. These are shown in Figure 2.6.1 and are collectively known as the 'big five' mass extinctions. In order of increasing severity, they occurred during the:

- Late Cretaceous (when about 60% of all known species died out)
  - Late Triassic (about 75% of species)
  - Late Devonian (about 80% of species)
  - Late Ordovician (about 85% of species)
  - Late Permian (about 95% of species).

The graph also shows that each of the big five extinctions marks the end of a geological period. This is no coincidence, because periods were originally defined on the basis of their fossil faunas according to Smith's principle of faunal succession. Indeed, the recognition of sudden mass extinctions led the great biologist and palaeontologist Baron George Cuvier to propose his theory of catastrophism in the early nineteenth century. In the rest of this chapter we will look at the causes of three extinction events: the end-Permian (the largest) and end-Cretaceous (the best-known) mass extinctions and the recent regional extinction that affected the Australasian megafauna (the most recent). An interesting feature common to each of these extinctions is that palaeontologists have not been able to agree entirely on what caused them. You should also note the steep rise in species diversity that occurs after each of the mass extinctions: these are the 'explosive radiations'.

As an idea, catastrophism was based partly on the accounts of Noah's great flood that are set down in certain religious texts. Using the Noachian flood as an example, Cuvier suggested that the sudden extinction of faunas marked violent global events, or revolutions as he called them. These revolutions were also responsible for mountain building and explained why fossils of marine animals could be found on mountain peaks. Throughout the first half of the nineteenth century, catastrophism competed with James Hutton's uniformitarian model of gradual geological change.

> **phyla**
> major groups of anatomically related organisms; plural of phylum
>
> definition

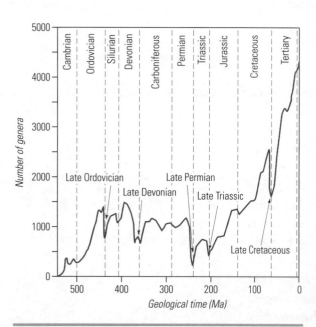

Figure 2.6.1 The diversity of marine animals during the Phanerozoic.

1

Outline how a mass extinction differs from a regional extinction.

2

How many mass extinctions have occurred since the beginning of the Cambrian? Name them, state when they occurred and indicate the percentage of the species that are thought to have become extinct when each of them occurred.

3

Investigate one of the mass extinction events not described in this book, that is, the Triassic, Devonian or Ordovician events. Briefly describe how this mass extinction was similar to or different from the K/T mass extinction.

THE END-PERMIAN MASS EXTINCTION

Over the course of a few million years at the end of the Permian, 95% of land-based and marine animal species became extinct. All groups of animals were affected, but the worst affected groups were bottom-attached marine filter feeders. This is the only mass extinction that has affected the hardy insects and it marks the end of the era of ancient life, the Palaeozoic. The reasons why the end-Permian mass extinction occurred are not clear, but indications are that a coincidence of several major whole environmental changes that affected the entire Earth was responsible. (See Figure 2.6.2.)

3 Enormous eruptions and carbon dioxide release trigger climate change

1 Drop in sea level

2 Oxidation of organic carbon

4 Return of the sea and flooding of low-lying lands

Figure 2.6.2 Causes of the end-Permian mass extinction.

A regression and transgression cycle

There was a sharp lowering of the sea level just before the end of Permian that dramatically reduced the amount of space on the continental shelves and decimated many marine communities. This particular lowering of the sea was the largest to occur during the 550 Ma of the Phanerozoic and was quickly followed by a transgression that raised the sea level back to close to where it had been.

The initial reduction in space resulting from the regression would have reduced the numbers of species able to live on the continental shelves. This would have particularly endangered species that live attached to the sea floor. Large-scale ocean currents would have been dramatically reorganised due to this regression and changed the locations where nutrients were brought to the surface from the deep sea. By itself, this change could have devastated previously productive shallow marine communities and caused the collapse of marine habitats. The rise of sea level that followed would have then destabilised the geography of the ocean's currents a second time. This would have placed new environmental stresses on those marine communities that were adjusting or establishing to take advantage of the new ocean currents.

A probable consequence of the regression–transgression cycle is global disturbance of terrestrial climates. The rainfall patterns across Pangaea would have changed, and so it is likely that land-based habitats and species were also affected by the regression. The following second major reorganisation of ocean currents due to the transgression that followed would have also completely changed rainfall patterns within Pangaea, possibly wiping out the survivors of the first set of changes.

It is easy to see why these two closely timed sea-level changes made the survival of bottom-attached filter feeders unlikely: their habitats were destroyed While such bottom-attached animals dominated Palaeozoic shallow marine faunas, Mesozoic faunas were dominated by mobile, active predators: a change in lifestyle that persists to the present day. Figure 2.6.3 shows the differences between the Palaeozoic and Mesozoic shallow marine communities.

A pulse of intense vulcanism

Massive eruptions of volcanic material in Siberia and Southern China took place right at the end of the Permian. They are thought to have initially caused substantial and sudden global cooling, accompanied by acid rains due to the particulates and sulfur compounds the eruption released into the atmosphere. Enough lava was erupted in about 600 000 years to bury all the Earth's land under 10 m of basalt.

The cooling possibly lasted for the duration of the end-Permian vulcanism. The ash and other particulates blasted into the upper atmosphere reflected so much sunlight that the amount of warming solar energy reaching the Earth's surface was greatly reduced. The cooling effect stopped after the eruptions finished as the particulates and sulfur compounds that were ejected into the atmosphere were rained out. The cooling may have been quickly followed by a more permanent global warming due to the massive amounts of carbon dioxide released during the eruptions. This would have caused an enhanced 'greenhouse effect' as the carbon dioxide prevented infra-red radiation escaping to space.

It is somewhat difficult to determine the temperature fluctuation due to this vulcanism, the time span involved, or the extent of the stress it had on the Late Permian ecosystems because of the incompleteness of the rock record. But it is not hard to understand that, if it did happen, only a very few of the species capable of tolerating a significant cooling event would also be able to tolerate the significant warming event that is thought to have followed.

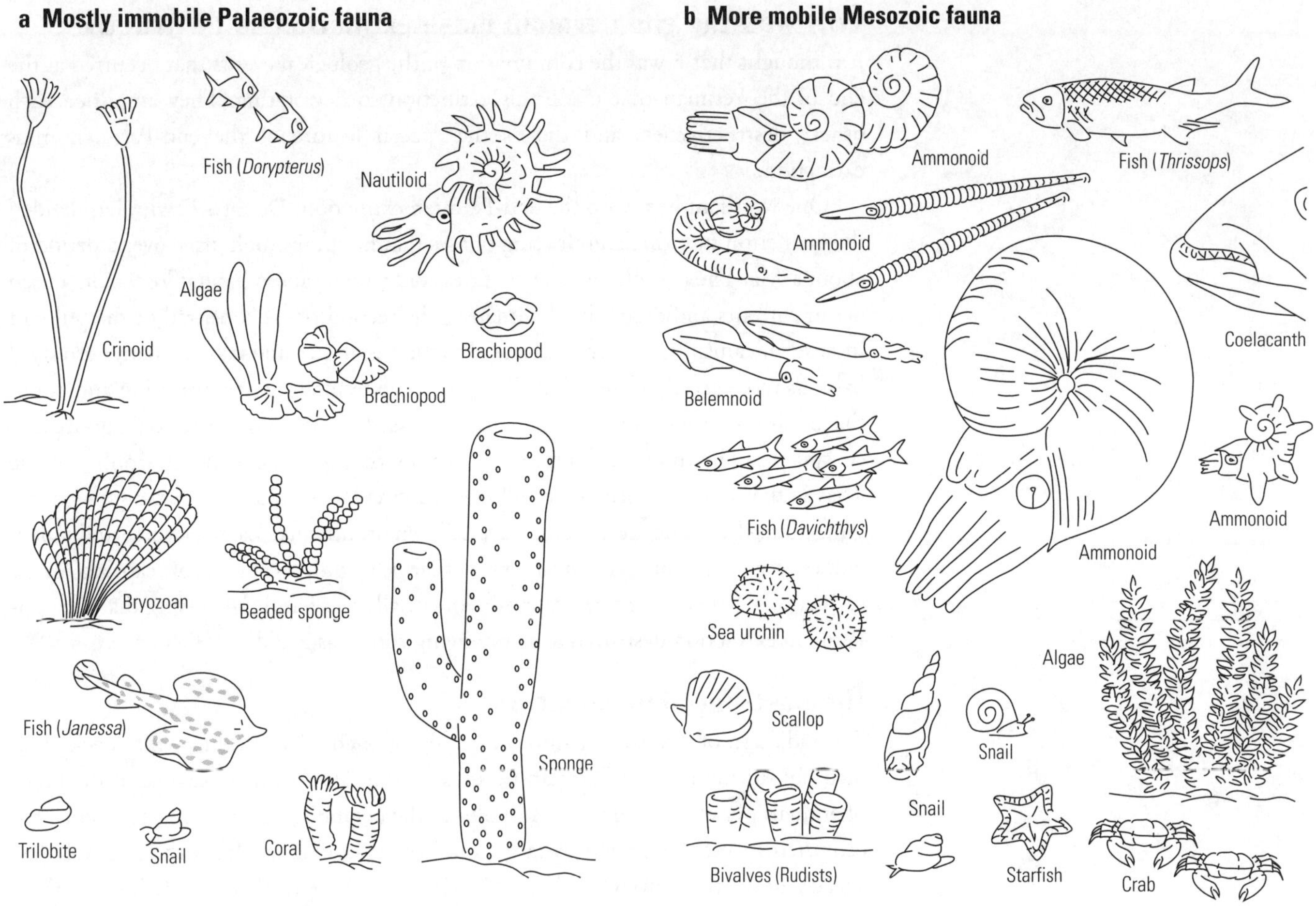

a Mostly immobile Palaeozoic fauna

Crinoid

Fish (*Dorypterus*)

Nautiloid

Algae

Brachiopod

Brachiopod

Bryozoan

Beaded sponge

Sponge

Fish (*Janessa*)

Trilobite

Snail

Coral

b More mobile Mesozoic fauna

Ammonoid

Fish (*Thrissops*)

Ammonoid

Coelacanth

Belemnoid

Fish (*Davichthys*)

Ammonoid

Ammonoid

Sea urchin

Algae

Scallop

Snail

Snail

Starfish

Crab

Bivalves (Rudists)

Figure 2.6.3 Marine faunas became more mobile after the end-Permian extinction.

Creation of an anoxic ocean

One possible consequence of the sea-level lowering that occurred at the end of the Permian is that a vast amount of the organic material buried in the shallow marine sediments was oxidised after being exposed to the atmosphere by weathering and erosion. Much of this material would have been turned into petroleum if the regression had not occurred but, instead, it was released back into the atmosphere as carbon dioxide. Similarly, a number of buried coal deposits that had formed at the margins of Pangaea during the Carboniferous and Early Permian were also exposed, eroded and weathered during the regression. So, the amount of oxygen in the latest Permian atmosphere would have decreased, while the amount of carbon dioxide present in it increased.

This should have had two major affects on the global marine environment. First, the weathering of organic material eroded from the shelves and transported into the oceans would have removed much of the ocean's dissolved oxygen, causing **anoxia**. This would have made it very difficult, if not impossible, for many animals to respire, especially those that lived attached to the sea floor because bottom waters would have been the first to become anoxic. Secondly, the release of a vast amount of carbon dioxide would have probably caused global warming and climate change, which can be a driving force that causes habitat loss and extinction. Furthermore, carbon dioxide is a potent poison for many marine invertebrate animals, especially those that are sedentary, thick shelled or have poor circulation. Overexposure to carbon dioxide induces a condition called acidising narcotosis, in which semisedation is quickly followed by death.

anoxia
not oxygenated

definition

Why was the end-Permian mass extinction so destructive?

It is thought that it was the combination of the geological events that occurred at the end of the Permian that made this extinction so devastating. They amplified each other's destructiveness and this is the special feature of the end-Permian mass extinction.

One world authority on the end-Permian extinction, Douglas Erwin, has divided the extinction into three reinforcing phases that he thinks took place over a period of about 1 Ma. The first phase was the regression, which caused habitat loss, reorganised ocean currents and destabilised climates. The second phase consisted of the pulse of intense vulcanism that increased the climatic instability and caused many terrestrial habitats to be greatly changed or disappear. The resultant global warming and fall in the amount of atmospheric oxygen also increased the stress on terrestrial and marine ecosystems. The third and final phase was the transgression at the beginning of the Triassic that once again reorganised oceanic circulation and climate patterns, at the same time drowning near shore terrestrial environments. Each of these individual phases was a major geological event that was quite capable of causing many extinctions. When the three occurred sequentially it is hardly surprising that the end-Permian extinction destroyed as many living things as it did.

The post-Permian radiation

The radiation of new species into the ecological niches left vacant by the species that died out during the end-Permian mass extinction led to a new balance in the kinds of animals present on the Earth. Initially, at the beginning of the Triassic it seems that ecosystems were dominated by large numbers of individuals from the relatively few species (5%) that survived. About 5 Ma after the end-Permian extinction these survivor species had radiated into a new and diverse fauna. The changes in species abundance and the newly dominant types of animals mark the major differences between the Palaeozoic era (ancient life) from the Mesozoic era (middle life).

The general nature of sea life changed from being dominated by bottom-attached filter feeders to being dominated by more active, crawling and swimming predators. On land, the extinction of the species that had been dominant, such as amphibians and mammal-like reptiles, gave other reptile groups the chance to become the dominant land animals. These included the dinosaurs and a number of other groups of reptilians that evolved and became successful after the end-Permian extinction. These include now-extinct groups that moved back into the sea (ichthyosaurs, pliosaurs and plesiosaurs), some that took to the air (pterosaurs, a group that includes the largest and most efficient flying animals to have ever lived) and some that lived in freshwater (the crocodilians, ferocious predators that still exist).

REVIEW ACTIVITIES

1

List and briefly describe the possible causes of the end-Permian mass extinction.

2

Explain why the end-Permian mass extinction was so severe and discuss why it is difficult to identify a single cause for this event.

EXTENSION ACTIVITY

3

Research the radiation of dinosaurs during the Triassic and Jurassic using sources available on the Internet, and in libraries and encyclopedias. Explain this radiation in terms of the exploitation of ecological niches left empty by the mass extinction event that preceded the radiation. Present your findings as a talk or illustrated essay.

THE END-CRETACEOUS MASS EXTINCTION

The end-Cretaceous mass extinction is the smallest, but most recent, of the big five extinctions. Hence, it is the best studied of the mass extinctions and the best understood. It is commonly referred to as the 'K/T event' or the 'Cretaceous–Tertiary boundary extinction'. Because it happened only 65 Ma ago there are many well-preserved, well-exposed and easily studied sequences of terrestrial and marine sedimentary rocks that span the period of time during which it occurred.

The K/T event has captured the public's imagination because it killed off the dinosaurs. Other groups of animals with which you may be familiar that died out at the Cretaceous–Tertiary boundary include the ammonites (free-swimming molluscan predators that resemble the nautilus and are related to squid), the plesiosaurs and the pterosaurs. There was a major decrease in the diversity of land animals as well as many groups of marine invertebrate animals at this boundary. Affected groups include the echinoderms (sea urchins and starfish), corals, most groups of molluscs and the foraminifera (unicellular marine animals). The mass extinction was worse in the marine environment than it was on land. Mammals, birds and many other groups of plants and animals survived the end-Cretaceous mass extinction and then evolved into the wide variety of modern creatures that we know today.

Causes of the K/T event

The scientists that study the causes of the K/T event have divided into two opposing groups. One group, which is the more publicly known of the two, favours a sudden, catastrophic extinction caused by a large extraterrestrial object colliding with the Earth. This object was either an asteroid or a comet and is commonly called a **bolide**. (See Figure 2.6.4.) The other group favours a more gradual extinction during the last 2 or 3 Ma of the Cretaceous. The gradualists suggest that the K/T event was caused by a regression of the sea in conjunction with global climate change that occurred due to vulcanism. This is very similar to the explanation used to account for the end-Permian extinction.

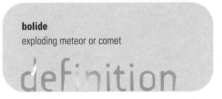

bolide
exploding meteor or comet

definition

Figure 2.6.4 The bolide impact.

Evidence for the bolide impact theory

The evidence for a bolide impact right at the Cretaceous–Tertiary boundary is overwhelming and includes the following three pieces of information as well as others.

THE IMPACT CRATER

There is a large impact crater located beneath the Yucatan Peninsula, which juts out into the Gulf of Mexico, that is consistent with the impact of a comet or asteroid about 10 km in diameter. This is the so-called Chixulub Crater and it has been dated to be about 65 Ma old, precisely the age of the K/T event.

THE BOUNDARY LAYER

A special layer of clay that can be found at many sites around the world separates the older Cretaceous rocks from the younger Tertiary rocks. This clay is generally green in colour and is known as the boundary layer because it occurs right on the Cretaceous–Tertiary boundary. The clay contains two very unusual features that are typically associated with bolides and their impact craters, providing strong evidence for the bolide impact theory. First, it contains a variety of quartz that only forms at extremely high pressure and is only naturally formed in meteorite impact craters. The quartz also usually displays fractures that are characteristic of impact shock. The second feature associated with the boundary clay is unusually high concentrations of the element iridium. Such high levels of iridium are typically found in meteorites, comets and asteroids.

THREE UNUSUAL DEPOSITS AT THE CRETACEOUS–TERTIARY BOUNDARY

Tsunami (tidal wave) deposits and bolide impact debris (called impact ejecta) are found at many places near the site of the Chixulub Crater, while soot is often observed in the boundary clay. The deposition of these three unusual materials confirms that a large bolide impact occurred at the Cretaceous–Tertiary boundary. The tsunami deposits suggest that a massive tidal wave swept through the area due to the bolide impact, while the boundary layer that occurs in the area relatively near the crater often contains special impact debris called tektites. Tektites are small droplet-shaped fragments of glassy rock that usually form during meteorite impacts. (See Figure 2.6.5.) They form when globs of rock melted by the impact fly through the air and freeze during transit. Large amounts of soot are found in the boundary layer from many sites around the world. This suggests that many of the world's forests were burnt as a consequence of the impact.

Other indications of global wildfires at the Cretaceous–Tertiary boundary are the presence of fullerenes in the boundary clay. These are intriguing molecules that only form during very high temperature combustion. Another indication is the 'spike' of fern spores in the fossil record just above the Cretaceous–Tertiary boundary deposits. This indicates that ferns almost totally replaced trees in the flora for a period of about 100 000 years after the impact.

Assessing the evidence

The sorts of stresses that are thought to have resulted from the Chixulub impact can be very destructive of habitat and cause widespread extinction of species. High-resolution cores drilled through the boundary layer by Ocean Drilling Program scientists show that the K/T event was as near to instantaneous as it is possible to measure geologically. Some palaeontologists support the alternative view that the bolide impact may have amplified a global extinction that was already occurring due to more normal geological processes, such as vulcanism and regression. While this may be true, it is very likely that the direct consequences of the Chixulub impact provided sufficient ecological stress to cause the K/T event by itself.

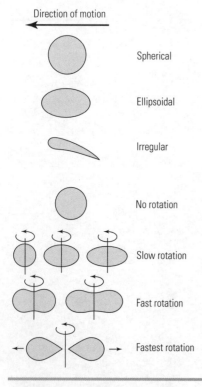

Direction of motion

Spherical

Ellipsoidal

Irregular

No rotation

Slow rotation

Fast rotation

Fastest rotation

Figure 2.6.5 Tektites acquire different shapes depending on how they move through the air.

The post-extinction Tertiary radiation

One important consequence of the end-Cretaceous extinction was the emptying out of a vast number of ecological niches. Few animals over 10 kg in weight survived the K/T event. The radiations that followed included a major diversification of bird and mammal species on land when these creatures took over from the dinosaurs that had occupied them prior to the catastrophe. The demise of marine reptiles led to the re-establishment of the fishes as the predominant marine predators. Mammals also moved into the marine realm, with whales and dolphins evolving during the Early Tertiary. Much later, our own species evolved as a consequence of these radiations.

REVIEW ACTIVITIES

1

Describe the geological feature that marks the Cretaceous–Tertiary boundary at sites all around the world.

2

List the special geological deposits that occur around the Chixulub site at the Cretaceous–Tertiary boundary.

3

What was the most likely cause of the K/T event? Outline the evidence that supports this explanation.

EXTENSION ACTIVITY

4

Find and evaluate some of the evidence for the end-Cretaceous bolide impact gathered by the Ocean Drilling Program (ODP). (*Hint*: Download the Blast from the Past poster from the ODP's website <http://www.oceandrilling.org>.)

EXTINCTION OF THE MEGAFAUNA

Over the last 60 000 years many large land mammals, large land reptiles and large flightless birds disappeared from Australia, New Zealand, North and South America and Madagascar. These large animals are collectively known as **megafauna** and an interesting aspect of the extinction of the megafauna is that our species, *Homo sapiens*, arrived in all of these places just before the extinctions occurred.

Australia's first human inhabitants are thought to have walked here through Indonesia and Papua New Guinea via land bridges, but they must have travelled from Bali to Lombok in small boats across a deep-water strait. It was possible to travel here largely by walking at various times during the last ice age because the sea level was lower than it is today, and so what is now submerged continental shelf was then exposed as dry land. The oldest widely accepted time for the arrival of the first humans in Australia is 60 000 years ago. The Americas were settled about 14 000 years ago when the sea level was at it lowest level at the peak of glacial conditions. New Zealand and Madagascar were much harder to get to because they were isolated by many hundreds of kilometres of deep ocean and required ocean-going vessels to make the trip. Madagascar was settled about 2000 years ago by people who sailed from Southern Asia while the Maoris are thought to have arrived in New Zealand about 1000 years ago.

The Australian megafauna consisted of land animals weighing more than 100 kg. All these mammals, reptiles and flightless birds had become extinct by about 30 000 years ago, a relatively short time after the first human settlers arrived here. They included the giant flightless birds, such as *Genyornis*; giant wombats, such as *Diprotodon* (see Figure 2.6.6, page 112); giant kangaroos, such as *Procoptodon*, which

megafauna
giant animals, generally extinct

definition

Figure 2.6.6 *Diprotodon.*

was up to 3 m high and weighed about 250 kg; as well as giant goannas, such as *Megalania.* In addition to these large animals, another six of the seven groups of land animals weighing between 45 and 100 kg also became extinct at about the same time. The surviving animals are members of the family that includes kangaroos and wallabies.

The exact timing of the megafauna extinction is currently the subject of robust debate because it is difficult to establish. It is very difficult to reliably date bone by carbon 14 methods because the ages involved are close to or over the limit that the technique is able to measure and because the bone is often contaminated with younger material. An extinction date of 50 000 years has been suggested. This is based on ages determined by other dating techniques for the sediments in which the youngest partially complete skeletons of megafauna animals have been found. Partially complete skeletons are thought to give a more reliable age because they record the burial of a newly dead animal, whereas individual bones may be fragments of a much older animal that were washed in to the site of deposition. The age of 50 000 years is confirmed by the age of the last eggs that were laid by *Genyornis.* These egg fragments have been dated by methods other than carbon 14 dating.

What caused the megafauna extinction?

It is not only the timing of the megafauna extinction that is hotly debated. Three different ideas have been proposed to explain the disappearance of these animals. They are:

- Their habitat was destroyed due to climate change.
- They were hunted to extinction by humans. This is the so-called blitzkrieg hypothesis, which suggests that these animals might have had no fear of humans or that they might have been too slow to escape human hunters.
- Their habitat was changed and destroyed due to human activity, especially the use of fire in hunting. The introduction of feral animals and plants must also have played a role in habitat modification.

All these ideas are plausible explanations for the disappearance of the megafauna, and the available data support all three proposed explanations to varying degrees.

The difficulty scientists have in agreeing on a single cause is partly because there is some doubt about the exact time that the megafauna animals died out. However, the main sources of disagreement come from:

- uncertainty about the time that Australia became drier, especially the interior
- uncertainty about the time when humans occupied Australia.

As far as we can tell, humans populated the Australian continent at about the same time as the extinction.

A solution to the conflict over the Australian megafauna extinction probably lies in examining the megafauna extinctions that occurred in other places in the world. It turns out that the timing of the extinctions elsewhere provides much stronger support for the blitzkrieg hypothesis than the climate change hypothesis. In the Americas the last megafauna extinctions took place about 11 000 years ago, which was less than 3000 years after the first arrival of humans there. People arrived in Madagascar about 2000 years ago, and the 500 kg elephant bird and the giant lemurs that lived there died out a few centuries later. In New Zealand a recent, detailed study has shown that the extinction of the moas (eleven different species of flightless birds weighing from 20 to 250 kg) occurred about 1300 AD, less than 100 years after Polynesian people arrived there. Similarly, then, it seems likely that the Australian megafauna extinction occurred relatively soon after the arrival of humans.

Each of these events is associated with a different climatic regime and a different pattern of climate change and consequent habitat change. During the Australian event, the continent was drying out and becoming cooler. During the Americas' event, global climate was warming and becoming wetter. The Madagascan and New Zealand events are not associated with large-scale climate change. However, the four megafauna extinction events do have one aspect in common: the arrival of humans. So, there is a strong indication of guilt by association and repetition that human beings (or the domesticated and feral species they brought with them) were the prime culprits for the megafauna extinctions.

REVIEW ACTIVITIES

1
Name four different geographical areas of the world where megafauna extinctions occurred and indicate the time when these extinctions occurred.

2
What does the term 'blitzkrieg hypothesis' mean?

3
Draw a time line to show the development of our knowledge and understanding of the universe.

4
What was the megafauna and when did it go extinct in Australia?

5
Outline two possible explanations for the megafauna extinction that do not involve direct human action.

6
Why can palaeontologists and anthropologists dispute the causes of the Australian megafauna regional extinction?

EXTENSION ACTIVITIES

7
Find images in books or on the Internet that show what the Australian megafauna animals and their habitats looked like.

8
Find out about the habits and habitat of an Australian megafauna animal and present this information in a talk to your class.

ARE WE CAUSING A NEW MASS EXTINCTION?

There is a great deal of agreement among biologists, ecologists, environmental activists, the general community and even our politicians that some of Australia's natural habitats and ecosystems have been dramatically changed since the arrival of Europeans in 1788. In 1996 a government report into the state of the Australian environment was published. The report indicated that, of those species that were alive at the time of settlement, 5% of plants, 23% of mammals, 9% of birds, 7% of reptiles, 16% of amphibians and 9% of freshwater fish had become extinct or were threatened with extinction. At 9%, the actual rate of extinction of Australia's native mammals was the world's highest. Overexploitation of our ecosystems by land clearing, land salinisation, overfishing, mechanised agriculture and river damming, as well as other factors (such as pollution and urbanisation) have all contributed to these extinctions. These activities have mostly caused habitat change, and our activities on this continent have undoubtedly greatly reduced the area of natural habitat that was present in 1788.

We have already seen that one of the main consequences of habitat destruction is a corresponding loss of species, that is, the extinction of species. As the human population increases, there will be further exploitation and the area of natural habitat left will therefore decrease. It follows that more and more species will become extinct in the future as a consequence of habitat destruction.

The global picture

There is no general agreement on the actual increase in the current rates of species loss due to human population growth and increasing exploitation of the planet. This is because there just aren't enough biologists to identify the vast variety of plants and animals on the Earth. But the experts in this field indicate that the present rate of extinction of species on the whole planet is somewhere between fifty and 1000 times the background rate of extinction. The present-day global extinction rate is therefore directly comparable to the rate of extinction during mass extinction events. Some predictions have been made that more than 30% of all the Earth's flowering plants and more than 25% of all the Earth's animals will become extinct due to habitat destruction and overexploitation during the next 100 years or so. If these predictions come true, the current global biodiversity crisis will meet the criteria for a mass extinction, that is, 30% of all species in a geologically short time.

REVIEW
ACTIVITIES

EXTENSION
ACTIVITIES

1

What percentage of the terrestrial plants and major vertebrate animal groups will have gone extinct in Australia since European settlement if the currently endangered species in those groups die out?

2

Explain why it is possible to view the current global biodiversity crisis' as being a mass extinction event.

3

Consult the *2001 State of the Environment Report: Australia* at the Environment Australia website <http://www.ea.gov.au/soe/2001/> and find out if any progress has been made in fixing Australia's biodiversity crisis.

4

Find out what is meant by the term 'background rate of extinction'. Compare and contrast this background rate to the estimates of recent extinction rates both globally and in Australia.

SUMMARY

A regional extinction occurs when species that are unique to a particular geographical area become extinct at about the same time.

A mass extinction occurs when more than 30% of species die out suddenly at different sites all around the world.

There have been five mass extinctions. From smallest to largest, they were the end-Cretaceous (about 60% of all known species died out), end-Triassic (75%), late-Devonian (80%), end-Ordovician (85%) and end-Permian (95%).

Explosive radiations of new genera have always followed mass extinctions as new species evolved to fill empty ecological niches.

A single cause for the end-Permian mass extinction has not been identified, but this catastrophe is thought to have occurred when several major geological crises occurred together.

The end-Permian crisis was a drastic climate change that occurred at the same time as a large fall, and then a large rise, in sea level.

The end-Cretaceous extinction is generally thought to have occurred almost instantly due to a comet striking the planet.

The end-Cretaceous comet impact probably started an intense global bushfire that completely destroyed the planet's habitats and ecosystems.

Megafauna are several different groups of giant animals that were unique to specific geographical areas of the world and became extinct over the last 60 000 years.

There is no agreed cause for each of the individual megafauna extinctions.

Climate change, and predation by humans and/or domesticated animals are the likely causes for the extinction of the individual megafaunas.

The present-day rate of species extinction is similar to the rate during mass extinctions.

Extinctions and geological boundaries

This exercise will give you a better understanding of the relationship between the geological boundaries of the Phanerozoic and mass extinction events.

Procedure

1

Use the material in this chapter, along with resources on the Internet or in your library, to research land-based and marine vertebrates of the Permian, Triassic, Cretaceous and Tertiary.

2

Locate diorama images or other pictorial reconstructions of typical vertebrates of these times and compare them with one another.

3

Find tree of life type diagrams that show the evolutionary relationships of the dominant groups of vertebrate animals that lived during the Phanerozoic.

4

Identify several groups of vertebrates (for example, dinosaurs, marine reptiles, flying reptiles and the many different types of mammals) on the vertebrate evolutionary tree. For each group, determine when it:

a first appeared

b diversified and became abundant

c declined

d went extinct.

5

Organise your findings into a table under the following headings:

- name of group
- time of appearance
- time of abundance
- time of decline
- time of extinction.

ACTIVITIES

1

Name some groups of vertebrates that declined or died out at the end of the Permian.

2

When did the dinosaurs, flying reptiles and marine reptiles die out?

3

When did most groups of mammals appear or diversify?

4

Imagine that you are in charge of excavating a rich and diverse vertebrate fossil deposit that spans the Cretaceous–Tertiary boundary.

a Describe how the appearance of vertebrate fossils should change across the Cretaceous–Tertiary boundary.

b Explain how you could accurately determine the layer that marks the boundary between the Cretaceous and Tertiary.

5

Outline and explain the relationship between the timing of the extinction and radiation of two particular groups of vertebrates with the timing of the big five mass extinctions.

Caring for
the country

3

In this section we look at how human activities have affected the environment and the steps that are being taken to restore damaged environments. A damaged environment can usually be restored. However, it may be impossible to restore a damaged environment to its original condition if its biodiversity has been permanently lost and its topography has been changed due to mineral extraction, improved drainage or levelling. With our greater understanding of environmental processes and systems, after restoration an environment may be better than the original. It may have a wider biodiversity or a more varied habitat. This can be seen where poor-quality farmland has been mined and after rehabilitation the area has become a series of water bodies and wetlands.

CONTENTS

3.1 Weathering and erosion of Australia's land surfaces

OUTCOMES

At the end of this chapter you should be able to:

describe the low fertility of Australian soils in terms of:
- slow rate of soil formation
- long period of depletion of nutrient ions
- stability of Australian continent in terms of low relief

outline a cause of soil erosion in NSW due to:
- an agricultural process
- a natural process
- urbanisation

and identify a management strategy that prevents or reduces each of these three causes of soil erosion.

THE LOW FERTILITY OF AUSTRALIAN SOILS

Australian soils may be up to 100 Ma old. They are described as very old, highly weathered and mostly infertile. The character and properties of Australian soils reflect unusual continental stability. Australia has experienced little effect from the soil-rejuvenating impact of uplift and mountain building, vulcanism and glaciation. The absence of these processes has preserved an ancient land surface and soil over much of the continental area. Many areas show the effects of deep weathering, duricrusting of **silicrete** and **laterite**, leaching and extreme nutrient depletion. These effects can be seen in the Australian environment in features such as a flat terrain; evidence of **palaeodrainage**; **etch plains**; **duricrusts**; and deep, weathered profiles.

The nature and distribution of the present-day soils in many areas is linked to the **geomorphology**. Australia is generally very flat, with most mountain ranges on the east and west coasts. The flat terrain is the result of the weathering agents of wind, water and heat having eroded ancient mountain ranges over millions of years. The eroded material was deposited in low-lying areas, which now form vast expanses of flat terrain.

Since the arrival of Europeans in the nineteenth century, soil degradation has accelerated rapidly, and in some areas it has reached crisis point. Degradation of soil can be in the form of nutrient loss, **acidification**, salinisation, erosion, and decline in soil structure and biodiversity.

Most of Europe and North America were stripped of their former soils by the icesheets of the Pleistocene age (1.8 Ma to 11 000 years ago), and soil formation restarted on fresh rock surfaces or on the glacial deposits about 10 000 years ago. By contrast, in Australia, apart from the very small areas that were glaciated in the southeast and Tasmania, the soils formed on land surfaces that have been continuously exposed to weathering, probably since the Late Tertiary (65–18 Ma ago). As a result, the soils are ancient and have deeply weathered structures. Where soil formation occurs, it is a slow process; 1 cm of soil may take hundreds of years to form. These ancient, weathered soils dominate the soil pattern in many areas and, due to the

definitions

silicrete
a silica-rich, solid layer in soil

laterite
a soil residue composed of secondary iron oxides

palaeodrainage
ancient drainage system

etch plain
flat ground marked by wind and water erosion

duricrust
a solid layer on the surface of the soil

geomorphology
the manner in which the land surface has been altered by erosion and deposition of material

acidification
the process by which soil becomes increasingly acidic

intense weathering, it causes plant nutrition problems that are not encountered in younger soils.

The productivity of Australian soils is largely determined by the supply of water. Only about 10% of the continent receives sufficient rainfall for agricultural plant growth, and some of this falls or drains onto soils too steep and stony, or too elevated and cold, to support crops. Where moisture is continuously or seasonally abundant, such as in the southern parts of the continent, productivity is governed largely by the availability of **trace elements**. (See Table 3.1.1.) There is an almost universal need to enhance crops by using phosphate fertilisers as well as a widespread need for sulfur and potassium. Much of Australia has low and unreliable rainfall, and in these areas little arable (cropping) agriculture or sown pasture production is possible.

Table 3.1.1 Trace elements needed by plants

Element	Signs of deficiency	Soil type where deficiency commonly occurs
Iron (Fe)	Nematode attack on roots	Soils high in calcium, manganese, phosphorus or heavy metals (such as copper and zinc) and with poor drainage, high pH or oxygen deficiency
Boron (B)	Failure to set seed; internal cellular breakdown; death of top buds	No specific soil type
Zinc (Zn)	Reduced leaf size, distorted or puckered leaf margins	No specific soil type
Copper (Cu)	Small size of new growth; wilting	Some peat soils
Manganese (Mn)	Brown spots on leaves, producing a chequered appearance	Found in acidic soils
Molybdenum (Mo)	Twisted leaves	No specific soil type
Chlorine (Cl)	Leaves that wilt, become bronze, then die; club roots	No specific soil type

Since the formation of the soils, the nutrient levels have changed drastically. Over a long period of time, rain has leached nutrients out of the soil. As a result, most nutrients have disappeared. Nutrients such as phosphorus and nitrogen are particularly deficient.

Soil biodiversity is essential to maintain a fertile, healthy soil. The animals, plants and micro-organisms living in topsoils are both numerous and highly diverse. In $1\,cm^3$ of soil, there can be more than 6 billion micro-organisms. Soil-dwelling species are possibly the most important factor in maintaining a healthy ecosystem. Soil micro-organisms can be divided into a number of groups dependent on their function in the soil. The roles include breaking down, absorbing, recycling and releasing nutrients for animals and plants.

Australian soils are slowly becoming more acid. The rate of acidification in bushland is very slow. Farming speeds up the process by increasing the rate of two processes that produce acidity: the amount of nitrogen that flows through the soil; and the amount of nutrients absorbed from the soil and exported in produce. Most nitrogen that accumulates in Australian soils comes from the nitrogen-fixing bacteria found in **legumes**. These processes are not acidifying in themselves. Instead, when plants die and are broken down in the soil, organic nitrogen is converted into acidic nitrates. The acidic nitrates can be taken up by plants and other organisms or **denitrified**. By these two processes, acidic nitrates are recycled and do not affect the levels of acid in the soil. If, however, the acidic nitrates in solution move down through the soil profile below the level of plant roots then the process of acidification will take place.

trace elements
elements that are required in very small amounts by plants for normal growth

legumes
members of the Leguminosae plant family, which includes plants such as beans, gorse and peas

denitrified
having nitrogen removed by a process

One of the major effects of soil acidification is its impact on the availability of individual nutrients. The greater the acidity, the lower the amount of nutrients, especially nitrogen, phosphorus and potassium. A second effect is the reduction in trace elements. For example, an increase in acidification leads to a reduction in the availability of the element molybdenum, a trace element vital for plant reproduction. Thirdly, increasing acidification leads to the development of aluminium and/or manganese toxicity in plants, which results in reduced growth and, eventually, death. The fourth major effect is that leguminous plants fail to produce the root nodules that contain the nitrogen-fixing bacteria that convert nitrogen gas into nitrates, which are vital for plant growth.

Climate change has also led to a drying out of the continent since Tertiary times. This has led to an accumulation of soluble salts in soil and ground water and a spread of low-nutrient sands in dune fields and sand sheets. Rainfall is limited or highly seasonal and, for crop **irrigation**, farmers depend on water stored in dams, ground water or water from major rivers. Around two-thirds of irrigated land in Australia is located in the Murray-Darling Basin.

Australia has old, fragile soils that are undergoing change. Since European settlement, the soils have been progressively affected by introduced organisms. There was an approximate equilibrium between the soils and the organisms that were present when the soils first formed. Introduced organisms are widely different from the original organisms and, as a result of their introduction, the soils changed and continue to change. The extent to which the soils are eroding shows that a new equilibrium has not yet been achieved.

The distribution of soil organisms is complex and dependent on soil type. For instance, waterlogged soils are often lacking in oxygen because oxygen diffuses 10 000 times slower through soil than through atmospheric gases. Therefore, waterlogged soils have a different group of micro-organisms than dry soils. This means that waterlogged soils have entirely different soil conditions and, therefore, plant species and related organisms than dry soils. A severely degraded soil will have lost most of its biodiversity, making the soil difficult to rehabilitate.

definition

irrigation
the supplying of land with water from artificial channels to promote vegetation

REVIEW ACTIVITIES

1
Outline why the character and properties of Australian soils are unusual.

2
Describe how the Australian topography shows the effects of continental stability.

3
What human activity speeds up the process of soil acidification?

4
Outline what effect climate change has had on Australian soils.

5
Explain why the soil is eroding faster since European settlement.

6
Explain why plants living in waterlogged soils often suffer from a lack of oxygen.

EXTENSION ACTIVITIES

7
Investigate what steps are being carried out to reduce and prevent soil acidification.

8
Research where and how nitrogen is fixed by bacteria in legumes.

CAUSES OF SOIL EROSION

Soil erosion is the movement of particles of soil, surface sediments and rocks by the action of water, glaciers, winds, waves and so on. Soil erosion processes include wind erosion, sheet erosion, rill erosion, gully erosion and mass movements. Erosion is greatest in arid or semiarid regions and where soil is poorly developed and receives little protection from vegetation. Human land use also causes erosion because it disturbs the soil. The tendency of a soil to be eroded is dependent on its characteristics. The characteristics influencing erosion include soil texture, organic matter, calcium carbonate content and rock content.

Work carried out on soils in New South Wales and Victoria has found that there are significant differences between the rate of erosion and the rate of soil formation. The trial plots were based in temperate lowland areas and it was found that erosion rates may increase with changes in climate or altitude.

Rates of soil erosion can be affected by a number of factors, such as:
- size and velocity of raindrops
- **permeability** of the soil
- soil particle size and shape
- slope angle
- exposure
- vegetation cover.

An average-sized raindrop of 5 mm diameter falls through still air at 32 km/h, and this speed increases if the raindrop becomes larger or there is an increase in wind speed. The raindrop hits the soil's surface like a miniature bomb and displaces surface soil particles. With an increase in the velocity of the raindrop, there is also an increase in soil particle disruption.

If the soil has low permeability, the water is unable to infiltrate the soil profile and surface run-off increases. This generally results in an increase in the transportation of soil particles.

The size of the soil particles is an important factor in the ease with which a soil is eroded. The smaller the size of the soil particles, the easier the soil can be eroded by water or wind. The shape of the soil particles can also affect whether the soil is likely to be eroded. Irregular-shaped particles may increase the likelihood of the soil particles clumping together. This may reduce soil erosion; although excessive clumping may also cause a decrease in soil permeability due to a reduction in air spaces within the soil.

Soils on slopes greater than 20° are more likely to erode than soils on lesser slopes. Steep slopes are also more prone to mass movements and erosive surface run-off.

The amount of exposure of soil to the erosive agents of water and wind affects the degree of erosion. Small fields or paddocks surrounded by trees or other forms of barriers will suffer less wind erosion than large fields or paddocks. Intensive European agricultural practices (such as the removal of natural vegetation, ploughing and leaving fields and paddocks **fallow**) expose the soil surface to the erosive agents of wind and water.

Figure 3.1.1 A raindrop hits the soil's surface like a miniature bomb.

1

Recall what factors result in soil erosion.

2

Work carried out on trial plots in New South Wales and Victoria found that erosion rates could be affected by what changes?

3

The shape and size of soil particles are very important factors as to whether soils may or may not be eroded. describe why these factors are so important.

4

Explain why small fields or paddocks are generally less susceptible to wind erosion than large fields or paddocks.

5

Design an experimental method to test the statement: 'Soils on slopes greater than 20% are more likely to erode than soils on lesser slopes'.

6

Research why the removal of natural vegetation results in an increase in soil erosion.

PROCESSES OF SOIL EROSION

Soil erosion is a natural process. The rate at which it occurs depends on land use, geology, geomorphology, climate, soil texture, and the nature and density of vegetation in the area. Soils generally are more prone to erosion as particle size decreases and are particularly susceptible to erosion under heavy summer rainfall, when vegetative cover is low.

Wind erosion

The ability of the wind to carry soil depends on conditions such as climate, plant cover and surface roughness. Wind erosion occurs when soils are bared of vegetation or are exposed to high-velocity winds. When the velocity of the wind overcomes the gravitational and cohesive forces of the soil particles, the wind will move the soil and carry it away.

The wind is able to move soil particles in several ways. If the soil particle is less than 0.1 mm in diameter it can be picked up and carried through the air. A particle that is larger than 0.1 mm can be pushed along on the ground. This occurs in one of two ways. If the particle is 0.1–0.5 mm in diameter, the wind pushes it along on the ground in a hopping or bouncing fashion. This is called **saltation**. If the particle is greater than 0.5 mm in diameter, the wind rolls it along on the surface. This is called **soil creep**.

The most easily recognisable form of soil erosion occurs when soil is carried up into the air, forming a dust storm. The wind sorts the soil particles. Finer material, containing organic matter and clay and silt particles, is removed and carried through the air. Slightly heavier material is sorted when it is pushed along the surface by the wind. Deposition of this slightly larger material occurs at obstructions, such as fence lines, tree lines and roads.

Following wind erosion, soil may look lighter because organic matter, clays and iron oxides have been removed. Also, the soil surface may be smooth, with little sign of previous cultivation. In areas with sandy soils, particles may be sorted, leaving the surface covered with a coarse, sandy layer. The impact of wind erosion is most severe

saltation
occurs when the wind blows soil particles (0.1–0.5 mm in size) in a bouncing manner across the ground

soil creep
occurs when the wind blows soil particles (more than 0.5 mm in size) in a rolling manner across the ground

definitions

when strong winds combine with drought periods to create dust storms that carry fine soil particles, or fractions, hundreds of kilometres. An example is the 1983 dust storm that engulfed Melbourne.

The coarser, less fertile material is left behind and the productive capacity of the soil is reduced because most soil nutrients are attached to the soil particles that have been carried away. Over a long period, the physical structure of the soil is altered because the topsoil is lost, exposing the subsoil.

Wind erosion mainly affects inland farming areas, especially those where crops are planted, as there is a cycle of ploughing, growing and harvesting. This cycle increases the opportunity for wind erosion, which occurs most frequently when the soil is ploughed and after harvesting when the soil has lost its plant coverage. Soils that are prone to wind erosion are predominantly sandy soils with low levels of organic material. (See Figure 3.1.2.)

Wind erosion can be prevented or reduced by carrying out a number of strategies:
- avoiding ground cover of less than 30% on sandy soils
- reducing wind access to the soil—usually through stubble or mulch retention at the soil surface or by maintaining soil roughness—and controlling wind speed over a property by planting lines of trees to act as windbreaks
- giving special attention to slopes with a north-western aspect as they are often the driest and exposed to harsh northerly winds
- modifying fallowing practices and machinery use by employing minimum tillage and direct-drilling techniques
- reducing grazing or stocking rates.

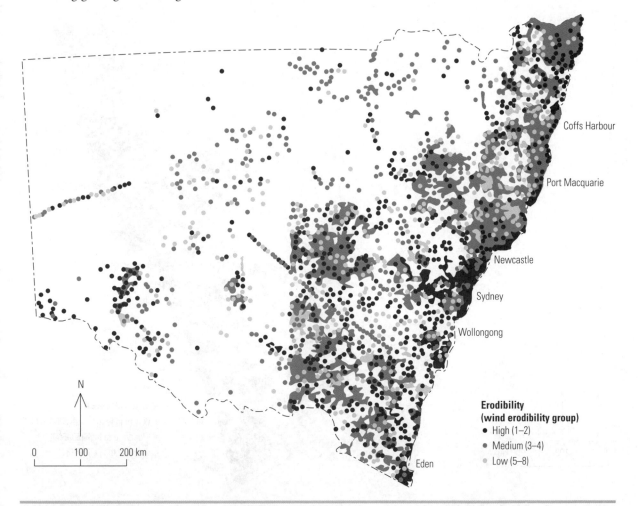

Figure 3.1.2 Areas of New South Wales that are prone to wind erosion.

Sheet erosion

Sheet erosion occurs when the uniformly thin layer of soil is removed by the splash of raindrops or water run-off. This thin layer of topsoil is lost gradually, and because the damage is not immediately obvious it makes it difficult to monitor.

When raindrops hit the surface of the soil they break up soil clumps, dislodge soil particles and compact the soil surface. If the amount of rainfall exceeds the amount of rain that the soil can soak up, a surface film of water forms, building up into flows that are 2–3 mm deep. Continued rainfall causes turbulence within the flow, increasing the erosive power of the water by up to 200 times.

Sheet erosion removes the topmost layer of the soil, which is composed of the finest particles and holds the bulk of plant-available nutrients. Organic matter adheres to the topsoil, but when erosion occurs the organic material is removed and the vegetation decreases. As a result, many soils begin to form a crust, which air and water can no longer penetrate. The loss of the topsoil reduces the productivity of the land. This form of erosion may also result in the removal of seeds or seedlings, which reduces the ability of the soil to store water. The soil that has been removed may cause crop and pasture damage, a deterioration of water quality and an increase in sedimentation in dams, lakes and watercourses.

Soils that are particularly vulnerable to sheet erosion include those that are repeatedly cultivated, are left fallow or are bare due to overgrazing by farm or pest animals. Red-brown duplex soils on slight slopes are particularly vulnerable to sheet erosion. (See Figure 3.1.3.)

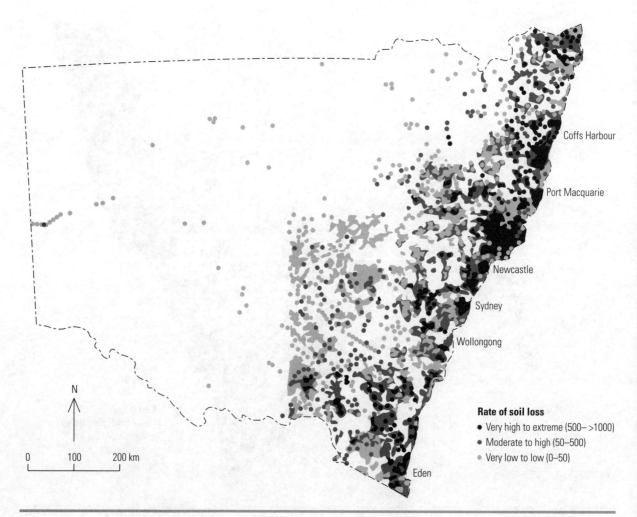

Figure 3.1.3 Areas of New South Wales that are prone to sheet erosion.

A number of actions can be taken in order to reduce or prevent sheet erosion. They include:

- retaining vegetative cover
- modifying fallow, cultivation and rotation processes of cropping land
- reducing the pressures from grazing
- improving vegetation and soil management to maximise rainfall infiltration and its use by the vegetation.

Rill erosion

Rill erosion often occurs with sheet erosion and is commonly seen in paddocks of recently cultivated soils following high-intensity rainfall. It is easily identifiable as a series of little channels called rills, up to 30 cm in depth. (See Figure 3.1.4.)

If rainfall exceeds the amount that can be taken up by the soil, a surface film of water forms. Rill erosion occurs when this surface water forms deeper, faster-flowing channels, which follow depressions or low points in paddocks. The velocity of this water can break off and pick up soil particles, making these channels the preferred routes for sediment transportation.

The loss of topsoil reduces the fertility of the remaining soil and, consequently, lowers its productivity. The transport of sediment along the rills causes sedimentation of watercourses, dams and lakes, as well as deterioration in water quality and damage to aquatic environments.

Figure 3.1.4 Rill erosion.

Rill erosion is common on agricultural land devoid of vegetation, as well as areas where overgrazing has occurred. Red-brown duplex soils are vulnerable to rill erosion. Strategies to prevent or reduce rill erosion include:

- avoiding extensive summer fallow periods
- modifying cultivation and rotation processes of cropping land
- controlling grazing pressures
- improving vegetation cover
- improving soil management practices to increase the amount of organic matter in the soil, and promote water uptake by the soil.

Gully erosion

When the small channels, or rills, are large enough to restrict vehicular access, the channels are referred to as gullies. Water can travel along these gullies at high velocity and, in the process, it can remove large amounts of soil. (See Figure 3.1.5.) This results in deep, wide gullies occurring along depressions and drainage lines. Lateral erosion may also occur in gullies when water undercuts the sides, causing them to slump. Undercutting of the gully bed occurs where there are waterfalls, and leads to a deepening of the gully.

Gullies can reach a depth of 30 m, which reduces the use of the land on which they are situated. The resulting deposition of the soil downstream reduces water quality and increases sedimentation in dams and reservoirs.

Figure 3.1.5 Gully erosion.

All soils are susceptible to gully erosion. However, those soils with more rounded particles, such as sand, have a greater susceptibility.

Gully erosion can be prevented or reduced by:
• reviewing current land use
• reducing the amount of water entering gullies by increasing plant uptake
• planting along drainage ways
• building diversion banks, detention dams, **groynes** and stabilising banks.

groyne
a jetty built out into a river to prevent erosion of a beach or bank

definition

Figure 3.1.6 Stream-bank erosion.

Stream-bank erosion

Stream-bank erosion is the direct removal of banks and beds by flowing water. This process is very similar to gully erosion, but gully erosion tends to be more seasonal or ephemeral. Stream-bank erosion usually occurs after heavy rains. If the catchment has poor plant coverage there is excessive run-off. The resulting increased volume of water and higher velocity of flow has a major impact further downstream, where it causes stream-bank erosion. (See Figure 3.1.6.) As the sediment load increases it has a greater potential to grind and excavate the banks and bed of the stream. Erosion of the bed and banks is exacerbated by a lack of stream-bank vegetation, which gives stock better access to the stream.

Stream-bank erosion can be reduced or prevented by the following actions:
• reducing run-off
• reducing the access of stock to stream banks
• improving vegetation cover on stream banks
• improving engineering measures, such as retention walls, weirs and dams.

Mass movements

In mass movements of soil, gravity is the principal force. This form of soil erosion occurs on slopes. Mass movements may take the form of slips, slumping, landslides and creep. Rapid movement of soil or rock that behaves differently from the stationary underlying material is described as a landslide. Creep tends to be a much slower, more long-term process where the upper layers of the soil or rock slide slowly downhill. Whether it is a landslide, slip or slump, a number of factors lead to the mass movement. These can include slope, geology, soil type and structure, vegetation type, water, loads and lateral support.

Generally, mass movements occur when the weight of the surface material on a slope exceeds the restraining ability of the underlying material. Mass movements most frequently occur when the slope has an angle greater than 25°, has little vegetation cover and is in an area where annual rainfall exceeds 900 mm^3.

Strategies to prevent or reduce the risk of mass movements include:
• diverting the water away from areas that are prone to land slippage
• revegetating the prone area with deep-rooted trees and perennial plants
• improving surface and subsurface drainage
• removing stock from the area.

Human action

Erosion is a natural process that can be modified by human action, including land clearance, agricultural processes (such as ploughing, irrigation and grazing), forestry, construction, surface mining and urbanisation. It has been estimated that 15%, or 2000 million ha, of the Earth's land surface has been degraded by such processes. About 50% of that total is due to human-induced water erosion and 33% has been affected by wind erosion. The remaining 17% of the soil degradation has been caused by the action of chemicals.

1
Define the term 'erosion'.

2
Explain how the rate of erosion can be affected.

3
Discuss why raindrops are a major erosive agent.

4
Describe the conditions that increase the susceptibility of soils to wind erosion.

5
Differentiate between the following types of erosion:
a rill erosion
b gully erosion
c stream-bank erosion.

6
Compare and contrast, with regard to their effects and the methods of reducing their effects, the following types of erosion:
a wind erosion   **d** gully erosion
b sheet erosion   **e** stream-bank erosion.
c rill erosion

IMPACT OF INDUSTRY, URBANISATION AND AGRICULTURE

The by-products of industry may destroy vegetation by increasing soil acidification. Soil acidification can be minimised by improving emission-control technologies, thereby reducing the amount of waste produced.

Parts of the Australian landscape were scarred during the nineteenth and early twentieth centuries by industry searching for raw materials. Over the latter part of the last century, industry generally adopted a more environmentally sensitive approach to its search for materials, exploration became more precise, and damaged areas were restored. Certain mining practices, such as the use of cyanide in the extraction of gold, once led to soil contamination through toxic-waste disposal. Legislation, better industrial processes and more responsible environmental attitudes have led to improvements in the minimisation, recycling, treatment and disposal of waste.

Urbanisation can increase soil degradation in a number of ways, such as by altering natural hydrological pathways, reducing soil cover and water infiltration, and causing compaction and contamination. It can alter the hydrology of a **watershed** or a specific drainage area in two ways. First, the construction of roofs, driveways, pavements and parking lots makes an increased percentage of the soil surface impervious to infiltration by water. It has been estimated that, in residential areas, 25% of the area of a 1400 m² property is impervious to infiltration. For smaller areas, say, 560 m², the impervious area is 80%. Impervious surfaces decrease infiltration and increase overland flow. This increases the frequency and height of flood peaks during heavy storms. More topsoil is washed away, erosion increases on the areas of open land and there is less **recharge** to ground water bodies. A second change caused by urbanisation is the introduction of stormwater drains. Water is taken directly by the drains to stream channels for discharge, with a shortened travel time and increased proportion of run-off. This also increases the chance of flooding.

During the process of urbanisation, large-scale land clearing takes place and this may cause soil erosion. Once the development has been built, the original native vegetation is replaced with introduced ornamental plants. These 'garden' plants may have shallow roots and there may be reduced ground cover, both of which result in reduced soil fertility and increased erosion rates.

watershed
a ridge or crest separating two drainage areas

recharge
where surface water enters the ground through the process of infiltration

1

Outline the three things that have led to improvements in the minimisation, recycling, treatment and disposal of waste.

2

Describe how urbanisation can have an effect on the hydrology of a watershed.

3

Explain why there has been an increase in the frequency and height of flood peaks.

4

Research how cyanide is used in the extraction and purification of gold. Explain the effect of cyanide on the environment.

5

Research how exploration for mineral resources has become more precise over the last fifty years.

MANAGEMENT STRATEGIES TO PREVENT AND REDUCE SOIL EROSION

The Australian landscape is facing a crisis and it affects all Australians, whether they live in rural, regional or urban areas. Salinity, habitat loss, soil degradation, and river degradation and pollution are clear warnings that the way we use and manage the land is not sustainable.

Over the last two decades, government and the agriculture sector have realised that soil erosion is a major problem in New South Wales and that statewide measures need to be carried out to reduce the problem. Soil degradation not only undermines rural viability but also the viability of tourism.

The NSW Department of Land and Water Conservation has legislative responsibility for the protection and **conservation** of the state's soil resources and the prevention of soil erosion. The state soils policy was passed by the state Cabinet in 1987 and covers all soils in New South Wales used by agriculture. The policy's goal is to achieve proper use and management of soils, thereby maintaining, and in many instances improving, their utility and productive capacity, with a consequent reduction in environmental damage.

The principles of the policy are outlined below:

• The use of soils should not lead to their degradation.
• Both the NSW Government and other interested parties have a joint responsibility to prevent degradation.
• A state soil conservation organisation with statutory powers is required to prevent further soil degradation and to protect and improve degraded soils.
• **Total catchment management** is required to conserve and rehabilitate soils.
• The fertility and structure of agricultural soils should be maintained and improved.
• Soil use should be minimised.

Agriculture is a crucial part of the landscape, and is critical for the future of Australia's economic and social wellbeing. The challenge here is to ensure we protect and sustain the soils on which the future of agriculture depends as we move to more sustainable agricultural practices. Management of soil to prevent water and wind erosion is based on sensible soil conservation practices. For effective erosion control it is important to maintain good soil structure, protect the soil surface by adequate crop cover and use erosion control practices where necessary. These practices often control both water and wind erosion. Each erosion problem can be remedied by choosing one or more remedial practices appropriate to the problem.

Good soil structure is a result of management systems that include the regular use of soil-improving crops (such as **forage**), the frequent return of organic matter in manure and residues from crops (such as stubble), and tillage practices (ploughing) that avoid unnecessary breakdown of soil structure. Soil loss from ploughed land can easily exceed 7 tonnes per hectare.

Growing crops on slopes increases the possibility of soil erosion. Soil erosion can be reduced by ploughing and planting crops across, rather than with, the slope. This can reduce soil loss by 25%. Another method of planting that reduces erosion is strip cropping. This method involves alternating hay (a forage crop) and grain strips across a slope.

Crop rotation improves the overall efficiency of nitrogen uptake and utilisation in the soil. If certain cover crops are planted in the winter, erosion and run-off is prevented when the ground thaws, and nutrients are trapped in the soil and released to the spring crops.

Constructing bench-like channels, or terraces, enables water to be stored temporarily on slopes, allowing sediment deposition and water infiltration. When trying to reduce the possibility of severe gully erosion, grassed waterways provide a helpful solution. They force stormwater run-off to flow down the centre of an established grass strip and can carry very large quantities of stormwater across a field without erosion. Grassed waterways are also used as filters to remove sediment, but may sometimes lose their effectiveness when too much sediment builds up in the waterways. To prevent this, it is important that crop residues, buffer strips and other erosion control practices and structures be used along with grass waterways.

Diversion structures are channels that are constructed across slopes to ensure water flows to a desired outlet. They are similar to grassed waterways and are used most often for reducing and preventing gully erosion.

Drop structures are small dams used to stabilise steep waterways and other channels. They can handle large amounts of run-off water and are effective where falls are less than 2.5 m. In channel stabilisation, drop structures (such as a straight-drop spillway) are constructed to direct the flow of water through a weir, into a stilling basin. There the energy of the water is dissipated before it flows into the channel below.

Riparian strips are buffer strips of grass, shrubbery and other vegetation that are grown on the banks of rivers and streams and in areas with water conservation problems. The strips slow run-off and catch sediment. In water with a shallow flow, they can reduce sediment and the nutrients and herbicides attached to it by 30–50%. Riparian strips host biodiversity, which benefits the environment.

conservation
preserving something of value

total catchment management
a management practice that deals with a river system's whole catchment

forage
fodder for farm animals

definitions

REVIEW ACTIVITIES

1
Which state government has legislative responsibility for the protection and conservation of the state's soil resources?

2
Outline the measures that need to be taken to ensure effective erosion control.

3
Explain how terracing can reduce soil erosion.

EXTENSION ACTIVITIES

4
Describe three methods that builders can use to reduce or prevent soil erosion on building sites.

5
Justify the use of riparian strips on agricultural land.

SUMMARY

Soil degradation can be in the form of nutrient loss, acidification, salinisation, decline in soil structure, biodiversity and erosion.

Soil erosion is the movement of particles of soil, surface sediments and rocks by the action of water, glaciers, winds, waves and so on.

Soil erosion processes include wind erosion, sheet erosion, rill erosion, gully erosion and mass movements.

Rates of erosion are affected by the size and velocity of raindrops, the soil's permeability, the size and shape of soil particles, the slope angle, and the amount of exposure and vegetation cover.

Wind erosion may cause saltation, soil creep and dust storms. It removes the finer material, containing organic matter and clay and silt particles.

Urbanisation increases soil degradation by altering natural hydrological pathways, reducing soil cover and water infiltration, and causing compaction and contamination.

The NSW Department of Land and Water Conservation has legislative responsibility for the state's soil resources.

PRACTICAL EXERCISE
Soil erosion

In this exercise you will design and carry out experiments to investigate the differences between compacted and non-compacted soils.

Procedure

1
Use two types of soil: a sand-based and a clay-based soil. Describe how you can produce a compacted sample of each soil.

2
Design and carry out an experiment to measure how much water is absorbed by both compacted and non-compacted soils.

3
Design and carry out an experiment to measure erosion rates using a stream tray.

4
Once you have completed the experiments, complete the following activities.

ACTIVITIES

1
Are both types of soil equally easy to compact?

2
Is it easier to compact dry soils or damp soils?

3
What percentage of water is absorbed by each type of soil?

4
Did the soils absorb water at the same rate?

5
Which type of soil eroded most quickly?

6
Research why one soil suffered greater erosion than the other.

7
How did the compaction affect erosion rates?

8
Erosion by visitors is a major problem in national parks. Outline some of the ways the National Parks and Wildlife Service has tried to reduce compaction of soils in fragile areas.

3.2 Salinity of soils and water

OUTCOMES

At the end of this chapter you should be able to:

- identify regions of Australia with naturally saline land

- examine the possible consequences for soil salinity of land clearing and irrigation and outline precautions that could minimise the problem in each case

- summarise a specific example of a successful government or community strategy employed to rehabilitate salt-affected land in New South Wales.

SALINISATION

Salinisation is the build-up of salt within a soil and occurs when salt is moved around and concentrated in the soil. Because salt dissolves, it is carried in water. When water evaporates it leaves the salt behind in the surface layers of the soil. If this process continues long enough, the concentration of salt becomes too high for most plants and they die. Salt is already present in our soil because much of Eastern Australia was once covered by the sea. However, most of the salinity has resulted from humans altering the way water moves through the environment. This is done by irrigation, alteration of vegetation, and fertiliser use.

There are five types of salinity: irrigation, dryland, urban, river and industrial.

Irrigation salinity

Irrigation salinity is caused by excess water from irrigation, which raises the **watertable**, bringing salt to the surface. This form of salinity is estimated to affect 320 000 ha, or 15%, of irrigated land in New South Wales. A further 70–80% of the state's irrigated land is threatened by rising watertables and associated salinity problems. (See Figure 3.2.1.)

Irrigation salinisation can be reduced by using less water on crops and by growing crops that require less water. Unlike dryland salinity, which takes a long time to appear, irrigation salinity occurs soon after an irrigation scheme is introduced. Some of the problems of irrigation salinity are understood and a number of them have been solved.

Dryland salinity

Dryland salinity affects 120 000–174 000 ha of land in New South Wales. Large areas of the Western Slopes, the Hunter Valley and the Sydney Basin have saline water just 2 m below the surface. The Murray-Darling Basin Salinity Audit has projected that by 2050 salinity will affect 2–4 million ha within the basin.

Dryland salinity is caused when land is cleared and the native vegetation is replaced with introduced crops, which use less water than the native vegetation. This results in more water seeping down through the soil to join the existing ground water, which causes the watertable and the salts to rise. When the watertable has risen to within 2 m of the surface, capillary rise transports the salts to the surface.

watertable
the position in the Earth's crust below which cracks and other openings are filled with water

definition

Figure 3.2.1 Damage resulting from irrigation salinity.

Replanting cleared areas with trees and deep-rooted grasses is often used to lower rising watertables. Replanting has to take place over large areas for it to have any use.

Urban salinity

Urban salinity occurs in the built environment and is a combination of irrigation salinity and dryland salinity. It results when there is both land clearing and rising watertables. In urban environments, the watertable will rise for a number of reasons: excessive watering of gardens and parks, leaking from sewers and drainpipes, and obstruction or modification of drainage paths.

Urban salinity in the south-west of New South Wales is estimated to cause $9 million worth of damage to roads annually. In Wagga Wagga alone, the cost of salinity over the next thirty years will be in the hundreds of millions of dollars.

River salinity

River salinity is caused by saline discharge into rivers from other salinity affected areas. As salinity increases, water quality decreases. At Narromine, the Macquarie River flows through the town, carrying 230 000 tonnes of salt each year. Research has shown that the Macquarie, Bogan, Namoi, Lachlan and Castlereagh Rivers all show a significant increase in salinity levels. These levels are expected to continue to increase in the foreseeable future.

Industrial salinity

Industrial salinity is the result of chemical, washing and cooling processes in manufacturing procedures and mining practices as well as cooling water from coal-fired power stations. Mining activities that took place prior to strict rehabilitation regulations have led to abandoned mines being a major source of salt in some water catchment areas.

REVIEW
ACTIVITIES

EXTENSION
ACTIVITIES

1
Outline why salt is already present within the soil.

3
Outline the causes of urban salinity.

5
Explain why irrigation salinity appears at a faster rate than dryland salinity.

2
Describe what the five types of salinity are.

4
How much salt is carried by the Macquarie River?

6
Describe the types of damage caused by urban salinity.

THE EFFECTS OF SALINITY

The problems caused by salinity are very serious. Increasing salinity levels in rivers will put pressure on water supplies for many towns. The increasing salinity will result in a need for more thorough treatment of water, resulting in an increase in costs and infrastructure.

Apart from the direct affect of salinity on agricultural crops, such as rice and cotton, agricultural costs have increased due to mitigation procedures and loss of agricultural land. The cost of dryland salinity to NSW agriculture has been estimated to be at least $130 million and is increasing each year. Adverse effects on soils include changes in soil composition and structure and an increasing content of toxic substances that limit plant growth. The breakdown in soil structure, together with the loss of plant cover, results in a greater rate of soil erosion. It can also have an affect on the manufacturing and food-processing industries, as a number of their processes require low levels of salt.

The increase in salt levels has had an adverse affect on the natural environment. It has resulted in a loss of biodiversity and long-term damage to the soil structure.

Some further effects of salinity are listed below:

- Unless changes occur, on current trends, the total area of (mainly farming) land in Australia affected by salinity is expected to rise from 2.5 million ha to over 15.5 million ha; an increase of over 620%.
- In south-west Western Australia, creeping salinity has placed literally hundreds of plant species at risk of extinction.
- Salinity levels at Morgan on the Murray River in South Australia, where Adelaide extracts its water, will exceed the World Health Organisation (WHO) drinking water guideline of 800 electrical conductivity (EC) units by 2020 in two days out of five. (See Figure 3.2.2.)
- In south-west Western Australia, one-third of the wheat belt will be salt-affected within twenty years, and one-half of the divertible water resources is already salt-affected.
- On current salinity trends, internationally significant Ramsar-listed wetlands (such as the Macquarie Marshes in New South Wales) will become saline wastelands within a few decades.

Table 3.2.1 illustrates the extent of current and projected damage to selected Australian assets.

A number of features identify salt-affected areas. They include increased soil erosion, waterlogged soil, dying trees, bare patches, salt scalds, damage to buildings and salt-tolerant plant species, such as native reed, spiny rush, bulrush, couch grass and swamp oak. (See Figure 3.2.3.) Clear water in dams and streams is another indication that an area is salt-affected. High salt concentrations cause clay particles to immediately drop out of suspension and sink to the bottom.

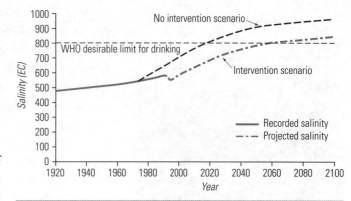

Figure 3.2.2 Salinity levels at Morgan on the Murray River in South Australia.

Figure 3.2.3 Salt-tolerant vegetation: sea barley grass.

Table 3.2.1 Australian assets at high from risk shallow watertables with a high salinity hazard				
Asset	**Unit**	**Year**		
		2000	**2020***	**2050***
Agricultural land	Hectares	4 650 000	6 371 000	13 660 000
Remnant and planted perennial vegetation	Hectares	631 000	777 000	2 020 000
Length of streams and lake perimeters	Kilometres	11 800	20 000	41 300
Railways	Kilometres	1 600	2 060	5 100
Roads	Kilometres	19 900	26 600	67 400
Towns	Number	68	125	219
Important wetlands	Number	80	81	130

*Projection

Figure 3.2.4 Dieback in trees caused by salinity.

As previously stated, increasing salinity levels result in higher rates of erosion and this can be seen in terms of sheet, rill, gully and wind erosion. (See Chapter 3.1.)

As the salt content increases, more salt-tolerant plants will replace the native vegetation. Salt-tolerant plants are more efficient at extracting water and are able to tolerate higher salt concentrations than more vulnerable species. As the watertable rises towards the soil surface, the airspaces in the soil fill with water. As a result, the roots of plants are starved of oxygen and the plants die. This process causes bare patches in pastures and paddocks. Trees are one of the first types of vegetation to respond to increasing salinity levels because their deep roots come into contact with the rising saline watertable before the root systems of other plants. This results in dieback of the trees. (See Figure 3.2.4.)

If salt levels are very high and the watertable has reached the soil surface, deposition of salt may occur on the surface and result in a salt scald.

High salinity levels can also damage houses, buildings and other structures. The salt is drawn up by capillary action from the soil and into brickwork. The salt dries out and expands, resulting in the breakdown of the bricks and mortar. This breakdown is called spalling. Salt also increases the corrosion rate of metals, especially those buried under the ground. Corrosion of metal reinforcements in structural concrete may cause foundations to shift or sink, leading to structural cracking, damage or collapse. The increasing height of watertables leads to the failure of septic tanks, resulting in environmental and health problems. Salt also breaks down bitumen and asphalt, which causes damage to roads and highways. Salt affects about 34% of the state's roads and 21% of national highways in south-western New South Wales.

REVIEW ACTIVITIES

1
Describe the features that can be used to identify a salt-affected area.

2
Explain why trees suffer from dieback and why it is one of the first symptoms of increasing salinity.

3
Define the term 'spalling'.

EXTENSION ACTIVITIES

4
Outline how salt causes corrosion of metal objects.

5
Go to the State of the Environment 2000 website <http://www.epa.nsw.gov.au/soe/soe2000/cl/cl_map_4.4php>. Locate map 4.4 (mapped, observed and measured soil salinity) and identify the main areas where salinity readings are very high and the main areas where salinity is very low.

SALINITY AND THE THREAT TO BIODIVERSITY

Salinity is a major threat to both aquatic and terrestrial biodiversity. High salinity levels kill both vegetation and animals living in the soil, leading to widespread loss of flora and fauna.

In Western Australia, 450 plant species are threatened with extinction from rising salinity. Many areas of remnant vegetation are directly threatened by salinity, particularly along creek and river frontages and floodplains. In some cases, entire plant communities (such as red gum, grey box and black box woodlands) are under serious threat in the medium to long term.

Many aquatic invertebrates are sensitive to increased salt levels. These include water snails, water fleas, dragonflies and some flies. These species are important sources of food for fish and other invertebrates. Aquatic plants are also salt-sensitive, particularly as seeds and seedlings.

Salinity is already reducing the diversity of aquatic life in rivers and wetlands. Salt sensitivity starts at concentrations as low as 800 mg of salt per litre, or about 480 EC units. Reproduction in many aquatic species declines significantly at salinity levels of 200 EC. Many NSW rivers already exceed these thresholds, including the Loddon (875 EC), the Avoca at Third Marsh (1440), the Bogan (730), the Castlereagh (640) and the Namoi (680).

To maintain salt levels below ecological thresholds in rivers it is necessary to ensure that the flow of water through rivers and wetlands is high enough to dilute the salt levels. Flow management is particularly important during spring, when fish and other aquatic life are breeding. Plant seeds and seedlings are also most sensitive in spring.

MANAGEMENT OF SALINITY

At a catchment level, steps are being taken to understand and manage the salinity problem. Information is required on:
- where and how much salt is stored in the landscape
- how subcatchments contribute to the whole
- how salt moves
- what affects the rate of movement of salt
- how changing something in one location affects the rest of the catchment or basin.

This information is sourced from catchment areas and passed on to the Centre for Natural Resources at the Australian National University in Canberra. The centre uses these data in a number of salinity computer models, such as PERFECT and CATSALT, to:
- evaluate where salinity is currently occurring in the landscape and where it may occur in the future
- predict the impacts of land-use changes on run-off, salt loads and salinities from catchments and streams
- evaluate the contributions of these changes to meeting local salinity targets.

The aim of the salinity modelling is to provide practical scientific and technical support on biophysical aspects of salinity management.

Households and communities can drastically reduce urban salinity by changing their gardening practices. These include:
- Water gardens only when it is necessary and use a sprinkler system with a timer.
- Mulch garden beds.
- Grow plants with low water needs, such as native trees and shrubs.
- Group plants of similar water usage.
- Test the salt load of the soil and use potted plants if the soil is too salty.
- Select salt-tolerant plants.
- Import fresh soil and use raised garden beds.

The salinity problem has become so large that both state and federal governments are taking measures to reduce or reverse the effects of salinity. The Federal Government has instigated two national audits on water quality and salinity levels. Two audits in 1996 and 2000 have provided the information on which Australian governments can develop their recent initiatives to deal with dryland salinity and water-quality problems. The Council of Australian Governments (CoAG) has agreed to a Commonwealth-initiated, $1.4 billion National Action Plan for Salinity and Water Quality. CoAG said the issues of salinity and deteriorating water quality are of major national significance and are best handled on a nationwide basis. The 2000 audit revealed that about 5.7 million ha in Australia are within regions mapped to be at risk from, or already affected by, dryland salinity. Ground water levels in many of these areas are still rising, so that in fifty years the affected area could increase to just over 17 million ha.

The NSW Government, after consultation with interested parties, produced a strategy aimed at dealing with the salinity problem. The NSW Salinity Strategy was released in October 2000 following summits in Dubbo and Wagga Wagga. Recommendations were gathered from the community, land managers, farmers, conservationists, representatives of Aboriginal communities, industries, local and state governments and scientists. The strategy set out to identify and explain the salinity problems being faced and how these problems can be managed.

The strategy identified that to slow the increase in salinity, a number of steps have to be taken. They include using less water and using water more efficiently so that less water enters the watertable. It also identified that protecting and managing our native vegetation is an essential step in managing the recharge of the watertable. The maintenance of existing native vegetation is both easier and less costly than revegetating an area.

In recharge areas the mix of vegetation has to be altered so that there is an increase in vegetation with long roots. It is necessary to change the way crops are managed and to increase tree planting. Taking some recharge areas out of agricultural use will slow the increase in salinity.

The efficient use of water is equally as vital in the urban situation. Local councils have to:

• Make better use of land affected by salt. It is important to change the mix of vegetation in recharge areas so that less water leaks through to the watertable.
• Use water more effectively and efficiently. There is a relationship between the available water and the quality of the water.
• Use engineered solutions. This is helpful in cases where other salinity management actions may actually increase the problem in the short term.
• Focus on landscapes at immediate threat from salinity. An understanding of the salinity features of a landscape allows efforts and resources to be concentrated on landscapes where high salinity may affect the productive capabilities of the land.

The Federal and NSW Governments have proposed spending over $3 billion in the foreseeable future on remediation work as well as preventive schemes to reduce the effects of salinity in New South Wales. A proposed breakdown of how the money will be spent is outlined in Table 3.2.2.

Salinity remediation measures

There are a number of ways that salinity can be combated. Some are accepted as being successful over a limited area, while others are being trialled or have been proposed.

Once an area has been identified as being salinity affected, remediation work can be carried out to improve the levels of salinity. The primary focus is to achieve a hydrological equilibrium between water supply and demand. This involves balancing water and salt removal and can be achieved by draining small areas with high watertables. Types of drainage include surface drainage, subsurface drains to divert ground water, and vertical drains in the form of pumping wells. Once it has been extracted, salt-laden water needs to be placed somewhere. It can be transported to evaporation basins, where the salt is harvested, or piped into the ocean.

To maintain lower salt levels it is necessary to reduce further recharge. This can be done by improving irrigation practices (such as lining or piping irrigation channels to reduce leakage), varying the irrigation patterns according to the growing season, replanting native vegetation in recharge areas or reducing ground water recharge rates.

Other salinity management procedures to address the problem include:

- introducing price incentives to encourage ground water extraction methods that control watertable levels
- intercepting saline ground water before it discharges into rivers or sensitive wetlands
- reusing slightly saline water for irrigation of salt-tolerant crops
- mixing saline and fresher water before reuse.

A more controversial procedure is to pump the saline water into deep aquifers, which could act as salt-storage facilities.

Table 3.2.2 Breakdown of proposed expenditure on the repair of salinity problems

Item	Extent	Percentage of total repair bill (approximate)
Farm, forestry and plantations for salinity management	16.5 million ha	41.9%
Revegetation mainly for biodiversity conservation (also meeting erosion control, river corridor and salinity objectives)	7.4 million ha	25.4%
Protection of river frontages	More than 500 000 km	2.2%
Improved land management, such as salt land agronomy and acid soils management	76 million ha	14.6%
Protection of rangelands with high biodiversity values	32 million ha	1.1%
Fencing of trees, remnant vegetation and so on	1.4 million km	5.0%
Environmental flow regimes for stressed rivers	15 major river systems	0.2%
Protection of valuable vegetation communities on private land	More than 5 million ha	1.0%
Other expenditure, including ongoing management, monitoring, information and education	–	5.2%

Information based on expenditure outlined by the NSW Salinity Strategy

REVIEW ACTIVITIES

1
Outline how salinity computer models, such as PERFECT and CATSALT, can be used to help solve salinity problems.

2
Recount how many hectares of land have been affected by dryland salinity according to the 2000 audit carried out by CoAG.

3
On whose recommendations was the NSW Salinity Strategy based, and what did the strategy set out to do?

4
Explain what steps have to be taken in order to slow down the increase in salinity.

EXTENSION ACTIVITIES

5
A number of salinity experts have proposed an increase in the planting of saltbush on farms affected by salinity. Saltbush can be used as an alternative fodder plant to grass. Research how the plant copes with high salinity levels.

6
Research the proposals that have been put forward by the South Australian Government to tackle the increasing salinity of the drinking water from the River Murray.

SUMMARY

- Salinisation is the build-up of salt within a soil. It occurs when salt is moved around and concentrated in the soil.

- Salt is already present in our soil because much of Eastern Australia was once covered by the sea.

- There are five types of salinity: irrigation, dryland, urban, river and industrial.

- Irrigation salinity is caused by excess water from irrigation, which raises the watertable, bringing salt to the surface.

- Dryland salinity is caused by land clearing.

- Urban salinity is a combination of both irrigation and dryland salinity and occurs in the built environment.

- River salinity is caused by saline discharge into rivers from other salinity affected areas.

- Industrial salinity is the result of chemical, washing and cooling processes in manufacturing procedures and mining practices as well as cooling water from coal-fired power stations.

- Features that identify salt-affected areas include an increase in soil erosion, salt-tolerant plant species, waterlogged soil, dying trees, bare patches, salt scalds, damage to buildings, and clear water in dams and streams.

- High salinity levels results in damage to houses and other structures caused by the deterioration of bricks, mortar, concrete and metal.

- Salinity is a major threat to both aquatic and terrestrial biodiversity.

- The Australian National University uses a number of computer models to evaluate and predict the potential effects of salinity.

- CoAG agreed to a Commonwealth-initiated, $1.4 billion National Action Plan for Salinity and Water Quality.

- To improve the levels of salinity, the primary focus is to achieve a hydrological equilibrium between water supply and demand. To achieve this equilibrium it is necessary to balance water and salt removal.

Cross-section of a watertable

The aim of this exercise is to investigate how changes in watertable depth affect soil salinity.

Background

A series of drillings was taken around the town of Gundoo to measure the depth of the watertable. The results are shown in Figure 3.2.6.

ACTIVITIES

1

The depth recordings shown in Figure 3.2.6 can be used to draw a cross-section of the watertable.

a Using graph paper, draw a cross-section of the surface of the land.

b Using the information from Figure 3.2.6, draw a cross-section of the watertable between points A and B at the correct depths below the surface of the land.

2

Arable farming has taken place to the east of the town during the last fifty years. However, recently the crops have been failing. Some of the farmers have attributed the crop failures to rising salinity levels. Describe the visual signs that could be used by the farmers to indicate whether salinity is the problem.

3

The town of Gundoo obtains most of its water from the Haygor River, which is becoming increasingly discoloured and saline. Research and then outline some of the alternative ways that could be used to obtain freshwater.

4

Use your cross-section to determine the areas that will be first affected as the watertable rises.

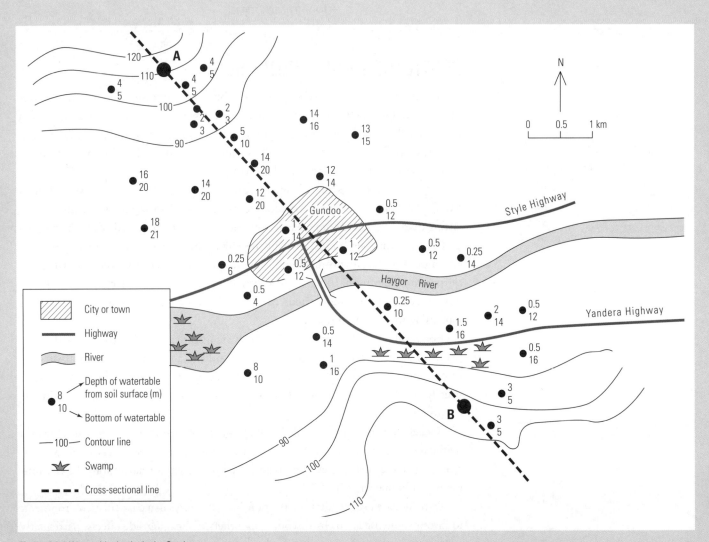

Figure 3.2.6 Watertable depths in the Gundoo area.

OUTCOMES

At the end of this chapter you should be able to:

assess alternative management practices that do not require the use of pesticides

discuss the effect of continually introducing new pesticides into the environment, including
– effect on non-target species
– accumulation in individuals (bio-accumulation) and magnification in animals higher up the food chain (biomagnification)
– human health impacts.

PESTICIDES AND THEIR USES

Using **pesticides** may be necessary at times, but in many cases there are alternatives that are often more effective in the long run and less harmful to the environment and humans. All pesticides are, by definition, toxic to some living thing: insecticides to insects, herbicides to plants, fungicides to fungi, and so on. In addition, they often have direct or indirect effects on other living things.

Synthetic pesticides are chemical compounds that have been invented in a laboratory. Some synthetic pesticides are more toxic than others, some are longer-lasting than others, and some release compounds that are more toxic than the original pesticides when they break down in the environment. Certain synthetic pesticides accumulate in the environment and cause harm far removed from the original site or purpose of application. Others, called persistent materials, just 'hang around' and do not break down for long periods of time. This results in possibly toxic chemicals remaining in the environment.

Botanical pesticides, almost all of which are insecticides, are derived directly from plants or animals. Some are more toxic than others; a few are even more toxic than some synthetics. However, all botanicals break down rapidly in the environment, usually in a matter of hours or days, and are not known to accumulate in the environment.

Rachel Carson's famous book *Silent Spring*, published in 1962, alerted the world to the dangers of pesticides. Supporters claim that pesticides are essential, whereas critics argue that many uses of pesticides are unnecessary or harmful to the environment and human health.

Pesticides are used to alter the relationships of species in a particular environment: to favour some life forms by discouraging others. Damaging side effects, particularly over the long term, can be subtle and, sometimes, profound.

definition

pesticides
include insecticides, herbicides and fungicides, which are designed to kill insects, weeds and diseases, respectively

Effects on non-target species

Pesticides can adversely affect wildlife through changes in the food web, direct and indirect poisoning, chemical **biomagnification**, and habitat changes. Their harmful effects may show up in animals that have no direct relationship to the original pest. In the 1960s, wildlife biologists were puzzled when the gannet (a seabird) population began to shrink. The shells of the gannet's eggs were too thin to protect the embryos. Only after the eggs had been analysed did the scientists identify the culprit: DDT, which had been biomagnified through higher levels of the food web. (See the DDT case study later in this chapter.)

Modern chemical pesticides are much less persistent than older pesticides, such as DDT, but they may be highly toxic to some non-target organisms. Phosphamidon, for example, which was intended to replace DDT in forest protection, killed large numbers of birds. The spraying of fenitrothion, which replaced both DDT and phosphamidon, has less severe effects, but it still remains a cause for environmental concern.

Pesticides also change wildlife habitat. For example, if a herbicide kills hardwood cover, birds and mammals that depend on that habitat can no longer live there. Nor can their predators. There are both winners and losers, however, since other animals may benefit from the changes. Regardless, the habitat changes caused by the herbicide affects the whole of the flora and fauna within the habitat.

Pesticides enter watercourses primarily through direct application, aerial spray drift and run-off from treated areas. Rinsing and filling spray equipment near streams can result in widespread contamination, killing fish and aquatic insects. Many pesticides enter aquatic sediments, where they can persist for long periods. Populations of aquatic insects, an important food source for fish, are drastically reduced after forest and agricultural spraying. Under some conditions, the spraying of insecticides such as deltamethrin and endosulfan may result directly in the death of fish. When herbicides enter watercourses, they may also affect vital plant life in aquatic systems.

It has been found that pesticides may not just affect the local area after application, but that they may drift in the air away from the point of application as pesticide drift. Examples of pesticide drift are:
- 200 m away, fish die within twenty-four hours
- 7.5 km away, 16% of the sprayed pesticide is still airborne
- 75 km away, minute amounts are measured in falling rain.

Ground water moves slowly, and once ground water supplies are contaminated they may remain so for decades.

> **biomagnification**
> the tendency of a pollutant to concentrate as it moves from one trophic level to the next, resulting in an increase in concentration of the pollutant from one link in a food chain to another
>
> definition

REVIEW ACTIVITIES

1
Define the term 'pesticide'.

2
Outline the main problems posed by pesticides.

3
Explain the advantage in using modern pesticides rather than older pesticides.

4
Describe the possible effects of pesticides entering watercourses.

EXTENSION ACTIVITIES

5
Pesticides are known to persist in watercourses both in the sediments and within the food chains. Research why pesticides persist in the sediments.

6
Read Rachel Carson's *Silent Spring*.

BIO-ACCUMULATION AND BIOMAGNIFICATION

bio-accumulation
the increase in concentration of a pollutant from the environment to the first organism in a food chain

Bio-accumulation

Bio-accumulation occurs when the concentration of a chemical in an organism becomes higher than its concentration in the air or water around the organism. Bio-accumulation is a normal process within organisms. All organisms bio-accumulate many important nutrients. For example, animals bio-accumulate vitamins A and D, trace minerals and amino acids. The problem occurs when the substances that are bio-accumulating are potentially harmful.

Bio-accumulation is the net result of the interaction of uptake, storage and elimination of a chemical. The process of bio-accumulation starts with a chemical passing from the environment into an organism's cells. These chemicals enter the cells by the process of diffusion, moving from an area of high concentration to an area of low concentration. Diffusion continues to operate within the organism, hindering the return of the chemical to the environment. Some chemicals are attracted to certain sites, where they bind to proteins or dissolve in fats. Heavy metals, such as mercury, and water-soluble chemicals bind tightly to specific sites within the body. For example, cobalt binds to sites in the liver. A number of chemicals are stored in the fat layers of an organism. There the chemicals are detoxified or rendered harmless until the fat layer is used up and the chemicals are recirculated into the body. If the fat layers are not used, bio-accumulation may continue.

The rate of bio-accumulation is affected if the organism can break down or excrete the chemical. A chemical that dissolves readily in fat but not in water tends to be more slowly eliminated by the body and will therefore have a greater potential to accumulate in the body. Bio-accumulation varies between individual organisms as well as between species. Large, fat, long-lived individuals or species with low rates of metabolism or excretion will bio-accumulate more than small, thin, short-lived organisms.

Biomagnification

Biomagnification follows bio-accumulation. It is the increase in concentration of a bio-accumulated chemical as it moves through a food chain. A large vertebrate, such as a frog, may eat thousands of insects through the course of its lifetime. By eating the insects, the animal will be exposed to insecticide doses many times the amount it absorbs directly. A single bird may eat dozens of these bio-accumulated animals, multiplying its own pesticide exposure with each feeding. (See Figure 3.3.1.) In this way, the levels of pesticide biomagnify up the food chain. US studies around Lake Michigan have shown carnivorous birds having 7000 times greater chemical residue than the lake mud.

In order for biomagnification to occur, the pollutant must be long-lived, mobile, soluble in fats and biologically active.

If a pollutant is short-lived, it will be broken down before it can become dangerous. If it is not mobile, it will stay in one place and is unlikely to be taken up by organisms. If the pollutant is soluble in water it will be excreted by the organism. Pollutants that dissolve in fats, however, may be retained for a long time. It is traditional to measure the amount of pollutants in fatty tissues of organisms such as birds or fish. In mammals, we often test the milk produced by females, since the milk contains a lot of fat and because the very young are often more susceptible to damage from toxins (poisons). If a pollutant is not active biologically it may biomagnify but it is not a cause of great concern as it is unlikely to cause any problems.

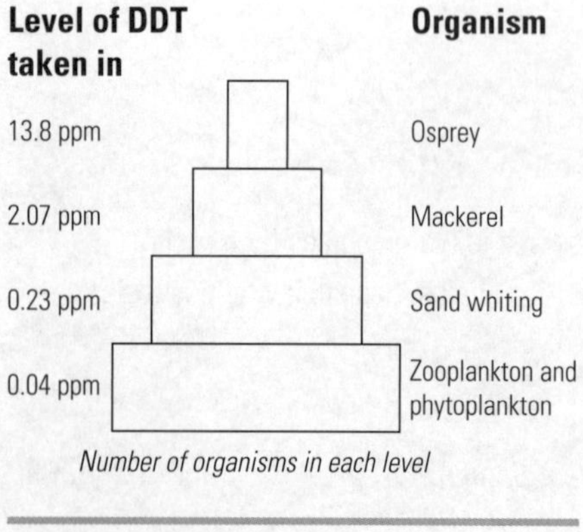

Level of DDT taken in	Organism
13.8 ppm	Osprey
2.07 ppm	Mackerel
0.23 ppm	Sand whiting
0.04 ppm	Zooplankton and phytoplankton

Number of organisms in each level

Figure 3.3.1 Biomagnification.

Adverse effects of bio-accumulation and biomagnification

Bio-accumulation and biomagnification cause concern because, together, they allow even small concentrations of chemicals in the environment to find their way into organisms in high enough dosages to cause problems.

A group of pollutants with a high rate of biomagnification due to their **lipophilic** nature are polychlorinated biphenyls (PCBs). PCBs are persistent organic pollutants. They have a long half-life and are slow to degrade, biologically, chemically and physically. PCBs are easily transported by wind, ocean currents and biota. There are many health problems associated with PCBs, among which are reproductive difficulties, immune system dysfunction, developmental difficulties (specifically neurological ones) and an increased risk of cancer. Since 1967, monitoring of certain organochlorines, such as PCBs, has been conducted. The Laboratory for Analytical Environmental Chemistry found that while the PCB levels have decreased since 1967 because of environmental protection programs the rate of decline indicates continued pollution. The levels are also such that they still pose both environmental and human health threats.

Table 3.3.1 lists some of the common substances that have the potential to biomagnify.

lipophilic
attracted to lipids, including fats

definition

Table 3.3.1 Common substances that biomagnify in the environment

Substance	Uses	Associated problems
PCBs	Insulators in transformers Plasticisers Fire retardant	Biomagnifies Impairs reproduction Widespread in aquatic systems
Polynuclear aromatic hydrocarbons (PAHs)	Component of petroleum products	Carcinogenic
Heavy metals: mercury, copper, cadmium, chromium, lead, nickel, zinc and tin (TBT or tributylin)	Mercury is used in gold mining and the other heavy metals listed are used in metal processing	Some may affect nervous system and reproduction
Cyanide	Leaching gold Fishing	Toxicity Concentrated by farming desert soils
Selenium	Used in solar cells and by the glass manufacturing industry	Reproductive failures Toxicity

REVIEW ACTIVITIES

1
Differentiate between bio-accumulation and biomagnification.

2
Recount the factors that make a chemical a likely candidate to biomagnify.

EXTENSION ACTIVITIES

3
Minamata disease is caused by the biomagnification of mercury. Research the cause and effect of this disease.

4
Explain why large, fat, long-lived individuals or species with low rates of metabolism or excretion will bio-accumulate more than small, thin, short-lived organisms.

PESTICIDES AND HUMAN HEALTH IMPACTS

Over the last thirty years there has been increasing evidence to link pesticides with illnesses in humans. Common pesticides used in the home and garden have been linked with a number of human health impacts. They appear to accelerate the ageing of the immune and nervous systems, resulting in a wide range of ailments, such as cancers and eye abnormalities.

Links have been identified between a number of specific pesticides and recognisable ailments. For example, contact with chlordane causes the formation of abnormal sperm that are incapable of movement. DDT has been linked with infertility in females and an increased risk of developing cancer. Lindane increases the risk of breast cancer, while organophosphates have been linked to tiredness, limb pain and suicidal tendencies.

Case study: DDT

DDT stands for dichlorodiphenyltrichloroethane. It is a chlorinated hydrocarbon, which is a class of chemicals that often fits the characteristics necessary for biomagnification. This pesticide was widely used to treat crops in many countries during the 1950s. In 1972 DDT was banned in developed countries, such as the USA and Australia. However, it is still widely used in the developing world, including parts of Africa and Asia. Most DDT in the environment is a result of past use. DDT is present at many waste sites, and releases from these sites may further contaminate the environment. DDT still enters the environment because of its current use in certain areas of the world.

While no human deaths have been reported from ingesting food sprayed with DDT, the bald eagle and osprey populations in the USA and the peregrine falcon population in Europe were pushed to the brink of extinction by 1970. DDT and dichlorodiphenyldichloroethylene (DDE) causes severe eggshell thinning in birds. The most severely affected are the birds of prey, or raptors, because they are at the top of the food chain. By continually consuming fish and small mammals that have low levels of DDT, which is not excreted or degraded, the toxin accumulates and becomes concentrated in the tissues of raptors. The offspring of contaminated birds have a very high mortality rate due to incomplete gestation in the thin eggshells. A connection was finally made between DDT and the rapidly diminishing populations of eagles, hawks and other birds of prey. The Environmental Protection Agency banned the use of DDT in the USA in 1972.

DDT is long-lived. It has a half-life of fifteen years, which means if 100 kg of DDT enters the environment, after 100 years over 1 kg of DDT would remain in the environment. (See Table 3.3.2.) By the processes of bio-accumulation and biomagnification much of the DDT would be present in organisms. Some organisms, including humans, are hardly affected by DDT. However, it is highly toxic to other organisms, including insects and birds.

The overuse of DDT produced two major effects. First, resistance formed in the insect population that survived the spraying. Secondly, DDT entered the air, soil and water.

Small amounts of DDT were released into the air when it was manufactured. DDT lasts for only a short time in the air, having a half-life of about two days, but in soil it endures much longer, possibly fifteen years. It attaches tightly to soil and is not leached out quickly. Large amounts of DDT were directly applied to the soil and some DDT may have entered the soil when it was stored or disposed of in waste sites. DDT in soil can be absorbed by some plants and by the animals or people who eat those plants.

Table 3.3.2 Half-life of DDT

Years from application	Amount of DDT remaining (kg)
0	100.00
15	50.00
30	25.00
45	12.50
60	6.25
75	3.13
90	1.56
105	0.78
120	0.39

DDT also entered surface water, either by direct spraying of the water during its application or indirectly when rain-washed soil containing DDT entered surface waters. In surface water, DDT may bind to soil particles mixed in water, which then settle to the bottom of the body of water. DDT in water can be absorbed by small aquatic organisms and become concentrated in the fish that eat those organisms. The levels of DDT in animals can be higher than in the surrounding environment because the fat cells of the organisms store DDT and because DDT persists for a long period of time.

During World War II, DDT was used extensively to reduce mosquito populations, and thus control malaria, in areas where troops were fighting, particularly in the tropics. It was also used on civilian populations in Europe, to prevent the spread of lice and the diseases they carried. Refugee populations and those living in destroyed cities would have otherwise faced epidemics of louse-born diseases. After the war, DDT became popular not only to protect humans from insect-borne diseases, but to protect crops as well. As the first of the modern pesticides, it was overused and soon led to the discovery of the phenomena of insect resistance to pesticides, bio-accumulation and biomagnification.

In the early 1950s, the Dayak People in Borneo suffered from malaria. WHO had a solution: it sprayed large amounts of DDT to kill the mosquitoes that carried the malaria. The mosquitoes died and the malaria declined. But there were side effects. The roofs of people's houses began to fall down because the DDT was also killing a parasitic wasp, which had previously controlled thatch-eating caterpillars. The DDT-poisoned insects were eaten by geckoes, which were eaten by cats. The cats started to die, the rats flourished, and the people were threatened by outbreaks of plague and typhus. To cope with these problems, which it had itself created, WHO was obliged to parachute live cats into Borneo.

REVIEW
ACTIVITIES

EXTENSION
ACTIVITY

1
Explain how raptor populations suffered due to the increasing levels of DDT.

2
Describe how DDT enters terrestrial food chains.

3
In a number of countries other pesticides have replaced DDT as a mosquito control. Research what has replaced DDT in this role.

EFFECTS OF INTRODUCING NEW PESTICIDES

Locust and grasshopper control is currently carried out with chemical pesticides. For many years, the product of choice was dieldrin, a persistent pesticide well suited to barrier treatment. However, concern about its negative impact on the environment caused it to be prohibited in most countries. Most modern pesticides replacing it are much less persistent and, therefore, have to be applied more frequently and in blanket treatments and larger volumes. So, even though they are less toxic than dieldrin, their environmental impact may well be worse. International bodies have become concerned about this and other pesticide issues and have initiated the development of alternative control methods. One example of an alternative method is the introduction of *Metarhizium anisopliae* var. *acridium*, which is an insect pathogenic (disease-causing) fungus that kills locusts and grasshoppers.

A number of pesticides have gone out of production because they pose environmental or health problems, and other pesticides are awaiting decisions as to their continued use. What can be used to replace the products being cancelled? Products from the pyrethroid family of insecticides are being widely used as replacements for the more acutely toxic insecticides. Pyrethrin is a natural insecticide that was originally extracted from the chrysanthemum flower. Pyrethroids are synthetic pesticides with pyrethrin-like compounds. They are fat-soluble, but are easily degraded and do not bioaccumulate. For the most part, consumers would not recognise the change.

ALTERNATIVES TO PESTICIDES

Modern pesticides, such as carbamates and organophosphates, are 'safer' in that they are not persistent, one of the requirements for biomagnification. However, they are more toxic, and insects are developing resistance to them. Pesticides are sometimes necessary to ensure a basic food supply and to protect human health. The concept of integrated pest management (IPM) has been developed to improve control of pests while decreasing the need for pesticides. IPM uses a variety of methods to control pests. These include pesticide use but also encompass biological controls and altering cultural practices, such as timing the planting and harvesting of crops to avoid periods of peak activity by pest species. It also includes scouting to determine how big a problem the pests are actually causing, rather than just spraying to prevent a problem that may never arise. Economics are watched closely; pesticides are never used if the cost of the pesticide would exceed the cost of the crops being grown. IPM relies heavily on information, and the Internet is being used extensively.

Sustainable agriculture is being introduced into the farming systems of the world. This type of farming integrates natural processes (such as nutrient cycling, nitrogen fixation, soil regeneration and pest predators) into food production processes. Over the last century, pesticides, inorganic fertilisers, animal foodstuffs, energy, and tractors and other machinery have become the primary means of increasing food production. These external inputs have replaced free, natural control processes and resources. Pesticides have replaced biological, cultural and mechanical methods for controlling pests, weeds and diseases. The basic challenge for sustainable agriculture is to maximise the use of locally available and renewable resources.

The replacement of pesticides by natural predators, habitat redesign, multiple cropping and the like has resulted in increasing yields by a small amount. A million wetland rice farmers in Bangladesh, China, India, Indonesia, Malaysia, the Philippines, Sri Lanka, Thailand and Vietnam have shifted to sustainable agriculture, where group-based farmer-field schools have enabled farmers to learn alternatives to pesticides while still increasing their yields by about 10%. Farmers in developed countries are finding that they can substantially reduce their inputs of costly pesticides (from 20–80%) and be financially better off. Yields do fall to begin with (by 10–15% typically), but there is compelling evidence that they soon rise and go on increasing. In the USA, for example, the top quarter of sustainable agriculture farmers now have higher yields than conventional farmers, as well as a much lower negative impact on the environment.

Barriers

In some instances, barriers can be used as an alternative to pesticides. Barriers do not kill pests; they simply keep them away from the places you do not want them. A good example is a screen door, which keeps flies out.

There are a number of different types of barriers. Floating row covers are thin, lightweight fabrics or plastics that are placed over growing plants. They allow light, air, and water to reach plants, but keep insects off. They are simply draped over the plants and secured on the sides with stones or soil. As plants grow, they push the fabric up. Netting is good for keeping birds off plants, especially while they come into fruit. Sheet copper can be cut to size (8 cm wide) and attached to raised beds or planters, keeping slugs out. A sticky barrier will prevent insects and mites from walking up trunks of trees or shrubs. Root weevils walk, rather than fly, to the leaves of their hosts. Sticky barriers around the trunks of shrubs and trees (as long as there are no branches touching the ground) prevent weevils from reaching the leaves.

Traps

Traps provide a further alternative to pesticides. All traps work by attracting a target pest into a container from which it cannot escape. Traps work best when there is not much competition. For example, a slug may smell a slug trap in the middle of the garden, but it will also smell, and eat, many other tasty things along the way. Sticky traps use a sticky barrier, with one or more attractants (such as colour, smell or shape) to bring the target pest in and keep it there. The type of sticky barrier shown in Figure 3.3.2 is similar to domestic flypaper. It is hung among the foliage of plants and is attractive to insects, usually because of its smell. The barrier has a sticky surface, preventing the escape of any insect that lands on it. Yellow is a colour commonly used since many insects associate yellow with flowers and, hence, plants. Beer-filled traps attract slugs, which then drink the beer, fall into the trap and drown.

Genetic engineering

Genetic engineering is one of the latest tools for controlling pests. It involves the artificial manipulation, modification and recombination of DNA or other nucleic acid molecules in order to modify an organism or population of organisms. The plants or animals modified in this way are called transgenic. The techniques of genetic engineering can be used to manipulate the genetic material of a cell in order to produce a new characteristic in an organism. Genes from plants, microbes and animals can be recombined to create recombinant DNA and introduced into the living cells of any of these organisms. It is now possible to transfer a gene from the DNA of one species to the DNA of another species. For cases in which scientists know exactly what a gene does and exactly how it does it, it is now possible to express that function in another species. This is genetic engineering.

Figure 3.3.2 A sticky barrier.

When a crop is sprayed with conventional insecticide, the harmful insects are not the only victims. Predatory insects may also be wiped out. Without any predators available, the pest populations can recover quickly, so that a second application of pesticide is required, which also kills the insect predators. This vicious cycle will be broken when predatory insects that are genetically engineered to tolerate the pesticide become available.

Farmers, for very obvious reasons, would prefer not to use pesticides. They are expensive and are dangerous to use. In Australia, cotton farmers are plagued by various insect pests, which are controlled using chemical insecticides. But there is a natural insecticide available and it has been used by organic farmers for almost a century. It is a bacterium called *Bacillus thuringiensis* (Bt). The bacterium produces a toxin that is deadly to caterpillars, but harmless to almost everything else except insects of the order Lepidoptera: butterflies and moths. So genetic engineers transferred the gene for Bt toxin from Bt to cotton. Then the cotton plants, which could make Bt toxin, were crossbred with other varieties in the old-fashioned way. Today, much of the US cotton crop is genetically engineered for the Bt toxin trait. As a result, the use of chemical insecticides in the cotton belt has declined dramatically. Since the Bt toxin is inside the plant instead of sprayed onto the plant, the only insects it can harm are those that eat the plant.

Reducing or stopping the spraying of cotton and replacing it with the new strains of cotton mean that cotton pesticides can also be replaced. Cotton pesticides are extremely damaging to the environment. They kill all insects in a cotton field, harmless or not, thus depriving insectivorous birds of their food. There is no way to prevent these pesticides from entering streams and rivers, where they are a serious hazard to aquatic life.

The gene for Bt toxin has been transferred into several other crops, including potatoes and corn. Approximately 30% of US corn is now **transgenic**, and the most popular transgenic varieties contain the Bt gene.

Other pesticide alternatives

Some of the garden's best friends are natural enemies to pests. Biological controls can destroy insects, slugs, mites and other unwelcome visitors. (For more information on biological controls see Section 4.) Some are very specialised and may be specific to individual target species, while others may be less specific.

Many materials have been used as alternatives to pesticides. Some have been effective, while the efficiency of others is questionable. These materials include lead arsenate, salt, copper, soap, sulfur, oil and wood ash. All have been used as pesticides throughout the ages, with varying levels of success and varying rates of toxicity and persistence. Most of those listed here are still used today, fortunately with a better understanding of their effects.

REVIEW
ACTIVITIES

EXTENSION
ACTIVITY

1
Describe how the IPM approach is used to fight pests.

2
Recount how genetic engineering can be used to make plants pest-resistant.

3
Describe the advantages of making plants pest-resistant.

4
List the measures that can be taken to replace or reduce the use of pesticides.

5
Explain what is meant by the term 'biological control'.

6
An alternative to the use of pesticides is the introduction of large numbers of sterilised males into a wild population. Explain how this may affect the pest population.

SUMMARY

Pesticides include insecticides, herbicides and fungicides, which are designed to kill insects, weeds and diseases, respectively. All pesticides are toxic to some living thing: insecticides to insects, herbicides to plants, fungicides to fungi, and so on.

Some synthetic pesticides release compounds that are more toxic than non-synthetic pesticides when they break down in the environment. Some accumulate in the environment and cause harm far removed from the original site or purpose of application.

Pesticides can adversely affect wildlife through changes in the food web, direct and indirect poisoning, chemical biomagnification and habitat changes.

Bio-accumulation is the increase in concentration of a pollutant from the environment to the first organism in a food chain.

Biomagnification refers to the tendency of a pollutant to concentrate as it moves from one trophic level to the next, increasing the concentration of the pollutant from one link in a food chain to another.

In order for biomagnification to occur, the pollutant must be long-lived, mobile, soluble in fats and biologically active.

Pesticides have been linked to serious human conditions, such as cancers, infertility and suicidal tendencies.

Biomagnification of the pesticide DDT caused the eggs of raptors to become too thin to protect the embryos.

Most modern pesticides replacing dieldrin are much less persistent than older forms and, therefore, have to be applied more frequently and in blanket treatments and larger volumes. So although they are less toxic than older pesticides, their environmental impact may well be worse.

The concept of IPM has been developed to improve control of pests while decreasing the need for pesticides. IPM uses a variety of methods to control pests.

PRACTICAL EXERCISE
Alternatives to pesticides

In this exercise you will design and carry out an experiment to measure the effectiveness of pesticide alternatives, such as different types of barriers.

Procedure

1
Explain what plants you would choose and give reasons for your choice.

2
Detail the variables, both dependent and independent, that have to be taken into account in order to obtain accurate sets of data.

3
Give a suggested time line for the experiment.

4
Detail the precautions that would need to be taken to ensure accuracy and safety.

5
Describe the changes that should be made to the original experiment in order to improve the results.

OUTCOMES

At the end of this chapter you should be able to:

assess management strategies and technologies that can be used to assist in the maintenance of natural processes in surface water,

including:
– drip versus overhead irrigation
– licensing irrigation/bore water users
– stormwater treatment technologies
– provision of environmental flows from dams.

WATER RESOURCES

Most of Australia's river systems are relatively short, confined to the coast or ephemeral (only running in the wet season). The exception is the Murray-Darling River system, which drains about one-seventh of the Australian landmass. The Murray-Darling Basin is the prime food production region of Australia.

The largest user of water resources in Australia, accounting for about 70% of the water consumed, is agriculture (including forestry and fishing), followed by domestic households (using approximately 10%) and manufacturing (about 5%). Whether used by industry or in the home, about 90% of the water used is eventually returned to the environment, where it replenishes water sources and can be used for other purposes.

Extraction of water in ever increasing amounts has led to river levels dropping and some wetlands either being lost or severely affected. **Environmental flows** have been introduced to counter the reduction in water levels and the threat to wetland destruction. The purpose of the artificial environmental flow is to improve the deteriorating health of rivers, sustain biodiversity and maintain flow patterns by stimulating or replacing chemical, geological and/or ecological processes.

Water extraction, combined with river regulation and land management practices, have significantly affected river ecosystems and water quality. When the balance between water extraction and environmental flows is altered in favour of water extraction, major problems occur. This is typified by the problems encountered by the Snowy River.

Case study: The Snowy River

In 1967 the Snowy River was dammed to provide a source of water for a hydro-electricity scheme and irrigation. Prior to the damming it was a huge, fast-flowing river, recognised for its trout fishing, abundant wildlife and clear water. After the damming, 99% of the water that used to flow along the river was sent westwards to the hydro-electricity scheme. (See Figure 3.4.1.) Only 1% of water was allowed to

environmental flows
artificial inputs of freshwater into rivers and other areas where natural flow levels have been drastically reduced because of upstream damming

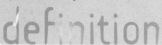

flow along the river below Jindabyne Dam. This resulted in a near empty riverbed, poor water quality, an increase in blue-green algae blooms, a reduction in biodiversity, an invasion by the weeping willow (an environmental weed), and an increase in soil erosion and salinity levels. It also caused economic damage to the people living below Jindabyne Dam.

Parties that were affected by the reduced flows undertook a campaign. Eventually they won an undertaking by the Victorian and New South Wales Governments to restore the Snowy River's flow to a level of 28% of the original flow. Under the deal, 21% of the river's flow will be restored within ten years and the remaining 7% will be restored at a later date. Funding totalling $300 million has been allocated to capital projects to boost water use efficiency and to fund water purchases to offset the environmental flows.

Figure 3.4.1 The flow of the Snowy River has been reduced dramatically.

Achieving sustainable environmental flows

To achieve sustainable water resource management, it is necessary to sustain and, where required, restore ecological processes and the biodiversity of aquatic ecosystems. Providing water for the environment is one component of a water allocation process in which the right to the use of water is distributed among various users. In the past, these have included only consumptive users, such as agriculture and industry. Today, it is recognised that there is a need to provide water for non-consumptive users. Environmental, recreational, navigational and hydro-electric requirements must be considered during the water allocation process. In many cases, the requirements of the non-consumptive users may be compatible with the needs of the environment. Where they are in conflict, priority should be given to the needs of the environment.

The provision of water for the environment should be enough to maintain the existing flora and fauna and their habitat. Where the environment can be restored to its original, pristine condition it may be possible to provide water to restore the aquatic ecosystem. Habitat restoration works could also be used in tandem with environmental flows to allow maximum environmental benefit from the environmental water provided. Any future development should be based on a full scientific assessment of the environmental water requirements of downstream aquatic ecosystems. These should be built into the design, operation and costs of the new development. If a scientific assessment of the environmental requirements shows that these cannot be met, then the development should not proceed.

Where it is shown that waste water is at least of the ambient quality of water within the stream, it may be possible to include waste water as part of an environmental water provision. This should be encouraged, provided that the quality of waste water is carefully monitored and contingency arrangements are in place in case of failure.

Water for consumption may be managed in more environmentally friendly ways while still meeting the needs of primary clients. There is scope to manage storages in 'smarter' ways and this should be encouraged.

It is essential to monitor the impacts of environmental water provisions to determine whether the water provision is adequate and to test the effectiveness of methods of determining environmental flows.

1

How much water is used by Australian agriculture?

2

Recall the consequences of ever increasing water extraction.

3

Explain how sustainable water resource management can be achieved.

4

Describe why it is essential to monitor the impacts of environmental water provisions.

5

Justify why rivers such as the Snowy River and wetlands should be maintained.

IRRIGATION

Throughout the world, irrigation is probably the most important use of water (except for drinking). Almost 60% of the world's usage of freshwater goes towards irrigation. Large-scale farming could not provide food for the world's large populations without the irrigation of crop fields by water taken from rivers, lakes, reservoirs and wells. But of the water used for irrigation, only about one-half is reusable. The rest is lost by evaporation into the air, by transpiration from plants or during transit due to leaking pipes. It is therefore important that any irrigation system is as efficient as possible.

The most common types of irrigation methods are flood irrigation, drip irrigation and spray irrigation.

Flood irrigation

The simplest method of irrigation is to cover the whole field with water by flooding it. This system is called flood, or furrow, irrigation. Water is pumped or brought to the fields and allowed to flow along the ground among the crops. This method is simple and cheap, and is widely used by the developing world. The disadvantage of this system is that about one-half of the water used does not reach the crops and is wasted.

Farmers who use flood irrigation have made it more efficient by employing the following techniques:

- *Levelling of fields.* Flood irrigation uses gravity to transport water and, since water flows downhill, it will miss any raised part of the field. To combat this problem, farmers use levelling equipment, some of which is guided by a laser beam, to scrape a field flat before planting. This allows water to flow evenly throughout the fields.
- *Surge flooding.* Traditional flooding involves releasing water once onto a field. Surge flooding, however, releases water at prearranged intervals, reducing unwanted run-off.
- *Capture and reuse of run-off.* A large amount of flood irrigation water is wasted when it runs off the edge of the fields. Farmers capture the run-off into ponds or dams and then pump it back up to the front of the field, where it is reused for the next cycle of irrigation.

Drip irrigation

Drip irrigation is a very efficient method of irrigating fruits and vegetables; it is much more efficient than flood irrigation. Water is sent through a network of plastic pipes. The pipes are laid along the rows of crops or buried along their roots. The pipes contain very small holes, through which water drips out. Drip irrigation minimises water wastage by delivering water directly to the point of uptake and ensuring that crops only receive the amount they need. It reduces the amount of water that enters the watertable and the possibility of soil erosion by run-off. Evaporation is far lower and up to one-quarter of the water used is saved.

Spray irrigation

Spray, or overhead, irrigation is very common. This method is similar to the way you may water your lawn at home: stand there with a hose and spray the water in all directions. Large-scale spray irrigation systems are in use on large farms today. These systems have a long tube fixed at one end to the water source, such as a well or bore. In spray irrigation, water flows through the tube and is shot out by a system of spray guns.

A common type of spray irrigation system is the centre pivot system. (See Figure 3.4.2.) This system involves the use of a number of metal frames on rolling wheels. Each frame holds a water tube out into the fields. The tube at the centre of the circle is fixed to the water source. The water sprays out from holes in the tube and, at the same time, squirts out from a water gun at the end of the tube. Electric motors move each frame in a big circle around the field.

Using overhead irrigation systems has several disadvantages. It can result in stormwater run-off, which may contain such contaminants as pesticides and nutrients. The run-off can also lead to soil erosion. Excess water may enter the watertable, increasing the risk of salinisation. Traditional spray irrigation basically involves shooting water through the air onto fields. This is inefficient for a number of reasons. Large volumes of water are used because the spray is indiscriminate. Also, the spray is relatively fine, which allows the water to be easily lost by the process of evaporation or dispersed from the target area by wind action.

An improved method involves spraying water gently from hanging pipes placed in close proximity to the crops. This allows more accurate targeting of the plants, meaning farmers can use, and lose, less water. This makes it a more efficient and cost-effective method. Replacing traditional spray irrigation with this method increases irrigation efficiency from about 60% to over 90%. Plus, less electricity is needed.

Figure 3.4.2 The centre pivot system is a common type of spray irrigation.

REVIEW ACTIVITIES

1
Outline the main methods of agricultural irrigation.

2
Explain the disadvantages of using overhead irrigation.

EXTENSION ACTIVITY

3
Propose reasons as to why many farmers prefer to irrigate their crops using spray irrigation rather than drip irrigation.

WATER MANAGEMENT ACT

In November 2001 the *Water Management Act* was passed by the upper house of the NSW Parliament. The purpose of this Act is to provide a framework for managing the water resources within the state and regulating the amount of water the various categories of users may be allowed to have.

The Act states that intensive livestock producers are entitled to use up to 10% of the total rainwater run-off that falls on their property. This 10% is called the

regulated rivers
those waterways with dams and other structures on them that regulate or restrict their natural flow

unregulated rivers
those waterways where the flow is not regulated or restricted

harvestable right. It can be used for any purpose on the property and covers water obtained from rivers as well as from bores. The inclusion of bores is important because excessive extraction of water from the watertable would result in changes in the subsurface flow within the watertable. If water consumption exceeds the harvestable right, a water licence has to be obtained. This licence allows the holder to be allocated a further defined amount of water.

A significant change introduced by the Water Management Act is that any land-holders in possession of a water licence must also obtain a water use approval. This is a site-specific licence, allowing a land-holder to use a certain amount of water at a particular location. The water use approval is not tradeable between landowners, unlike the more general water licence that was granted prior to the 2001 Water Management Act. The aim is to prevent the application of water on land not suitable for cropping.

Rivers are divided into two main categories: regulated and unregulated. For **regulated rivers** the Act provides town water, stock and domestic usage licences with the highest priority classification in terms of access to water in times of severe water restrictions and rationing. The type of licence determines the security of access to a water supply. For **unregulated rivers**, there are several classes of licence. Each class allows holders to pump water from their allocation once the flow of the watercourse reaches a specific level.

REVIEW ACTIVITIES

1
What does the 2001 Water Management Act set out to do?

2
Explain the difference between regulated and unregulated rivers.

EXTENSION ACTIVITY

3
Does the amount of water extracted by users affect whether a river is regulated?

STORMWATER TREATMENT

Stormwater is pure rainwater plus anything that the rain carries along with it. In urban areas, stormwater runs off paved areas, such as roads and driveways. Animal faeces and litter in the form of paper, plastic bags and bottles are washed into the drains, along with oil and rubber from roads and driveways. Green waste and garden chemicals enter the stormwater drains after being washed in from gardens. Unlike sewage, stormwater is not treated, but it may be filtered using pollution traps to remove large pollutants, such as plastic bags and bottles. Most pollutants, however, enter the waterways.

Impact and causes of stormwater pollution

Stormwater pollution has a major impact on the environment, flora, fauna and humans. Sediments in the water reduce light, which affects the photosynthetic processes in plants. If the stormwater contains decaying green waste, oxygen levels become depleted and this results in a lack of oxygen for plants and animals. Soil washed into waterways by run-off clouds the water and clogs the gills of fish. Litter clogs waterways, causing toxicity as it breaks down. Stormwater pollution results in an increase in bacteria, posing a health risk to humans and animals swimming after rain. Stormwater pollution also results in a loss of visual amenities.

Common causes of stormwater pollution from domestic premises include:

- using detergent to wash cars and allowing the detergent to run down into street drains
- repairing cars on the street and letting oil and other liquids enter drains
- allowing the accumulation of garden waste in gutters
- dropping litter
- cleaning out dirty paint brushes
- hosing down driveways
- neglecting to pick up dog faeces.

Commercial companies also carry out activities that can lead to stormwater pollution. These may include:

- failing to clean out restaurant grease traps, allowing grease to enter drains
- allowing chemicals, such as oil and other wastes, to enter drains from motor vehicle repairers or printers
- neglecting to use sediment traps on building sites, allowing material to enter drains.

Types of stormwater pollution traps

There are numerous types of pollution traps. They include artificial wetlands, oil and litter booms, trash racks, gross pollution traps and sediment traps.

Artificial wetland systems are both economical and environmentally friendly. There are two types of artificial wetlands: free water surface systems and subsurface flow systems. Free water surface systems require a substantial amount of land. They consist of channels and shallow basins with an impervious subsurface of clay to prevent seepage and a layer of soil to support the roots of aquatic plants. Subsurface flow systems consist of trenches with an impervious lining and/or a layer of soil for plants. These systems can be designed to allow stormwater to flow horizontally through the root system. This maximises filtration and take up by the soil. It also maximises nutrient uptake by plants and micro-organisms and encourages the breakdown of nutrients by microbes. Harvesting of the plants within the wetland will remove the accumulated nutrients.

Oil and litter booms are floating barriers that are placed in moving water. They are designed to collect surface oil and floating litter. Trash racks are grates that are usually located towards the end of stormwater drains, where they collect large items of rubbish from the stormwater.

Gross pollution traps are formed by a series of chambers and separation screens. (See Figure 3.4.3.) It operates in a similar way to a domestic swimming pool's skimmer box. Water enters a large chamber whose walls are made from a metal screen that is perforated to let the water pass through. Any large solids are trapped in the chamber and then sink into a catchment sump at the base of the chamber. Sediment traps slow the water flow, allowing the suspended sediments to drop out of suspension.

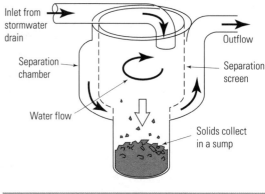

Figure 3.4.3 A gross pollution trap.

REVIEW
ACTIVITIES

EXTENSION
ACTIVITY

1
Define the term 'stormwater'.

2
Describe the sources and types of stormwater pollution.

3
Explain the processes used to treat stormwater.

4
There are a number of different types of surface and subsurface flow systems available for stormwater treatment. Research and outline the structure of two examples of each system.

SEWAGE TREATMENT

Humans produce a vast amount of waste water as a result of industrial, agricultural and domestic processes. If this water returns to rivers and oceans untreated it has the potential to severely damage those habitats. Industrial processes produce waste water that contains heavy metals and possibly toxic organic compounds that may cause long-term, irreparable damage to the environment. The Environmental Protection Authority and local water authorities set strict guidelines as to what materials can be discharged into sewers and what materials are totally banned from being disposed of through the sewers.

Domestic sewage and some industrial waste water is carried via sewage systems to a sewage treatment plant, where the water can undergo up to three levels of treatment in order to ensure the end product is clean water. (See Figure 3.4.4.) The three levels of treatment are primary, secondary and tertiary.

Primary treatment is mainly involved in the removal of solids from sewage water. In primary treatment the waste water initially passes through a series of screens whose mesh size decreases through the series. This removes 65% of solids. Once the sewage has passed through the series of screens, the water enters large settling tanks. Substances that are less dense than water, such as oil and scum, float to the surface and are skimmed off. Other substances are more dense than water and sink to the bottom, where they form a sludge. This sludge can be used on fields as a fertiliser after further treatment. The remaining liquid is chlorinated in order to kill off any harmful pathogens. At this point the water can either be returned to the environment or passed to the secondary level of treatment.

Figure 3.4.4 Sewage treatment plant.

Secondary treatment is primarily focused on the treatment and removal of pathogenic micro-organisms, the breakdown of organic matter and purification. Beneficial bacteria are used to kill and remove pathogenic bacteria. In a simple sewage treatment process, micro-organisms are encouraged to grow on stones over which the sewage is trickled. The micro-organisms, which need oxygen to thrive, feed on the bacteria in the sewage and purify the water. These treatment units are called percolating filters.

This process can be sped up by blowing air into tanks of sewage, where the micro-organisms float freely and feed on the bacteria. These treatment units are called aeration tanks. Following this step, the waste water is settled in tanks to separate the biological sludge from the purified waste water. The final step in secondary treatment is to chlorinate the water, which kills any micro-organisms in the water. After this treatment the water can be returned to the environment or passed on to the tertiary level of treatment.

The tertiary level of treatment is not as commonly used in Australia as it is in other countries. Its primary function is to remove high levels of nutrients (such as nitrates and phosphates) and heavy metals (such as mercury and lead) from the treated water. Various methods may be used, including sand filters, reed beds or grass plots. Disinfection, using ultraviolet light to kill bacteria, is another method, and is being used in a number of coastal sewage treatment schemes.

The first step in tertiary treatment is to add alum to the water, which causes **colloidal particles** to clump together and settle out. This is aided by passing the water through activated charcoal, which removes dissolved organic compounds. Lime is then added to the water, which causes the phosphates in the water to precipitate out

colloidal particles
particles of between 1 and 1000 mm in size

definition

and then settle. The next step is to carry out ion exchange. An ion is an electrically charged atom, formed by the loss or gain of one or more electrons. Both types of ions, cations and anions (positively and negatively charged atomic ions, respectively), can be exchanged between the surface of an organic resin and the water undergoing treatment. It improves the quality of the water by removing calcium. This softens the water and helps remove other mineral ions, including heavy metals.

The final step in tertiary treatment, and one that is not always carried out, is to apply reverse osmosis to the treated water. Reverse osmosis uses water under high pressure that flows out of a high ionic strength solution into a lower ionic strength solution on the opposite side of a semipermeable membrane. This removes more calcium and magnesium ions, helping to further soften the water. At the end of this process all organic material, 99% of unwanted ions and most microbes have been removed. The water is now fit to be released into the environment.

All cities produce great quantities of sewage and stormwater. Coastal cities are able to reduce the cost of sewage treatment by discharging this water into the sea. Originally the water received primary treatment and then was discharged off the coast in relatively shallow water. This led to high levels of pollution on coastlines, adversely affecting water quality and coastal ecosystems. With the implementation of deep-ocean outfalls, the level of beach pollution has been drastically reduced but not eradicated. The deep-ocean outfalls discharge the waste up to 3 km off shore, increasing the dilution of the discharge. As yet, we are not sure what effect this discharge has on ocean ecosystems.

Non-coastal communities have had to find alternatives to deep-ocean outfalls. These include such methods as lagoons, land filtration and artificial wetlands. These methods rely on natural processes, such as using ultraviolet radiation to disinfect sewage or break it down.

REVIEW ACTIVITIES

1
Describe why waste water is potentially harmful to the environment.

2
Differentiate between primary, secondary and tertiary sewage treatments.

3
Outline the steps in sewage treatment. Compare the methods used for treating stormwater and sewage.

4
Explain the advantage of deep-ocean outfalls.

EXTENSION ACTIVITIES

5
The CSIRO has developed an alternative process for the secondary treatment of waste water called Sirofloc. Research and evaluate this process.

6
Research how the methods of lagoons, land filtration and artificial wetlands are able to break down and deal with sewage.

SUMMARY

Agriculture (including forestry and fishing) is the largest user of water resources in Australia, accounting for about 70% of the water consumed. It is followed by domestic households (using approximately 10%) and manufacturing (about 5%).

Environmental flows are artificial inputs of freshwater into rivers and other areas where natural flow levels have been drastically reduced because of upstream damming. They have been introduced to improve the deteriorating health of rivers, sustain biodiversity and maintain flow patterns by stimulating or replacing chemical, geological and/or ecological processes.

Water allocation must include both consumptive users (such as agriculture and industry) as well as non-consumptive users, including environmental requirements, recreation and navigation.

Irrigation (water for agriculture, in particular crops) is probably the most important use of water (except for drinking). Almost 60% of the world's usage of freshwater goes towards irrigation.

The most common type of irrigation methods are flood, or furrow, irrigation; drip irrigation; and spray, or overhead, irrigation.

In 2001 the NSW Parliament passed the Water Management Act. The Act is a framework for managing the water resources within the state and regulating the amount of water the various categories of users may be allowed to have.

Stormwater is pure rainwater plus anything that the rain carries along with it. In urban areas, stormwater runs off paved areas. Animal faeces, litter, and oil and rubber from roads and drives are washed into drains.

The different types of pollution traps include artificial wetlands, oil and litter booms, trash racks, gross pollution traps and sediment traps.

Sewage can receive three levels of treatment: primary, secondary and tertiary.

Primary treatment is the removal of solids from sewage water. Secondary treatment involves the removal of pathogenic micro-organisms, the breakdown of organic matter and purification. Tertiary treatment removes high levels of nutrients and heavy metals from the treated water.

PRACTICAL EXERCISE
Sequencing and designing a flow chart

The aims of this exercise are to learn the steps in sewage treatment and practise the use of flow charts to show information.

ACTIVITIES

1

Using the information in the box, reorganise the steps in each of the three stages of treatment (primary, secondary and tertiary) into the correct sequence required for the treatment of sewage water

2

Using the boxed information in the correct sequence, draw a flow chart to show how sewage water may be treated.

A The remaining liquid is chlorinated in order to kill off any harmful pathogens.

B Substances that are less dense than water float to the surface and are skimmed off.

C The water can either be returned to the environment or passed to the next level of treatment.

D Substances that are more dense than water sink to the bottom and form a sludge.

E The waste water passes through a series of screens.

F The water enters large settling tanks.

G The water can now be returned to the environment or passed to the next level of treatment.

H The water is chlorinated to kill any micro-organisms in the water.

I The waste water is settled in tanks to separate the biological sludge from the purified waste water.

J Micro-organisms are encouraged to grow on percolating filters or in aeration tanks, where they feed on bacteria.

K The water may be passed through activated charcoal, which removes dissolved organic compounds.

L Lime is added to the water, causing the phosphates to settle out.

M Alum is added to the water to make the colloidal particles settle out.

N Reverse osmosis is carried out to remove calcium and magnesium ions

O Microbes and 99% of unwanted ions have been removed.

P The water is now fit to be released to the environment.

Q Ion exchange is carried out to remove calcium.

OUTCOMES

At the end of this chapter you should be able to:

summarise types of chemical reactions involved in the formation of greenhouse gases and acid rain from the burning of fossil fuels (word equations only)

analyse different scientific views on the causes of global warming to discuss predictions on the effects of global warming

outline the way in which chlorofluorocarbons and other halides can reduce the percentage of ozone in the stratosphere

analyse implications of national and international strategies related to maintaining and protecting the atmosphere and the hydrosphere

summarise the evidence for ozone depletion.

definitions

air pollution
events that add to, or subtract from, the usual constituents of air and alter its physical or chemical properties

ozone layer
a region in the outer portion of the stratosphere at an elevation of about 30 km, where much of the atmospheric ozone is concentrated

pollutants
usually considered to be those substances that are added to the air or water in sufficient concentrations to produce a measurable effect on living things or materials, such as stone or steel structures

industrial revolution
the period when changes were brought about by the introduction of technology and mass production methods. It began in the late eighteenth century.

incineration
combustion at high temperatures

combustion
a chemical reaction in which heat, and often light, is produced

particulate
composed of particles

AIR POLLUTION: AN OVERVIEW

Worldwide, rising **air pollution** has an adverse impact on human health, plants and soils, and the built environment. It also affects the global climate and depletes the **ozone layer**. Excessive heat and noise are also forms of air pollution. As air pollution knows no boundaries, national and international avoidance and control strategies are vital in lessening its impacts.

There are two principal sources of air pollution: natural occurrences and human activities. Naturally occurring air pollution has existed since the planet formed, and arises from sandstorms, bushfires and volcanic eruptions. The sources and effects of naturally occurring air pollution generally tend to be of a fairly short duration, whereas **pollutants** caused by human activities may last for decades. Human-induced pollution has accelerated rapidly since the start of the **industrial revolution** with the increase in the manufacturing of goods. The sources of human-induced air pollution include industrial processes, electricity generation, transportation, domestic heating and **incineration** of waste.

Types of air pollutants

There are two general groups of air pollutants: primary and secondary. Primary pollutants are those that are emitted directly from identifiable sources, such as domestic fuel **combustion**, transportation, solid waste disposal, industrial processes and agricultural burning. Animal production is another source of primary pollutants. During digestive processes pollutant gases, such as methane, are released. Primary pollutants come in a number of forms, including:

• suspended **particulate** matters, such as dust, soot, pollen, lead and asbestos
• oxides of carbon (carbon dioxide and carbon monoxide)

- oxides of sulfur and nitrogen (sulfur dioxide, sulfur trioxide, nitric oxide, nitrogen dioxide and nitrous oxide)
- hydrogen sulfide
- organic gases and volatile compounds, such as methane, benzene, formaldehyde and chlorofluorocarbons
- radioactive substances (radium -222 and strontium -90)
- gaseous halogens, hydrogen fluoride, chlorine and hydrochloric acid.

Secondary pollutants are those that are produced in the air by the interaction of two or more primary pollutants. The degree of their impact depends upon relative concentrations, humidity, temperature and solar radiation. Secondary pollutants include ozone, organic hydroperoxides, peroxyacyl nitrates, several aldehydes and a variety of free radicals.

REVIEW ACTIVITIES

1
Distinguish between naturally occurring and human-induced pollution.

2
Explain what a pollutant is.

3
Differentiate between primary and secondary pollutants. Give examples of each.

4
What affects the degree of impact caused by secondary pollutants?

EXTENSION ACTIVITIES

5
During winter, Sydney suffers from smoke carried by wind from slow combustion heaters burning fuels in outlying districts. What pollutants are released from these fires? Using data from the Environmental Protection Authority, show how these levels vary according to the seasons.

6
Research the effect of the pollutants hydroperoxides and peroxyacylnitrates on the environment.

AIR POLLUTION FROM THE BURNING OF FOSSIL FUELS

The remains of organisms embedded in the Earth have a high carbon and/or hydrogen content and can be used by humans as fuels, especially **coal**, oil, and **natural gas**. These fuels are referred to as **fossil fuels**. With the advent of the industrial revolution in the nineteenth century there was a major increase in the burning of fossil fuels. Reliance on fossil fuels continued to increase with the development of the internal combustion engine, which is used in most vehicles today. Most of the pollutants produced in the combustion of fossil fuels add to the **greenhouse effect** in the atmosphere, thus leading to **global warming**.

A variety of pollutants are released when gas in the form of natural gas, liquefied **petroleum** or coal gas is burnt in water heaters, boilers and furnaces. These pollutants include carbon monoxide, carbon dioxide, nitrogen oxides and soot particulates.

Oil is not as refined as gas. Consequently, the burning of oil (such as in oil-fired furnaces and boilers) creates less-efficient combustion and produces a wider variety of pollutants than the burning of gas. Pollutants emitted from the combustion of oil include sulfur dioxide, sulfur trioxide, carbon monoxide, carbon dioxide, nitrogen oxides, aldehydes, and particulates formed by soot and unburnt hydrocarbons.

The combustion of solid fuels (such as coal, coke or wood) also results in the release of a number of pollutants, including sulfur dioxide, carbon monoxide, carbon dioxide, hydrocarbons and soot.

coal
a rock composed of the compressed and heated remains of buried vegetable matter

natural gas
the gaseous component of petroleum

fossil fuel
a combustible material of biological origin that is found as altered carbon compounds within geological deposits

global warming
the significant rise in the global average temperature of the Earth's atmosphere. It is attributed to the greenhouse effect.

greenhouse effect
the increase in temperature in a greenhouse caused by the radiant heat from the Sun passing through the glass, which traps the heat in the greenhouse. The Earth's temperature is similarly increased when the atmosphere, acting like the glass of a greenhouse, traps the Sun's heat.

petroleum
liquid and/or gaseous hydrocarbons derived from the remains of microscopic plants and animals

definitions

Suspended particulates

Industrial and agricultural processes, transportation, urban development and various other human activities emit particles into the atmosphere. The main sources of particulate emissions are incinerators, open-air burning of bush and solid waste, and mining activities.

Particulates remain in the troposphere (about 11–30 km above the Earth's surface) for only a few days. However, if they are injected into the stratosphere (above the troposphere) they may travel around the world for several years. This may have a severe impact on global climate.

Some particulates are small in size and remain suspended in the atmosphere for a long period. They may be emitted directly into the atmosphere or may result from chemical reactions between pollutant gases. Suspended particulates less than 10 microns in diameter are small enough to be inhaled by humans and may lead to respiratory illness if the concentrations are sufficiently high. Particulates larger than 10 microns in diameter consist mainly of dust, coarse dirt and **fly ash** from industrial and erosive processes. These large particulates usually settle rapidly.

Sulfur dioxide

Sulfur dioxide is a dangerous primary pollutant. It causes harm to living things and building materials. Each year, 150 million tonnes of sulfur dioxide are discharged into the Earth's atmosphere as a result of the burning of fossil fuels, in particular coal. In Australia, most electricity generation uses coal. Some sulfur dioxide is released in the process of smelting during the production of zinc, lead and copper.

Sulfur dioxide in air is oxidised to sulfur trioxide, which then reacts with water to form sulfuric acid. The sulfuric acid in the air falls as **acid rain**. The overall oxidation reaction is summarised below:

Sulfur dioxide + water + oxygen → sulfuric acid

Certain coal processing methods can reduce the sulfur content of coal, producing a coal that releases less sulfur dioxide than coal processed by conventional methods. One of these methods involves treating the coal with 10% sodium hydroxide at a high temperature and pressure. The coal is then washed and dried. A second method is to add athracene to ground coal. These methods reduce the sulfur content of coal from more than 15% to less than 1%.

Sulfur dioxide emissions from industrial processes can be reduced by removing sulfur dioxide from chimney gases by a process called scrubbing. In this process, sulfur dioxide is heated as it passes through beds of calcium carbonate, or limestone. There is a chemical reaction between the limestone and the sulfur dioxide, which causes the limestone to absorb the sulfur dioxide. The waste can then be removed.

Nitrogen oxides

Nitrogen oxides are major air pollutants. They include nitrogen oxide, nitrogen dioxide and nitrous oxide. As a major greenhouse gas, nitrous oxide is dealt with in the next section on greenhouse gases. The major sources of nitrogen oxides include the combustion of fossil fuels in power plants and automobiles, and a variety of industrial processes. Nitrogen oxides react with hydrocarbons in sunlight to form **photochemical smog**. They also contribute to ground-level ozone pollution and acid rain. The nitrogen oxides, especially nitrogen dioxide, have a detrimental impact on human and animal health. They also cause fabrics to deteriorate, dyes to fade, and metal to corrode.

fly ash
the fine ash produced from pulverised

acid rain
highly acidic rain caused by air pollution

photochemical
things related to chemical reactions that involve light as an energy source

smog
a combination of smoke and fog

1

Compare the pollutants produced by gas heaters and oil boilers.

2

Explain the effects of different-sized particulates.

3

Outline how sulfur dioxide emissions can be reduced.

4

Outline different methods carried out by industry to reduce nitrogen dioxide levels.

5

Compare the working of wet and dry sulfur dioxide scrubbers.

GREENHOUSE GASES

Many chemical compounds found in the atmosphere act as greenhouse gases. These gases allow solar radiation (sunlight), which is radiated in the visible and ultraviolet spectra, to enter the atmosphere unimpeded. When it strikes the Earth's surface, some of the solar radiation is reflected as infra-red radiation (heat). (See Figure 3.5.1.) Greenhouse gases absorb the infra-red radiation as it is reflected back towards space, trapping the heat in the atmosphere.

Greenhouse gases comprise less than 1% of the atmosphere. Their level is determined by the balance between their sources and what destroys them, or their **sinks**. Humans increase greenhouse gas levels by introducing new sources or by interfering with sinks.

The major naturally occurring greenhouse gases in the atmosphere are carbon dioxide, methane, nitrous oxide, water vapour and ozone. Human-made greenhouse gases include chlorofluorocarbons (CFCs), hydrofluorocarbons (HFCs), perfluoro-carbons (PFCs) and sulfur hexafluoride. They are fluorine-containing halocarbons. The atmospheric concentrations of these human-made gases are relatively small in comparison with the natural greenhouse gases. However, they have the potential to greatly increase global warming due to their potency and extremely long atmospheric lifetimes. As they remain in the atmosphere almost indefinitely, concentrations of these gases will increase as long as emissions continue. Their potential to be signifi-cant and important contributors to global climate change make them an essential element of any greenhouse gas mitigation strategy.

Carbon dioxide

Concentrations of carbon dioxide in the atmosphere are regulated by numerous processes, collectively known as the carbon cycle. (See Section 3 of *Earth and Environmental Science: The Preliminary Course*.) The movement of carbon dioxide between the atmosphere and carbon sinks on land and in the oceans is dominated by natural processes, such as photosynthesis and respiration. Human activities globally produce 3.3 billion tonnes of carbon dioxide per year. Since the beginning of the industrial revolution there has been a 30% increase in atmospheric carbon dioxide. In 1900 the amount of carbon dioxide in the atmosphere was around 290 parts per million (**ppm**), but by 1998 this had risen to 366 ppm. There are seasonal variations in carbon dioxide concentrations, and this is mainly caused by fluctuations in photosynthesis in mid-latitude forests.

The main source of **anthropogenic** carbon dioxide is from the burning of fossil fuels, such as coal, petroleum and natural gas. Half a kilogram of coal generates enough electricity to light a 100 watt bulb for ten hours, but the combustion of coal to create the required electricity produces 1.5 kg of carbon dioxide. Carbon dioxide

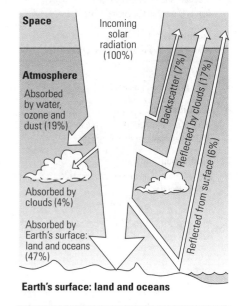

Figure 3.5.1 The balance between incoming and outgoing radiation.

sink
occurs where atmospheric carbon dioxide is absorbed from the air and converted into other substances. Sinks include seas and forests, and occur where large-scale photosynthesis takes place.

ppm
a way to describe how much of a substance is contained in a sample. In atmospheric sampling, it represents parts per million parts of air.

anthropogenic
derived from human activities

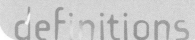

is produced by combustion because of the oxidation of carbon and can be shown by the following equation:

Carbon + oxygen → carbon monoxide

This is immediately followed by a second reaction:

Carbon monoxide + oxygen → carbon dioxide

Methane

Methane is a colourless, odourless, flammable hydrocarbon gas. It is the product of **decomposition** of organic matter and the **carbonisation** of coal. It is often called swamp gas because it is abundant around freshwater, especially swamps. Bacteria are involved in decomposition and cause methane to be released. Bacteria involved in the breakdown of plant material are found in cows, sheep, goats, buffaloes, camels and termites. A single **herbivore**, such as a cow, releases approximately 250 g of methane per day. Annually these animals produce vast amounts of methane. Methane is also released from natural **reservoirs** of fossil fuels, such as coal seams and oil **reserves**, and from the incomplete combustion of fossil fuels.

As a contributor to the greenhouse effect, methane is second only to carbon dioxide. Since the start of the industrial revolution, atmospheric methane levels have doubled and are expected to double again by 2050. Each year 300–500 million tonnes of methane enter the atmosphere from livestock, coal mining, drilling for oil and gas, rice cultivation, and the decomposition of organic garbage in **landfill sites**. Methane has a short life expectancy of ten years in the atmosphere before it breaks down. However, it traps twenty times more heat than carbon dioxide.

Nitrous oxide

Nitrous oxide is a greenhouse gas with both human and natural sources. Over the last 2000 years, nitrous oxide concentrations have been steady at 280 **ppb**. However, there has been a dramatic increase since the beginning of the industrial revolution, and the level continues to increase by about 0.3% per year. Nitrous oxide has a greater impact on the greenhouse effect than carbon dioxide, being approximately 310 times more powerful in trapping heat in the atmosphere. It also has a greater greenhouse potential due to its long atmospheric lifetime, approximately 120 years.

Agricultural processes, fossil fuel combustion, solid waste disposal and industrial processes (such as nitric acid production) are the primary cause of nitrous oxide emissions. Nitrous oxide is a product of the reaction that occurs between nitrogen and oxygen during fossil fuel combustion. The volume emitted varies with the fuel type, technology or pollution-control device used, as well as maintenance and operating practices.

Nitrous oxide is produced naturally in soils and water through the microbial processes of **denitrification** and **nitrification**. However, natural emissions of nitrous oxide are increased by human intervention. Examples include the use of synthetic and organic fertilisers, production of nitrogen-fixing crops, cultivation of soils with a high organic content, and the application of livestock manure to croplands and pasture.

HFCs and PFCs

HFCs and PFCs are used in a variety of industrial processes, including semiconductor manufacture, refrigeration, and fire protection and retardation. These gases also are inadvertently emitted during some manufacturing processes. For example, HFC-23 is emitted during the production of HCFC-22, and PFCs are emitted during the aluminium smelting process.

Sulfur hexafluoride

Sulfur hexafluoride is an extremely potent and persistent greenhouse gas. It has an atmospheric lifetime of 3200 years. The gas accumulates in an essentially undegraded state for many centuries. As a result, a relatively small amount of sulfur hexafluoride can have a significant impact on global climate.

The primary user of sulfur hexafluoride is the electric power industry. Other industrial applications include use in magnesium melting operations and as a gas tracer. Because of its non-toxic, inert properties, sulfur hexafluoride is also used in a few highly specialised medical procedures.

REVIEW ACTIVITIES

1
Recall the names of the main greenhouse gases and the effect they have on global warming.

2
Account for the increase in carbon dioxide levels since the start of the industrial revolution.

3
Describe how herbivores contribute to atmospheric methane levels.

4
Explain what HFCs and PFCs are.

EXTENSION ACTIVITIES

5
Explain why the human-induced greenhouse effect has only occurred since the start of the industrial revolution.

6
Which gas is regarded as being the main cause of human-induced global warming? Justify the decision that it is the main greenhouse gas.

GREENHOUSE GASES AND GLOBAL WARMING

Causes of global warming

If the Earth had no atmosphere, its average surface temperature would be a very low −30 °C rather than the comfortable 20 °C found today. The current surface temperature is due to a suite of gases called greenhouse gases, which affect the overall energy balance of the Earth-atmosphere system by absorbing infra-red radiation. In its natural state, the Earth-atmosphere system balances absorption of solar radiation by emission of infra-red radiation to space. Due to greenhouse gases, the atmosphere absorbs more infra-red energy than it re-radiates to space, resulting in a net warming of the Earth-atmosphere system and an increase in surface temperature. This is the natural greenhouse effect. When more greenhouse gases are released into the atmosphere due to human activity, more infra-red radiation is trapped in the Earth's surface. This contributes to the enhanced greenhouse effect.

An increase in concentration of greenhouse gases causes a reduction in outgoing infra-red radiation. The Earth's climate must change in some way to restore the balance between incoming and outgoing radiation. This climatic change will include a global warming of the Earth's surface and the lower atmosphere. A small rise in temperature will induce many other changes, for example, cloud cover and wind patterns will increase in some regions and decrease in others. Some of these changes may act to increase the warming (positive feedbacks), others to reduce it (negative feedbacks).

Effects of global warming

It has been forecast that the mean global surface temperature will rise by 5.8 °C by the end of 2100. This projection takes into account the effects of aerosols (which tend to cool the climate) as well as the delaying effects of the oceans (which act as heat reservoirs and have a large thermal capacity). However, there are many uncertainties associated with this projection, such as future emission rates of greenhouse gases, climate feedbacks, and the size of the ocean delay.

If global warming continues to takes place, the sea level will rise due to two different processes. First, warmer temperatures cause the sea level to rise due to the thermal expansion of sea water. The melting of glaciers and icesheets in Greenland and Antarctica would also add water to the ocean. It is predicted that the Earth's average sea level will rise by 0.09 m to 0.88 m between 1990 and 2100.

Over half of the human population lives within 100 km of the sea. Most of this population lives in urban areas that serve as seaports. A measurable rise in sea level would have a severe impact on low-lying coastal areas and islands. There would be an increase in beach erosion rates along coastlines. Fresh ground water would be displaced for a substantial distance inland. The rise in sea level would also result in the loss of coastal wetlands, which would reduce fish populations, especially shellfish. Increased salinity in estuaries could reduce the abundance of freshwater species, but increase the presence of marine species.

Global rainfall is likely to increase because of global warming as there will be increased evaporation from water bodies. However, it is not known how regional rainfall patterns will change. Some regions may have more rainfall, while others may have droughts. There would be a change in soil moisture contents over various continents. Furthermore, higher temperatures would probably increase evaporation. These changes would probably create new stresses for water management with regard to water demand and supply. Higher concentrations of carbon dioxide would allow plants to grow bigger and faster.

Global warming potentials

In order to assist policymakers to measure the impact of various greenhouse gases on global warming, the concept of global warming potentials (GWPs) was introduced by the Intergovernmental Panel on Climate Change in its 1990 report. The panel assigned each greenhouse gas a GWP rating based on its relative contribution to global warming. It took into account the differing atmospheric lifetimes of the various gases as well as the differences in their ability to absorb radiation.

REVIEW ACTIVITIES

EXTENSION ACTIVITY

1
Explain the process involved in the greenhouse effect.

2
Briefly describe the trends that are occurring in the quantities of greenhouse gases in the atmosphere.

3
List the possible effects that global warming will have on the planet.

4
Define the term 'global warming potential'.

5
Human-induced global warming is a contentious issue. Some governments and scientists reject the idea of human-induced global warming. Outline the main arguments put forward by each side.

ACID RAIN

Causes of acid rain

Acid rain is caused by air pollution in heavily populated or industrialised areas and is the result of sulfur dioxide and nitrogen dioxide mixing with water droplets in the atmosphere. Acid rain can form as a result of two processes. In some cases, hydrochloric acid can be released into the atmosphere as a result of industrial processes. More commonly it is due to the pollutants that form from the oxidation of nitrogen oxides or sulfur dioxide gases, which have been released into the atmosphere from fossil fuel combustion and other sources. It may take several days for these gases to be converted into acidic compounds. They may be carried many kilometres from their point of origin. Acid rain formation can also take place at the Earth's surface when the nitrogen oxides and sulfur dioxide settle on the landscape and mix with dew or frost.

Sulfur dioxide undergoes the following reactions to produce acids:

Sulfur dioxide + water → sulfurous acid

$$SO_2 \text{ (g)} + H_2O \text{ (l)} \rightarrow H_2SO_3 \text{ (aq)}$$

Sulfur dioxide can be gradually oxidised to sulfur trioxide, which is quickly converted in moist air to sulfuric acid. This reaction can be expressed as follows:

Sulfur trioxide + water → sulfuric acid

$$SO_3 \text{ (g)} + H_2O \text{ (l)} \rightarrow H_2SO_4 \text{ (aq)}$$

Acids of nitrogen form as a result of the following atmospheric chemical reaction:

Nitrogen dioxide + water → nitric acid + nitrous acid

$$2NO_2 \text{ (g)} + H_2O \text{ (l)} \rightarrow HNO_3 \text{ (aq)} + HNO_2 \text{ (aq)}$$

Scientists in North America and Europe have found that rain and snow in those continents are becoming increasingly acidic. This appears to be linked to the vastly increased emission of greenhouse gases, such as sulfur and nitrous oxides. These gases dissolve in water to form strong acids.

Pure water has a pH of 7, showing that it is a neutral solution. A solution with a pH less than 7 is regarded as being acidic. Rain has a pH range of 6.0–6.5. The pH level of pure water is lower than that of rain because as rain falls it picks up carbon dioxide. In so doing, it forms a dilute solution of carbonic acid. In polluted regions of Europe the pH of rain has been recorded as low as 2. The normal pH range of acid rain in Australia is 4.0–5.6. However, in the lower Hunter Valley, where fossil fuels are burnt in the production of electricity, a pH as low as 3.6 has been recorded.

Effects of acid rain

Although acid rain appears to pose no threat to human health, it can do considerable damage to human structures and equipment. (See Figure 3.5.2.) More importantly, it has serious implications for ecological systems, ranging from changes in the rate at which nutrients are leached from plants and soil, through to acidification of lakes and rivers and effects on the metabolism of organisms. In Southern Norway, for example, acid rain has altered the acidity of certain streams to the point where salmon eggs can no longer develop and salmon runs have vanished.

The threat of acid rain is not as high in Australia as in some of the more industrialised countries, due to our relatively low levels of industrialisation. Despite this, certain areas experience a localised affect of acid rain. Western Sydney is one of these areas because of the effects of the prevailing wind currents and the industry located there.

Figure 3.5.2 Acid rain can damage statues and buildings.

EFFECT ON SOIL AND ROCKS

Acid rain has a number of effects on soil and rocks. When acidic rainwater comes into contact with soil and rock, the sulfate ions and calcium ions in the rainwater crystallise. The effect is pronounced in rock crevices and cracks. As the crystals grow and expand within the crevices, they cause the rock to crumble. Acid rain can also dissolve the cementing agent between sand particles in sandstone, especially when the cementing agent is calcium carbonate. This results in chemical weathering of the rock and it crumbles. Acid rain may also react with or dissolve minerals in a soil, resulting in an increase in heavy metals within the soil.

EFFECT ON THE URBAN ENVIRONMENT

Acid rain may react with rock-based building materials in the above-mentioned ways. Statues and building facades made of marble, which is a **metamorphic** form of calcium carbonate, suffer badly from acid rain. The marble is easily converted by the acid rain to gypsum, which is a softer material and is easily weathered. The gypsum readily binds with dust and soot particles, disfiguring the buildings or statues. Acid rain also reacts with metal building materials, increasing their corrosion rate. Metal building materials can be protected from corrosion, usually by applying a protective layer, such as paint.

metamorphic
the structure or composition of a rock or mineral that has been changed due to heat or pressure

definition

REVIEW ACTIVITIES

1
Explain why rain and snow are becoming increasingly acidic.

2
Describe the possible consequences of increasingly acidic rain.

EXTENSION ACTIVITY

3
Discuss the role of nitrogen and sulfur oxides in the environment.

THE DEPLETION OF THE OZONE LAYER

The ozone layer plays an important role in reducing the amount of ultraviolet radiation able to penetrate the atmosphere and reach the Earth's surface. (See *Earth and Environmental Science: The Preliminary Course*, Chapter 1.4.) Recently, it has been discovered that this layer is under threat from a group of compounds called chlorofluorocarbons (CFCs).

Between the 1930s and 1980s CFCs were used extensively in a variety of manufactured goods. As a result, large amounts of this group of compounds were released into the atmosphere. CFCs remain in the stratosphere for a long time and this leads to the destruction of ozone.

CFCs are unaffected by ultraviolet radiation in the troposphere, but are attacked by ultraviolet radiation in the stratosphere. When CFCs are attacked in the stratosphere, chlorine atoms are released. The chlorine atoms react with methane in the air and produce hydrogen chloride, which causes acid rain. When a free chlorine atom reacts with an ozone molecule it converts the ozone molecule to an oxygen molecule by removing one of the oxygen atoms. The chlorine atom then combines with the free oxygen atom to form chlorine monoxide. The free oxygen atom combines with the ozone molecule to form two molecules of oxygen. (See the following equations.)

Trichlorfluoromethane + ultraviolet radiation → dichlorofluoromethane + chlorine

CCl_3F (g) + ultraviolet radiation → CCl_2F (g) + Cl (g)

Chlorine + ozone → chlorine monoxide + oxygen

Cl (g) + O_3 (g) → ClO (g) + O_2 (g)

Chlorine monoxide + oxygen → oxygen + chlorine

ClO (g) + O (g) → O_2 (g) + Cl (g)

Ozone + oxygen → oxygen

O_3 (g) + O (g) → $2O_2$ (g)

It became apparent to scientists that the ozone levels were being depleted because more ozone was being destroyed than created. When ozone is lost, the ozone layer becomes thinner and this results in an increase in ultraviolet radiation reaching the Earth's surface. The stratosphere cools because less radiation is absorbed, and the temperature of the lower troposphere layer increases.

The thickness of the ozone layer varies naturally over time. Ozone levels drop in winter and at night, and increase when there is more sunshine. This makes it difficult for scientists to precisely determine how much of the ozone loss can be attributed to human activity rather than natural depletion. However, they are able to map the areas of the layer that are damaged. In 1985, members of the British Antarctic Survey reported that they had discovered a hole in the ozone layer over Antarctica. Ozone is produced in the tropics and transported to the poles, where it is more prone to thinning. The air over the poles is very cold and chlorine atoms are released from chlorine nitrate, which react with the ozone as the air temperature rises in spring.

Since the discovery of the Antarctic hole, a similar, but larger, hole has been discovered over the Arctic. Because it is bigger, the southern hole may pose a greater threat with regard to the penetration of ultraviolet radiation.

The uses and replacement of CFCs and other halides

CFCs are examples of a group of compounds called halogenated alkanes. CFCs were originally developed during the 1930s to replace ammonia as a refrigerant. They had several advantages over ammonia: they were unreactive, had low toxicity and low flammability, and were easily liquefied when compressed. These properties meant that by the late 1980s CFCs were widely used in a variety of household appliances, such as airconditioners and refrigerators. CFCs were also used as propellants in common household spray cans (such as deodorants and insecticides), as solvents, in electronic circuits, in the dry-cleaning industry, and in the making of expanded polystyrene and other plastics.

Halides called bromofluorocarbons were mainly used in firefighting systems. Despite most countries banning the use of this group of chemicals by the end of the 1980s, the levels of bromofluorocarbons in the atmosphere increased by 25% between 1988 and 1998.

Once governments were convinced that CFCs and other halides were having negative effects on the atmosphere, a race started to develop suitable harmless compounds to replace them. Two replacement compounds in current use are hydrochlorofluorocarbons (HCFCs) and hydrofluorocarbons (HFCs). Unlike CFC and other halides, these replacement compounds are not harmful to the ozone layer. HCFCs break down in the troposphere due to their increased reactivity. HFCs do not contain chlorine atoms and, therefore, do not affect ozone.

1

Recall what the letters 'CFC' stand for.

2

Summarise how CFCs affect the ozone layer.

3

Recount why the ozone layer is important to the Earth's stable conditions.

4

Explain what bromofluoro-carbons were used for.

5

Use the Internet or library to find photographs that show changes in the size of the hole in the ozone layer over Antarctica. Describe how the hole has changed over five years. If possible, describe how it changes during the course of a year.

6

Research how HCFCs break down in the troposphere.

STRATEGIES TO REDUCE POLLUTION EMISSIONS

Measures are being taken by many governments to reduce the levels of some primary pollutants. Sulfur dioxide emissions have been reduced by using coal that is cleaner or has been treated to reduce its sulfur compounds. A number of industries, such as the power generation industry, have introduced scrubbers to remove sulfur compounds from their emissions. These scrubbers also reduce particulate emissions. The use of lead, which has been identified as particularly harmful to young children, has been greatly reduced. Government legislation and positive action by the petroleum industry have seen leaded petrol replaced with non-leaded petrol. Also, lead-based paints have been banned and replaced with lead-free products.

Australian strategies, such as the *Clean Air Act 1990*, have been implemented to remove CFCs and other harmful fluorocarbons from industrial and domestic use. They are being replaced with non-harmful compounds, including HFCs and PFCs. For example, HFC-134a is a substitute for CFC-12.

The Montreal Protocol and the Kyoto Agreement are two international strategies that have been formulated to tackle the problems associated with air pollutants.

The Montreal Protocol

The 1987 Montreal Protocol resulted from an international conference held to discuss the control of substances that deplete the ozone layer. The Montreal conference was preceded by the 1985 Vienna Convention for the Protection of the Ozone Layer. The Vienna convention outlined the responsibilities that each country has to protect human health and the environment against the effects of ozone depletion. This provided the framework for the Montreal Protocol.

The Montreal conference resulted in a series of rigorous meetings and negotiations. The outcome was a landmark international agreement designed to protect the stratospheric ozone layer: The Montreal Protocol on Substances That Deplete the Ozone Layer. At the conference it was agreed to phase out ozone-depleting chemical compounds, such as CFCs and other halides and carbon tetrachloride, by 2000. It was also agreed to phase out methyl chloroform by 2005. It set the elimination of ozone-depleting substances as its final objective. Convincing countries to agree to the protocol proved to be difficult, but this was eventually achieved on 16 September 1987.

The delicacy of the negotiations is shown in the final agreement, which contained clauses to cover the special circumstances of several groups of countries, especially developing countries, which did not want the protocol to hinder their development. The protocol is flexible: it can be adjusted as the scientific evidence strengthens, without having to be completely renegotiated. This has been done a number of times since the protocol was signed in 1987. The protocol came into force on 1 January 1989, by which time it had been ratified by twenty-nine countries and the European Economic Community (EEC), representing approximately 82% of world users.

Later conferences finetuned the Montreal Protocol, increasing the list of controlled substances and defining timetables for the elimination of those substances. At a meeting in London in 1990, the parties to the Montreal Protocol agreed to a specific timetable for the phasing out of controlled substances. These controlled substances include CFCs and other halides, carbon tetrachloride, methyl chloroform, HCFCs, HBFCs and methyl bromide. Another meeting of the parties was held in Copenhagen in 1992, where it was agreed to accelerate the schedules for phasing out the controlled substances. The amendments arising from this meeting were to freeze HCFC production by 1996 and totally phase out HCFCs by 2030.

The production and consumption of ozone-depleting substances has been reduced more rapidly than is required by the Montreal Protocol. Focused efforts by industry have enabled their progress in phasing out ozone-depleting substances to surpass national and international efforts.

The Kyoto Agreement

In 1997, 160 countries signed the Kyoto agreement on global warming. The agreement bound all signatories to reduce their annual greenhouse gas emissions by an average of 5.2% relative to 1990 emissions. Annual emissions produced by Australia, Norway and Iceland can increase by 8%, 1% and 10%, respectively. Included among the signatories is the USA, which is the world's largest polluter; each year the USA produces twice as much pollution as Japan and three times more than France. Many commentators argued that the USA got off lightly by agreeing to make a reduction of just 7% of greenhouse gas emissions, while lesser polluters, such as the members of the EEC, agreed to an 8% reduction.

The most disputed provision of the deal is the US proposal to permit a trade in emission quotas. This would allow richer countries to 'buy' the right to pollute the environment from those countries that do not use their full emission quota. In its current form, the Kyoto Agreement does not cover developing countries, such as India and China. The US Senate has warned that it will not ratify the agreement unless developing countries are bound by it. Although the USA has signed the agreement, the US Government cannot implement it until it has been passed by the US legislature. The delegates at the climate conference in Kyoto established a framework for an agreement to fight global warming, but many details are still rather sketchy and need to be finalised.

REVIEW
EXTENSION

ACTIVITIES

ACTIVITY

1
Critically evaluate the importance of the Montreal Protocol and the Kyoto Agreement.

2
Explain the differences between the Montreal Agreement and the Kyoto Agreement.

3
Explain why the Kyoto Agreement allows specific countries to have different reduction rates.

SUMMARY

- Air pollution has an adverse impact on human health, plants and soil. It also affects the global climate and depletes the ozone layer.

- Air pollution may be defined as events that add to, or subtract from, the usual constituents of air and alter its physical or chemical properties.

- There are two principal sources of air pollution: natural occurrences and human activities.

- There are two general groups of pollutants: primary and secondary. Primary pollutants are those that are emitted directly from identifiable sources. Secondary pollutants are those that are produced in the air by the interaction of two or more primary pollutants.

- Many chemical compounds found in the atmosphere act as greenhouse gases. These gases allow sunlight to enter the atmosphere unimpeded. Greenhouse gases absorb infra-red radiation as it is reflected back towards space, trapping the heat in the atmosphere. This results in a net warming of the Earth-atmosphere system and an increase in surface temperature. This is the natural greenhouse effect. The release of greenhouse gases to the atmosphere due to human activity is called the enhanced greenhouse effect.

- The major natural greenhouse gases in the atmosphere are carbon dioxide, methane, nitrous oxide, water vapour and ozone. Human-made greenhouse gases include CFCs, HFCs, PFCs and sulfur hexafluoride.

- GWPs reflect the relative strength of individual greenhouse gases with respect to their contribution to global warming.

- Acid rain is caused by air pollution in heavily populated or industrialised areas and is the result of sulfur oxides and nitrogen dioxide mixing with water droplets in the atmosphere.

- Acid rain appears to pose no threat to human health, but can do considerable damage to human structures and equipment and has serious implications for ecological systems.

- CFCs are ozone-depleting compounds that were used extensively between the 1930s and 1980s.

- Bromofluorocarbons were mainly used in firefighting systems.

- The Montreal Protocol is a landmark international agreement to control ozone-depleting substances in order to protect the stratospheric ozone layer.

- The Kyoto Agreement was signed by 160 countries and binds them to reducing their annual greenhouse gas emissions by an average of 5.2%.

CFCs and other halides

In this exercise you will undertake research to identify the uses of CFCs and other halides, how they are being replaced and the role they have had in the depletion of the ozone layer.

ACTIVITIES

1

Make a list of the household items, such as furnishings and household appliances, in which CFCs were used. For each item, identify how the CFC was used.

2

Give examples of two other halogenated alkanes that have been replaced. Describe their uses.

3

For each use identified in activities 1 and 2, name the chemical that is now used as a replacement and identify any possible problems associated with these chemicals.

4

Outline the international and Australian restrictions that have been placed on the use of CFCs and other halides.

5

Identify how long the levels of CFCs and other halides are expected to remain at danger levels in the atmosphere.

OUTCOMES

At the end of this chapter you should be able to:

- define the qualities of geological features that need to be considered in selecting areas for waste dumps

- evaluate the methods currently used for the disposal, treatment and/or recycling of both solid and liquid waste

- assess attempts at mine-site rehabilitation and current methods of rehabilitating mined areas.

WASTE MANAGEMENT

Waste can be categorised in many different ways. It can be categorised as solid waste or **liquid waste** or it can broken down into waste sectors (that is, the sectors of society or industry that produce the waste) or types (such as paper, glass and packaging). Waste sectors include the building industry, households, offices and power plants. Each sector generates a specific type of waste. The waste generated by the building sector includes bricks, concrete, plastics, paints and timber, while offices generate such waste as paper, packaging and plastic.

Significant volumes (14 million tonnes annually) of solid waste are being produced from all sectors of the Australian community. In many Australian cities, landfills are now full to overflowing, toxic waste dumps are contentious community issues and waste disposal costs to taxpayers are rising rapidly. In response to these problems, the federal and state governments proposed a National Waste Minimisation Strategy to address the problems of waste generation and landfill disposal. In 1995 the NSW Government passed the Waste Minimisation Strategy. The strategy aimed to reduce the total municipal waste going to landfill by 50% by 2000. The target was based on the weight of waste per capita in 1990.

The government recognised that policy initiatives were needed if this goal was to be achieved. One policy initiated was the national kerbside recycling strategy. This strategy aimed to achieve the following recycling targets by 2000: 25% of plastic containers, 45% of glass, 65% of aluminium cans, 40% of steel cans, 20% of liquid paperboard containers, 40% of newsprint and 71% of paper packaging.

The Waste Minimisation Strategy established a waste management hierarchy to provide an order of preference for selection of appropriate waste management techniques. This order is based on the apparent effectiveness of each technique in conserving resources and protecting the environment against pollution. The components of the waste management hierarchy, in order of decreasing preference, are reduction, reuse, recycling and removal.

Reduction

Of all the forms of waste management, waste reduction has the greatest potential for conserving resources and protecting the environment. It is aimed at avoiding the production or purchase of superfluous items, and is based on the simple assumption that items no longer required or used sooner or later become waste.

Two essential management issues need to be addressed when considering waste reduction solutions for community waste. These are:
- the most cost-effective means of identifying and monitoring waste generation and the costs associated with waste
- the steps that can be taken at a local level to change consumer habits in order to decrease waste generation.

In recent years, many industries have adopted new technologies and management practices to reduce waste in the production and marketing of their products. Examples include the production of concentrated washing detergents, which require less packaging, and the glass industry reducing the weight of bottles. Despite these initiatives by producers, the volume of waste in most Australian cities continues to increase. This trend suggests that the consumer is a significant part of the problem and, to reverse this trend, more emphasis must be placed on educating the consumer.

Most people are unaware of how easily they can reduce their waste. For example, paint wastage can be avoided if the amount of paint required is calculated carefully before purchase. Having shorter showers, not running the tap while brushing teeth and reducing the number of times a toilet is unnecessarily flushed are all ways in which water use can be minimised, while also reducing the amount of waste water produced.

Reuse

Reuse usually refers to reusing a product in its original form without additional processing other than, for example, cleaning. There is increasing pressure on suppliers to take back, and reuse, the major parts of the packaging **waste stream**. This is most easily done with large packaging items, wooden pallets and the like. The consumer is able to minimise waste by reusing wrapping paper, plastic shopping bags and disposable cups. The retreading of tyres is an example of reuse because the tyre wall and basic frame are used and full reprocessing is not required.

A major difficulty with reuse is recognising opportunities to reuse products that are traditionally regarded as waste. Another waste reuse problem frequently arises in the home. It is the reuse of bottles as containers for pesticides and other hazardous materials. This is unsafe because the reused containers do not have appropriate identification and warning labels, creating the risk of accidental contact with or ingestion of the hazardous material. Such reuse should be actively discouraged.

waste stream
the different types of waste that are disposed of

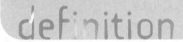

Recycling

Recycling has become one of the most accepted means of reducing solid waste generated by the community. It involves diverting material from the waste stream for reprocessing into similar products. Many types of waste can be recycled, including glass, metals, plastics, rubber, leather, textiles, wood, and food and garden wastes. Some processing is required to create new products from these types of waste. Milk cartons and aluminium cans need to be reprocessed into new containers before they are refilled. However, milk bottles are not subject to processing, other than cleaning, so they are considered to be reused rather than recycled.

Recycling waste can make good environmental and economic sense because it:
- conserves natural resources
- sometimes saves energy in production and transport
- reduces the direct risk of pollution and may reduce the cost of pollution control
- reduces the demand for landfill space
- reduces removal costs.

A number of recycling programs are used to collect household wastes. These include kerbside residential recycling. (See Figure 3.6.1.) Management issues requiring consideration in recycling from residential areas include:

- determining which waste components will be separated
- selecting the type, number and capacity of containers to hold recyclables
- paying for the containers, equipment and facilities
- deciding on the frequency of collection
- changing waste storage habits
- securing generator (that is, public) cooperation and participation
- determining market capacity for the recovered and separated material.

The major problems associated with recycling are obtaining recyclables that are free of contaminants and separating recyclables from the waste stream. Some groups of recyclables, such as plastics, may have different chemical compositions and properties, which makes reprocessing difficult. Contaminants, such as ceramics in glass, prevent unlimited reprocessing of some recyclables and eventually they have to be disposed of in landfills. Separation can be achieved at source, such as in the home, or at a community waste transfer station. In many large cities a combination of both forms of separation is used. Source separation is an effective way to improve the performance of all aspects of waste management, such as making recycling more effective, improving the quality of recovered materials and decreasing the costs of landfill.

Figure 3.6.1 Kerbside recycling service.

Removal

There are various methods of removing, or disposing of, liquid and solid waste. Liquid waste is usually diverted, via sewers, to a sewage treatment plant or to the ocean. Other liquid wastes, especially those containing toxic chemicals, are disposed of at specialist sites run by chemical recycling companies. Some commercial wastes (such as biomedical wastes from hospitals and chemical waste from the manufacturing and building industries) are considered to be hazardous. Whether solid or liquid, hazardous waste cannot be disposed of in the same place as non-hazardous domestic and commercial waste.

Solid waste can be removed in a number of ways. They include recycling, land-filling, incineration, **composting**, **anaerobic digestion**, **gasification** and **pyrolysis**. Land-filling, recycling, incineration and composting are the favoured methods in New South Wales. By most estimates, over 80% of all municipal solid waste ends up in landfills, or waste dumps. Of the remainder, approximately 10% is incinerated and about the same is recycled. These figures are expected to change radically by 2010.

composting
biological decomposition of organic material under aerobic (open air) and thermophilic (over 70 °C) conditions to create humus-rich garden compost

anaerobic digestion
a biological process in which organic material in waste is broken down by the action of micro-organisms

gasification
a process where carbon-based wastes are heated in the presence of air or steam to produce fuel-rich gases

pyrolysis
a process where carbon-based wastes are heated in the absence of air to produce a mixture of gaseous and liquid fuels and a solid, inert residue (mainly carbon)

definitions

Anaerobic digestion, gasification and pyrolysis are used overseas, but are not favoured in Australia because of the problems they present in regard to cost, environmental pollution, wastage of valuable resources and siting of plant to carry out the processes.

LANDFILL

Landfilling remains the major form of waste removal in Australia. A typical landfill site is shown in Figure 3.6.2. Most waste management organisations prefer this removal option because it is relatively cheap to operate, allows the disposal of most types of solid waste and, once complete, provides useable land on which to build homes or industrial units.

On the negative side, this option requires the replacement of the waste products. This uses valuable resources and exacerbates climate change because fossil fuel energy is used to replace the products through mining and manufacturing. Also, the decomposition of organic material in landfills produces methane, a potent greenhouse gas. Landfills cause various nuisances, including increased traffic, noise, odours, smoke, dust, litter and pests. Old landfill sites have created environmental problems because toxins and other chemical compounds have leached from the sites and entered the watertable. This has caused water pollution and contamination of land and has proved to be a health hazard to humans and animals alike.

Typical engineered landfills are now continually monitored to ensure that they are performing correctly. There are four critical components of a secure landfill:
• the natural hydrogeological setting
• an impermeable bottom liner to prevent ground water pollution
• collection systems for gases and **leachate**
• a cover.

The natural setting must be selected carefully to minimise the possibility of wastes escaping to the ground water beneath the landfill. The three other components must be engineered.

The geological features of the landfill site have to have two qualities. To prevent leakage into the watertable, the rocks need to be as impervious as possible. However, in case leakage occurs, the geology needs to be as simple as possible, with no fractures, dikes or other complexities. This makes it possible to easily predict where the wastes will go. Fractured bedrock is highly undesirable beneath a landfill because the wastes cannot be located if they escape. Mines and quarries should be avoided because they frequently are in contact with ground water.

Modern landfill sites have bottom liners, which are barriers that prevent water from the site entering the watertable or passing into the ground water. The liner effectively creates a bathtub in the ground. If the bottom liner fails, wastes will migrate directly into the environment. There are three types of liners: clay, plastic (flexible membrane liners or FMLs) and composite. Clay liners are not very effective because natural clay is often fractured and cracked. Also, some chemicals can degrade clay. A mechanism called diffusion will move organic chemicals, such as benzene, through a 1 m thick clay landfill liner in approximately five years. The very best landfill liners today are made of a tough plastic film called high-density polyethylene (HDPE). However, a number of household chemicals, including moth balls, will degrade HDPE, allowing leachate to permeate it. Other household items (such as margarine, vinegar, alcohol and shoe polish) can cause it to develop stress cracks. Reports show that all plastic liners will have some leaks. A composite liner is a single liner made of two parts: a plastic liner and compacted soil, usually clay. All liners are at least slightly permeable. Additional leakage results from defects such as cracks, holes and faulty seams. Studies show that a 4 hectare landfill will have a leakage rate of between 0.4 and 40 litres per day.

Figure 3.6.2 A typical landfill site.

leachate
water that is contaminated with either chemicals or micro-organisms by contacting wastes

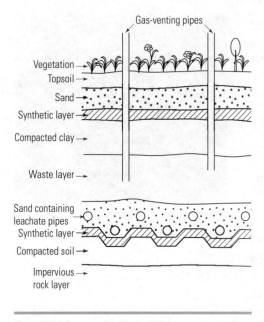

Gas-venting pipes

Vegetation →
Topsoil →
Sand →
Synthetic layer →
Compacted clay →

Waste layer →

Sand containing
leachate pipes
Synthetic layer →
Compacted soil →

Impervious →
rock layer

Figure 3.6.3 Cross-section of a landfill site.

Landfill sites must have some form of leachate collection system. As water passes down through the landfill site, soluble chemicals in the waste dissolve in the water and are carried away or leached. This water, or leachate, seeps to the bottom of a landfill, where it is collected by a system of pipes. (See Figure 3.6.3.) The leachate is then pumped to a waste-water treatment plant, where it is treated. If leachate collection pipes become clogged, leachate can build up in the landfill. The resulting liquid pressure becomes the main force driving waste out of the bottom of the landfill if the bottom liner fails.

As landfill sites contain high levels of organic waste, it is normal for them to produce large amounts of potentially dangerous methane. It is important that the site is vented to release the methane. It can be used as a source of energy or burnt off, rendering it harmless.

A cover, or cap, is an umbrella over the landfill to keep out water and thus prevent leachate formation. It will generally consist of several sloped layers: a clay or membrane liner (to prevent rain from intruding), overlain by a very permeable layer of sandy or gravelly soil (to promote rain run-off), overlain by topsoil in which vegetation can root (to stabilise the underlying layers of the cover). If the cover is not maintained, rain will enter the landfill. This will cause leachate to build up to the point where it overflows the sides of the landfill, allowing wastes to enter the environment. Covers are vulnerable to attack from at least seven sources:

- erosion by natural weathering
- penetration by roots
- burrowing or soil-dwelling animals
- sunlight, which dries out clay (permitting cracks to develop) or destroys membrane liners through the action of ultraviolet radiation
- subsidence, which can cause cracks in clay, tears in membrane liners and ponding on the surface
- rubber tyres, which float upward in a landfill
- human activities.

Due to the dangers of water infiltration, leaching of chemicals, methane production and breaching of the caps, landfill sites have to be closely monitored for a long period after they have gone out of commission. Landfilling is no longer a cheap option for disposal and requires consideration of many management issues. These include:

- justification of the need for landfills
- evaluation of site location, including environmental impacts, and public concerns and acceptance
- landfill design and cost-effectiveness
- management policies and regulations.

INCINERATION

Incineration is regarded as a very efficient form of waste disposal as most domestic waste can be incinerated. It greatly reduces the bulk of solid waste and produces an ash of a uniform composition. Trials have been carried out in using this ash as a building material to replace sand or as a component in the manufacture of besser blocks.

There are a number of problems with incineration. The main problem is air pollution and especially the release of dioxins, a highly toxic pollutant. Complete combustion of the waste requires very high temperatures (around 1000 °C). This produces 800 kg of waste gases for each tonne of rubbish. Also, for the method to be economical, the incineration plant must be situated close to the waste collection area.

When a proposed incineration site is planned it usually results in serious opposition from the local community on a number of environmental grounds.

Overall, this waste disposal method is not as popular as it could be. To make it more economical and attractive, incineration can be used to convert waste to energy. The waste-to-energy conversion process involves burning waste to extract energy and to reduce the mass and volume of waste going into landfills.

The Waverley-Woollahra incinerator in Sydney is currently the only reasonably large waste incineration plant in Australia. About 160 000 tonnes of waste is incinerated annually in this plant. This amounts to about 5% of the total quantity of waste disposed of in the Sydney region.

REVIEW ACTIVITIES

1
Outline the different methods of solid waste disposal.

2
Describe what geological features are required in order for a site to be used as a landfill.

3
Explain how leaching is prevented at landfill sites.

4
Outline the problems posed by leachate collection systems.

5
Describe the advantages and disadvantages of incineration as a solid waste disposal method.

EXTENSION ACTIVITIES

6
Describe how composting can be used as a waste disposal method.

7
Methane is one of the gases produced in waste dumps. This gas used to be burnt off as a waste gas. However, alternative uses have been found for the gas. Describe what uses can be made of methane.

MINE-SITE REHABILITATION

Some members of the community regard the mining industry as an exploiter and destroyer of the environment. However, little attention is paid to the effort, time and money being directed by the industry towards the rehabilitation of current and worked out mines. Mining in Australia is carried out within strict environmental constraints and is subject to constant monitoring. Today the mining industry is possibly one of the most environmentally conscious land users.

Common environmental problems resulting from mining include:
- soil erosion
- chemical pollution of soil, ground water and surface water
- alteration of the topography of the mine site (see Figure 3.6.4)
- acidification of soil and water
- siltation of streams and wetlands
- devegetation and introduction of noxious weeds
- presence of old mine workings or abandoned buildings and equipment.

In the nineteenth century it was common for mining companies to simply come into an area, clear a site and start mining. Once the mineral-bearing material had run out or mining it had become uneconomical, the company would abandon the site and move on. What was left was a cleared area of land open to soil erosion and littered with abandoned mine equipment. Unvegetated **spoil heaps** dotted the landscape. Chemical pollution would affect the environment due to toxic wastes leaching from spoil heaps or **tailing dams**.

Figure 3.6.4 An open-cast mine.

spoil heaps
waste materials from mining, excavating or quarrying activities

tailing dams
bodies of water containing mining residue held in suspension

definitions

Case study: Mt Lyell

The mine site at Mt Lyell in Tasmania provides an example of the adverse environmental effects of mining. However, it also provides an example of how to tackle the problems caused by mining. Mining at Mt Lyell caused the severe environmental damage of a 50 km² area of land. (See Figure 3.6.5.) For over 100 years, copper and gold was mined at this site. Smelting of the copper was done on site and sulfur dioxide emissions poisoned the vegetation. The sulfur-rich waste led to the rivers flowing through the site becoming highly acidic for up to 50 km downstream of the site. Tailings from the site caused severe sedimentation at the mouth of the rivers, in Macquarie Harbour.

The Tasmanian Government is studying the mine site in order to come up with an effective rehabilitation program for the repair of the site and its surrounds. At present it has identified the source of acid leakage and has capped some of the spoil heaps with an impervious clay cap to prevent run-off or leaching. Water movements in the local rivers have been monitored to measure and prevent further erosion, and copper and sulfur depositions in the rivers have been measured. Biological investigations have been carried out to determine the effects of the mine site on animals and plants. The program is likely to provide a role model for tackling other mine-affected sites.

Common rehabilitation techniques

Each mine site presents unique problems and, therefore, unique solutions in terms of rehabilitation. However, there are some techniques that can be carried out at all sites, although the methods employed may differ. These techniques include:

- removal, relocation or demolition of buildings and physical infrastructure
- closure of pits and shafts
- stabilisation of underground workings, soils and slopes
- treatment of waste, tailings and waste water
- revegetation of the land.

For further information refer to Section 6.

Figure 3.6.5 The Mt Lyell mine site.

ACTIVITIES

1

Explain why the mining industry has a poor reputation with some members of the community.

2

Outline how mining affects the environment.

3

What was mined at Mt Lyell?

4

Outline what escapes from spoil heaps, how it escapes and how its escape can be prevented.

5

Recall the common mine-site rehabilitation techniques.

ACTIVITY

6

Open-cast and underground mining are two different types of mineral extraction. Research in what ways mine-site rehabilitation may vary between the two.

SUMMARY

Waste can be categorised as solid waste or liquid waste or it can be broken down into sectors or types.

The components of the waste management hierarchy, in order of decreasing preference, are reduction, reuse, recycling and removal.

Waste reduction has the greatest potential for conserving resources and protecting the environment.

Reuse usually refers to reusing a product in its original form without additional processing other than, for example, cleaning.

Recycling diverts material from the waste stream for repro-cessing into similar products.

Methods of waste removal include dumping at sea, landfilling, recycling, pyrolysis, composting and incineration. Landfilling is the major form of waste disposal in Australia.

The critical components of a secure landfill are its natural hydrogeological setting, an impermeable bottom liner to prevent ground water pollution, gas and leachate collection systems, and a cover.

Old landfill sites have created environmental problems because toxins and other chemical compounds have leached from the sites and entered the watertable.

Common environmental problems resulting from mining include soil erosion, chemical pollution and acidification of soil and water, alteration of topography, siltation of streams and wetlands, devegetation, introduction of noxious weeds, and presence of old mine workings or abandoned buildings and equipment.

Common mine-site rehabilitation techniques are removal, relocation or demolition of buildings and physical infrastructure; closure of pits and shafts; stabilisation of underground workings, soils and slopes; treatment of waste, tailings and waste water; and revegetation of the land.

Mine rehabilitation

In this exercise you will conduct research on the Internet to find out how different types of mines have been rehabilitated and the intended uses of the rehabilitated sites.

ACTIVITIES

1

Locate information about the Cadia Hill mining operation by accessing the NSW Minerals Council website <http://www.nswmin.com.au/virtual_tours/option_9_7> and related websites and then complete the following activities:

a How long is the open-cast mine due to be operating for?

b How much waste rock is expected to be mined and how much gold will be produced?

c Explain what steps are being taken to rehabilitate the site.

d Why have the environmental staff decided to plant 10 000 to 20 000 trees per year rather than pasture plants, such as grass?

e Once the site has been rehabilitated what will it be used for?

2

Access the Alcoa website <http://www.alcoa.com.au/environment/miner.shtml> and associated websites and then complete the following activities:

a How much does Alcoa spend annually on environmental issues related to its mines?

b What is the ratio of tonnes mined to hectares rehabilitated?

c What is intended to replace Alcoa's Darling Range mine once mining has finished?

Introduced species and the Australian environment

4

Australia is unique as a continent because of the nature of its climate, topography and soils, its isolation from other major landmasses and its relatively short occupancy by Aboriginal peoples, who lived in harmony with their environment. All these factors resulted in a limited invasion by foreign living things and the formation of unique flora and fauna.

Major invasion by introduced species has only occurred in Australia since the arrival of European people in the late eighteenth century. They established towns and villages and employed European farming techniques, resulting in habitat change. They also brought a plethora of new foods and organisms, including plague rats, cockroaches and other vermin. Disease-causing viruses, bacteria and fungi were also introduced. This not only affected the health of humans, but also harmed crops and domestic animals. Eventually the introduced species escaped into the native populations of animals and plants.

Modern society is slowly learning the threat these introduced species pose to the Australian environment. Over the last 100 years, measures have been introduced to control these invasive species and regulatory bodies have been established. These include the Australian Quarantine and Inspection Service, which attempts to prevent further introduction of species. At the same time, in cooperation with other agencies, it attempts to understand the behaviour and life cycle of established invasive species in order to eradicate them. Eradication measures include biological control, culling and poisoning. Regulation of the discharge of ballast water is also employed in an effort to avoid the introduction of marine pests.

In this section we take a brief look at some of the introduced species that have had an adverse effect on the Australian environment. We also examine some methods of controlling introduced species.

CONTENTS

4.1 Survey of introduced species in Australia

OUTCOMES

At the end of this chapter you should be able to:

- define an introduced species as one that is not indigenous to a particular locality

- identify the criteria that can be used to classify introduced species

- discuss examples of introduced species to identify
  - plants or animals
  - food requirements
  - areas of invasion
  - aquatic or terrestrial
  - human mediated or non-human mediated

- discuss the reasons why different groups of people may have introduced plants and animals into the Australian environment

- discuss the reasons why different groups of people may have different opinions on the presence of an introduced organism as a pest, using an identified example such as the tourism value of water buffalo in the Northern Territory.

INTRODUCTION OF SPECIES TO AUSTRALIA

Humans are not the only species to have invaded the Australian continent. The arrival of humans in Australia has caused the introduction of at least 25 000 foreign animal and plant species.

Invasion can be defined as the successful founding of a colony that is able to produce viable young, where no previous colony existed. Invasion by **non-indigenous species** has been both accidental and intentional. Some species, such as strangler figs, may have arrived on Australian shores accidentally when seeds or other vegetative parts were carried by water currents from adjacent landmasses. Other species, such as dingoes and cats, have arrived in the company of humans.

Both aquatic and terrestrial ecosystems in Australia have been open to invasion by non-indigenous species. These non-indigenous species are called **introduced species**. Examples are salvinia, a highly successful aquatic plant, and European carp, both of which have altered Australia's aquatic environments.

Introduced species increase competition with native species and cause changes within ecosystems. This problem is not unique to Australia; all countries suffer from the effects of introduced species. Some of the Australian species that have been exported to other countries have had an adverse effect. For example, brush-tailed possums were exported from Australia to New Zealand, where they are now a major pest and have a bounty on their head. The Cootamundra wattle has been exported from Australia to Africa, Europe and the USA and it has seriously affected ecosystems there.

non-indigenous species
species that are not native to a particular environment

introduced species
a non-native species that is able to successfully establish a colony and produce viable young

definitions

As travel between countries has become easier for humans, there have been an increasing number of stowaways, ranging from microscopic pathogens through to insects, reptiles and small mammals. The Australian Quarantine and Inspection Service is at present attempting to prevent invasion by a number of introduced species. In Queensland, the devastating fire ant is being hunted and destroyed. In Tasmania, the hunt is on for stowaway foxes that have crossed the Bass Strait from Victoria, where they were introduced in the 1860s.

Feral animals are domestic animals, such as livestock or household pets, that have become established in the wild. They include feral cats, feral pigs, hares, feral goats and feral horses, or brumbies. Introduced animal species that did not originate from livestock or household pets are referred to as exotics, exotic wild animals or introduced wild animals.

Ecosystems are not only threatened by introduced animals. Non-native plants also have an adverse effect. Over the last 200 years, non-native plants have had a massive impact on Australia's biodiversity. Research has found that 24% of Victoria's flora is composed of non-native plants. (Currently there are no equivalent data on NSW flora.) These non-natives are made up of **exotics**, **naturalised aliens** and **noxious weeds**. They are mainly introduced species, although some of them are indigenous plants that have been introduced into New South Wales from Western Australia or Northern Queensland. As with introduced species from overseas, these organisms are able to outcompete natives and cause major changes in ecosystems. The methods of introduction of plants to the Australian environment are shown in Figure 4.1.1.

Under the NSW *Noxious Weeds Act 1993*, noxious weeds are placed in different categories according to the type and extent of the action required to control and/or eradicate them. These categories are as follows:

• W1 weeds must be reported to the local control authority and must be fully and continuously suppressed or destroyed
• W2 weeds must be fully and continuously suppressed and destroyed
• W3 weeds must be prevented from spreading and their numbers and distribution reduced
• W4 weeds must not be sold, propagated or knowingly distributed, and the occupier of the land on which these weeds exist must implement biological control or other control programs as directed by the local control authority.

Introduced animals and plants have had a great impact on the Australian environment and the productivity of many of its agricultural industries. The full impact of many species is still to be determined. However, sufficient is known for it to be a warning against any further uncontrolled introductions as they may result in the loss of biological diversity. Loss of biological diversity occurs when introduced species do not have the same natural controls in Australia as in their country of origin and can therefore displace Australia's native plants and animals.

The people who introduced these organisms into their new environment were generally lacking in concern about the consequences of their introduction. There was either no concern or no system for identifying the truly useful from the truly troublesome organism. Consequently, the introduction of a host of new animal species into Australia without considering their impact caused many unforseen results.

Introduction of organisms and lack of forethought continues to the present. A grass called *Hymenachne* was released in North Queensland in 1988 to improve the quality of pasture for cows in aquatic fringes and low-lying ground. Some experts objected strongly to the plant's introduction because it is highly invasive, chokes waterways and degrades water quality. Despite, these objections, *Hymenachne* was

feral animals
domestic animals that have become established in the wild, after escaping or being abandoned by their owners

exotics
non-native organisms that are not fully naturalised or fully acclimatised to the Australian environment

naturalised aliens
organisms that did not originate in Australia but have since established themselves here

acclimatise
to become habituated to a new climate or environment

noxious weeds
plants that, if left uncontrolled, will spread widely and cause serious economic loss to agriculture or have a detrimental effect on humans, animals, other plants or the environment

definitions

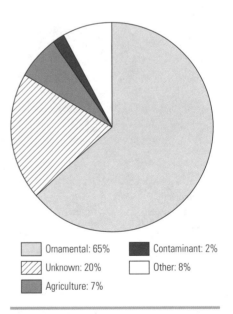

☐ Ornamental: 65%
▨ Unknown: 20%
▨ Agriculture: 7%
■ Contaminant: 2%
☐ Other: 8%

Figure 4.1.1 Methods of introduction of plants to the Australian environment.

introduced and it took just eleven years for this grass to become one of Australia's top twenty weeds.

Society is generally oblivious to or ignores the danger presented by most of these introductions. When measures are taken to remove some introduced organisms from the environment there may even be a public outcry. This was the case when deer were culled in the Royal National Park after the 1994 bushfires, and when brumbies were culled in the national park in the Casino area of Northern New South Wales.

METHODS OF INTRODUCTION

Non-human-mediated introductions

Non-human-mediated introduction of species is referred to as **colonisation** and it occurs without any intervention by human beings. Generally this form of introduction is accidental, and occurs in a variety of ways. Expansion by colonisation is normally halted by geographical barriers, such as mountain ranges, seas and altitude. Other barriers include lack of appropriate food sources, and excessive competition by other species for food and shelter. If these barriers can be overcome, the expansion of the species' range will continue. This was seen with the expansion of the human species from the confines of Africa millions of years ago. The expansion continued, overcoming barriers, until the species covered the world.

Some of the earliest species to attempt to come to Australia did so by crossing from Papua New Guinea to Australia across the land bridge that existed between the two landmasses before changes in sea levels covered it with what is now the Timor Sea.

The two main agents of dispersal and, hence, colonisation are wind and water. Spores of bacteria, moulds and simple plants, along with seeds of vascular or higher plants, can be carried thousands of kilometres in air currents. Larger organisms can also be carried by the wind. For example, spiderlings can be dispersed from their point of hatching by air currents. To increase their surface area, and therefore be blown further, they trail lines of fine threads. Flying insects can also be carried thousands of kilometres by air currents. In storm conditions, flying animals (such as insects, birds and bats) can be blown large distances off course. It is thought that the finches that populate the isolated Galapagos Islands were blown thousands of kilometres off course and that the original flock was flying close to the South American coast.

Seeds can be carried many kilometres by rivers and oceans. Some plants, such as coconuts, rely on these currents to transport their nuts to new areas where they can

definition

colonisation
the natural process by which a species expands its range

establish themselves. Other organisms can hitch a ride on floating debris and, if this debris is washed down a river into an ocean, can be carried long distances by ocean currents. Insects, reptiles and small mammals have been recorded as having colonised new areas by these means. It is not only terrestrial animals that can be dispersed in this way. Shallow water organisms (such as crabs, oysters and mussels) have been found attached to floating rubbish and other flotsam. Some of these organisms have survived crossing the Pacific Ocean, and once they have entered shallow water they can reproduce and start colonising the new shallow waters.

Some organisms hitch a lift with other organisms. Birds fly great distances and can carry hitchhikers that have become attached to their legs, beaks and feathers. These hitchhikers include seeds, some of which have barbs or hooks that allow the seeds to become tangled in the bird's feathers. Some animals (including leeches and small snails) attach themselves to the bird's feet or beak, and these hitchhikers drop off when the bird arrives at a new watercourse.

Non-human-mediated introductions usually end up in disaster for those organisms trying to colonise new environments. Most of these organisms, seeds or spores die long before they arrive at an area suitable for colonisation.

Human-mediated introductions

People have introduced organisms into Australia for various reasons. Many were deliberately introduced to the wild. Some came by accident, and others were escapees from aviaries, aquaria and zoos.

Introduced species were arriving in Australia long before the European settlers of the 1800s and 1900s. As you may know, the dingo arrived with the early Aboriginal settlers tens of thousands of years ago. Over the last several thousand years, fishermen from the island of Macassar were coming to Northern Australia to collect edible sea slugs. They brought tamarinds to eat and either discarded the seeds or deliberately planted them. Jequirity beans were brought to Australia as part of the goods traded by people from neighbouring countries. This bean originated in India and had spread through the tropical areas in the Southern Hemisphere. The beans grew into large creepers with red and black beans. Maccassan traders may have even introduced feral cats to Australia many years before Europeans did so.

An explosion of introduced species occurred during the end of the eighteenth century. In 1804, when Matthew Flinders visited Sydney, Flinders' botanist recorded over twenty non-native plants that had become weeds. These included plants such as plantain, nettle and scarlet pimpernel. Only a few weeks after the arrival of the First Fleet, a herd of cows escaped from the settlement at Sydney.

The early explorers deliberately set out to improve the newly discovered land. They felt it would benefit both Australian Aboriginals and European settlers if the countryside were full of animals and food plants that the Europeans would recognise. Captain Cook deliberately released both pigs and cows in Tasmania in the hope they would multiply. Captain William Bligh planted fruit trees in the wild, as well as oaks and firs. He also released chickens on Bruny Island off the coast of Northern Queensland. Today, parts of Australia possess wild food plants (such as olives, melons, blackberries, fennel, carrots and turnips) that were planted not so much by the early settlers but by the early explorers in order to 'improve' the countryside.

These early introductions by explorers were soon joined by new organisms introduced by settlers and, in particular, the newly formed acclimatisation society. Like the early explorers, the society's members wanted to 'improve' the environment and sought enjoyment by introducing animals and plants that they knew from their homelands. In 1862 the governor of Victoria addressed the inaugural acclimatisation society meeting in Melbourne. In his speech he advocated the introduction of

Figure 4.1.2 Stockmen used camels as a means of transport in Australia's arid interior.

monkeys to Victoria's forests. Acclimatisation society members were generally from the upper classes, accustomed to hunting and fishing, and so they were keen to introduce animals for these sports. This resulted in the introduction of salmon, trout, pheasants, deer, rabbits and foxes.

A wide range of animals and plants were introduced as ornamental or medicinal organisms. The blackbird, song thrush, goldfinch and Indian palm squirrels were brought in as domestic ornamental animals. Ostriches were originally introduced to supply feathers for the hat trade. Many introduced plants were imported as ornamental garden plants. The scotch thistle, Patterson's curse and the castor oil plant are examples. Some of Australia's noxious weeds were introduced as medicinal plants. These include St John's wort, which infests more than 360 000 ha in Eastern Australia. Other animals and plants became established in Australia after escaping from captivity.

Some animals (such as camels, horses and donkeys) were brought to Australia as beasts of burden. Camels were ideally suited as a means of transport in the arid central deserts of the continent. (See Figure 4.1.2.) Although they lost popularity with the advent of vehicles, today there is renewed interest in camels. We even export them to Saudi Arabia and other Arab countries.

New settlers preferred to eat food with which they were familiar. This resulted in the introduction of such species as pigs, oysters, Pacific rats and bees. The early Aborigines and Melanesians also brought dingoes and Pacific rats as sources of meat.

Some introduced plants required specialist pollination, and so it was important to also bring in some of the animals that specifically pollinated these plants. Animals that were introduced for this purpose include bumblebees and leaf-cutting bees.

Australia had a problem that settlers did not care for: snakes. Mongooses were introduced from Africa and India to help deal with the problem. Other biological controls that have been introduced include cane toads, mosquito fish, Indian mynah birds and some insects.

Other introductions were more accidental. The introduction of the dingo also brought in two types of tapeworm that were able to infest the kangaroo population. Rabbits brought in rabbit fleas and the black rat brought with it plague bacterium, which resulted in a series of plague epidemics in the overcrowded and dirty town of Sydney during the nineteenth century.

Introduced animal pests now affect almost every aspect of our community as well as the national economy. There are the obvious pests to agriculture (such as the European wild rabbit, the house mouse and the starling), but towns and cities are also affected. Many introduced birds and, in particular, rodents are serious pests in urban areas.

Although many of the introduced species were introduced in the nineteenth century through to the mid-twentieth century, organisms are still being introduced. This has occurred either accidentally, such as the fire ant outbreak in the Brisbane area, or deliberately. Unscrupulous collectors attempt to smuggle birds, reptiles and fish into Australia, and these organisms can carry disease or become established in the wild if they escape or their owners tire of keeping them. With the greater reliance on world trade and increased overseas travel, the opportunities for the accidental introduction of new species has increased rather than diminished.

The major exotic animal pests now cost Australia many millions of dollars annually and this is why continued quarantine vigilance is essential. The main method of preventing further entry of undesirable exotic animals is to prevent the importation of all plants and animals without proper quarantine and thorough biological screening procedures. The great challenge is to balance the desire to import new species with the risk of them becoming pests.

Australia now has the knowledge and experience to protect its native plants and animals and agricultural industries from introduced pests. Past experience has taught us that we must be vigilant in our approach to introduced organisms. The task cannot be left to governments alone. A responsible public is essential. Everyone eventually pays for the damage caused when introduced animals arrive in Australia and proceed to become a pest.

There are few cases of introduced wild animals that are not controversial or without problems today. Even those species that are considered desirable by some sectors of the community, such as trout, cause concern and their impact is still to be fully determined.

A commercial use has been found for some introduced species, such as rabbits, trout and carp. This complicates their management because these species are both pests and resources. Rabbits have been hunted and farmed for their pelts for the hat industry and for their meat. Commercial use of the rabbit has been sacrificed in the interest of reducing the damage they cause to agriculture. The trout continues to thrive in the wild. It causes severe damage to native aquatic insect and fish stocks, which it uses as food. Despite this, new stocks of trout are being released into the wild so they can be fished for sport. Carp are a major problem in the waterways of New South Wales. They have destroyed many freshwater ecosystems by increasing the turbidity of the water. Small industries that harvest the fish for food and plant fertiliser have developed. (See Figure 4.1.3.) At present these industries are very small, but they have the potential to be major export earners. Carp is a major source of proteins and other chemicals and could benefit poorer countries.

Figure 4.1.3 Charlie carp fertiliser.

REVIEW
ACTIVITIES

EXTENSION
ACTIVITIES

1

Define the term 'colonisation'.

2

Explain what the barriers are to the continuous expansion of a species.

3

Explain how wind and water act as agents of colonisation.

4

When was the dingo introduced into Australia?

5

Differentiate between human-mediated and non-human-mediated introductions.

6

a Outline the main reasons why humans introduced species to Australia.

b Give an example for each reason.

7

The strangler fig, a native of Australian rainforests, entered this country approximately 60 000 years ago. Carry out research into where this plant originated from.

8

Justify the introduction of the rainbow trout into Australian waters.

SUMMARY

- Introduced species are non-native species that are able to successfully establish a colony and produce viable young.

- Non-human-mediated introduction of species, or colonisation, occurs without any intervention by human beings.

- Humans have introduced many animals into the wild, either deliberately or by accident.

- Many introduced species were brought to Australia to be hunted or fished, to act as beasts of burden, as ornamental organisms, as a source of food or as pollinators.

- Like the early explorers, the acclimatisation society's members wanted to 'improve' the environment and sought enjoyment by introducing animals and plants that they knew from their homelands.

PRACTICAL EXERCISE
Modes of introduction of species

In this exercise you will carry out research to discover how a number of different introduced species have become established in Australia.

Procedure

1

Choose five plants and five animals that have been introduced into Australia and have become established here. (*Note*: Do not select any of the organisms discussed in Chapter 4.3.)

2

Conduct research into the organisms you have selected in order to complete the following activities. Complete the activities for each of your chosen organisms.

3

Once you have completed the activities, tabulate your answers. Your table should include all ten organisms.

ACTIVITIES

1
Where did the organism originate?

2
Why was it introduced?

3
When was it introduced?

4
To which part of Australia was it introduced?

5
How far has it spread since its introduction?

6
Why does it pose a threat?

7
What control methods are there for the organism?

4.2 Environmental impacts of introduced species

At the end of this chapter you should be able to:

- identify the biological and physical aspects of an environment

- explain why the physical aspects of Australian environments are so vulnerable to some introduced species

BIOLOGICAL AND PHYSICAL ASPECTS OF THE AUSTRALIAN ENVIRONMENT

An environment is the product of the interaction of numerous factors, all of which can be divided into biotic (biological) factors and abiotic (physical) factors.

Biotic factors are living features and include the:

- number of different species in the environment
- availability of individuals to mate with
- number of competitors competing for the same resources
- success rate of competitors
- number and type of predators
- availability of food
- number and type of **pathogens**.

Abiotic factors are those that can be described as non-living. They include the:

- availability of water
- type of water, such as freshwater, salt water, brackish, flowing, stagnant, deep or shallow
- availability and type of nutrients, such as in solution, simple compounds or complex chemical compounds
- availability and type of light, such as shade or full sun
- temperature variations over different time scales, for example, diurnal (daily) or seasonal
- availability and location of space, shelter and nesting sites
- air direction and movement, such as strong or sheltered
- altitude and topography
- oxygen concentration
- air and/or water pressure
- aspects of soil, such as soil type, texture, depth and pH (for example, acidic or alkaline)
- type of **substrate**, such as sand, mud or rock
- salinity levels.

An environment is shaped over millions of years by a combination of biotic and abiotic factors. If the biotic factors are altered, then it will cause the environment to change over time.

pathogen
a disease-causing organism

substrate
the material making up the subsoil

definitions

FEATURES OF THE AUSTRALIAN ENVIRONMENT VULNERABLE TO INTRODUCED SPECIES

We will now look at the physical aspects of Australian environments that are vulnerable to some introduced species.

Soils

Australia's soils are old and fragile and unable to support a wide range of plant life. When nutrients are removed by introduced plants, they are not replaced. The water requirements of native vegetation are different from those of thirsty introduced species. The greater water uptake by introduced species results in water being drawn up from the water table, which increases soil salinity.

Australian soils tend to be shallow and prone to soil erosion. The surface of Australia's fragile soils is disturbed by the hooves of animals, such as water buffalo and horses, which results in increased soil erosion. Predation by foxes and cats of ground-foraging birds (such as lyrebirds and scrub turkeys) have had major affects on vegetation patterns. This has resulted in increased soil erosion and altered water flow patterns. As the ground-foraging birds move around the forest floor they turn over the leaf litter, which encourages the germination of native trees. The removal of these birds results in reduced germination of native plants, allowing the more aggressive weeds to germinate in their place. The result is a loss of rainforests, which, in turn, results in changes in rainfall patterns and increased soil erosion.

Rabbits have two major effects on the Australian environment. Their habit of burrowing can result in soil erosion on sloping sites. They also change the nature of the habitat by eating germinating tree and shrub saplings. This results in a change from woodland to grassland habitat. It may also cause changes in watertable levels due to the change in vegetation type.

Climate

One major factor that has helped shape the Australian continent is its climate. Over millions of years, Australia has been slowly moving northwards and so over this time Australia's climate has changed little. This has allowed our indigenous species to adapt to the slow change. Unfortunately, the more rapid climate change that is currently taking place has allowed a number of introduced species to outcompete native organisms, which have been stressed by the climate change. This has resulted in alteration of certain habitats. In some areas, climbers such as lantana have replaced native rainforest trees, resulting in changes to soil structures and rainfall.

Australia's landmass is relatively small and is separated from polar regions by the Southern Ocean. It thus escapes the harsh polar air that invades Northern Hemisphere continents during winter. Northern Hemisphere continents have a greater temperature contrast between summer and winter. This means that the Northern Hemisphere species may be more active during the relatively warm winters in comparison to indigenous species. As a result, the species introduced from the Northern Hemisphere have a competitive advantage over Australia's native organisms.

Bushfires

Australia's geographical location and topography mean that almost all vegetation types in the country are fire prone. There are few high mountains and no truly alpine regions. Only the tropical rainforests of North Queensland can be said to be virtually fire-free.

Figure 4.2.1 Regrowth after a bushfire.

Fire has long been part of the Australian environment and has played an important role in shaping the flora and fauna. Climate change and human exploitation of the environment has resulted in more frequent and severe bushfires. A fire may be either beneficial or detrimental to individuals of a particular species, but the effect of a single fire is not as environmentally significant as a change to the fire regime. The kind of fuel available determines the kind of fire that burns, and the character of the fire helps shape the character of the fuel that is produced for future burns. In bushland and forested areas, where large numbers of introduced species may occur, the character of a fire may be affected by changes in the kind of fuel.

A number of Australian plant species require the action of bushfires for seed release or germination to take place. (See Figure 4.2.1.) Many Australian natives produce thick bark and large quantities of seeds in order to survive fires. However, the plants or seeds must not be exposed to too high a temperature, otherwise it may result in the native plants burning too severely for seeding or germination to take place. Gorse, a noxious weed found in many bushland areas in New South wales, burns very easily, but produces a great amount of heat. This causes the surrounding natives to burn too vigorously. A bushfire removes ground cover and young saplings (see Figure 4.2.2), allowing the introduction of weeds into the environment.

The cessation of annual burning by graziers in lands newly reserved for national parks is allowing open forests and woodlands to develop into closed forests. Open eucalypt woodlands on the rainforest fringes, which had been maintained by periodic burning, are now being invaded by rainforest species. These can be regarded as introduced species because they were not indigenous to the newly colonised areas.

The replacement of native plants by introduced species results in the loss of nesting sites. This leads to a reduction in pollinators and, in turn, a reduction of native plants. This process may result in a total change to a habitat.

The aquatic environment

The Australian continent has the lowest rainfall of the five continents. This results in relatively few rivers and lakes. Many of our freshwater bodies are under threat from introduced species.

Noxious weeds (such as water hyacinth and salvinia) may result in the destruction of a water body by killing off the living organisms underwater. The weeds deprive the submerged plants of sunlight, resulting in their death. This results in a degradation of water quality. Many aquatic introduced species grow rapidly and can clog waterways, resulting in reduced water quality and diminished water flow downstream. When the European carp stirs up the mud on the bottom of a water body, turbidity is increased and this results in a loss of biodiversity. The introduction of the weeping willow tree has also resulted in major changes in the nature of freshwater habitats. It has done so by outcompeting the native river red gum, which is used as a source of food as well as shelter by numerous native aquatic organisms. The result is that there is a loss in biodiversity, which reduces the water quality. Over a period of time, the watercourse will change in a number of other ways. For example, when the flow pattern is altered, organic matter will start to build up because there is a reduction in the number of organisms available to break down the organic matter.

Figure 4.2.2 A bushfire removes ground cover.

1

Distinguish between an environment's biotic and abiotic factors.

2

Describe how rainforests are affected by foxes and cats.

3

Explain how the replacement of river red gums by weeping willows affects the aquatic environment of a river.

4

Northern Hemisphere species are considered to have a competitive advantage over southern species. Explain why this may occur.

THREATS TO AUSTRALIAN FLORA AND FAUNA

The isolation of the Australian island continent from other landmasses for 55 Ma has created a sanctuary for flora and fauna. This isolation has resulted in Australia's biodiversity being vulnerable to introduced species. Marsupials were saved from competition with other species, although recent discoveries possibly show that marsupial mammals outcompeted early placental mammals to become the dominant group. Birds unique to Australia survived, and distinctive trees and plants developed. There are 700 bird species listed in Australia, and Australia has 20 000 species of plants. The main threat to the continuation of a number of native species in Australia is the introduction of alien species, which prey on and compete with native species for food and habitat.

Land clearing and the spread of towns and cities have destroyed many habitats. Fertilisers, such as superphosphate, have encouraged introduced pasture species at the expense of native grasses and the animals that feed on them.

Conditions in many parts of Australia have always been harsh. Our flora and fauna have lived through droughts, fire and flood and they have done this largely by surviving in refuge pockets where conditions are less severe than in other places. But, with European settlement, many refuges were destroyed and, in those that remained, feral animals moved in and killed, or competed with, native species.

Whether intentional or unintentional, the introduction of species into areas different from where they originate can have dramatic repercussions for the ecosystems and habitats where they take up residence. Other species are affected, flora and fauna are impacted, habitats are altered, and the ecosystems themselves may change in order to find a new biological balance.

In general, introduced species have a number of effects on the Australian environment. They threaten the survival of many native plants and animals because they:

- successfully compete for the available nutrients, water, space, sunlight, shelter and so on
- often survive better than native species as they may not be affected by the pests or diseases that would normally control them in their natural habitats
- reduce natural diversity
- replace or destroy the native plants that animals use for shelter, food and nesting.

The introduction of a species may have beneficial effects, such as a species acting as a controlling agent or providing a valuable resource. All too often, however, its introduction has a devastating outcome and starts a domino effect of problems. In most instances, species introduction has led to a serious loss of biodiversity where the existence of various native organisms is quickly threatened and often wiped out by the presence of the introduced species. It is in Australia that some of the most obvious and worst effects of species introduction can be seen.

Some species have been introduced as biological controls. The introduction of the moth cactoblastis to control the prickly pear cacti in Australia is an example of the successful introduction of a biological control. (It is discussed in more detail in Chapter 4.4.) By contrast, the mosquito fish (*Gambusia* sp.) is an example of a biological control gone wrong. (See Figure 4.2.3.) It was introduced to Australia to control mosquitos, but it prefers to eat native fish and frogs and their eggs.

Let's look at some other introduced species that have had detrimental effects in Australia.

The Indian mynah, a common bird in urban areas, displaces native species. Originally kept as a cage bird, it escaped and adapted well to the urban environment. There have been repeated sightings of groups of highly aggressive Indian mynahs ganging up against indigenous birds and driving them out of their territories. They are also known to take over native bird nesting sites. Pigeons, starlings and sparrows are poor competitors against native birds in the natural environment. However, in the urban environment these bird species outcompete the native birds for food and nesting sites.

Figure 4.2.3 The mosquito fish (*Gambusia* sp.): a biological control gone wrong.

Trout were introduced as a sport fish, but are highly predatory and have had disastrous effects on the stocks of native fish. However, trout are protected because of their value to the leisure fishing industry.

Rats and mice (such as the black rat, Norwegian rat and house mouse) are recent invaders from ships. They inhabit disturbed urban areas and displace some of the native species of rats in the bush surrounding urban developments. The black rat is a known carrier of pathogens and poses a risk to humans and other animals. Mice can reach plague proportions in rural areas and cause real problems for the agricultural industry when they eat crops.

Water buffalo were imported from South-East Asia as beasts of burden, but were turned loose when they were replaced by tractors and trucks. They now pose a major threat to the wetlands of Northern Australia as their weight and hooves breaks up the delicate soil crusts, resulting in soil erosion. However, water buffalo are a major money earner for the tourist safaris that take tourists to view them. The authorities have to balance the detrimental effects of the animals with the economic advantages of retaining them.

Goats were originally domesticated, but some either escaped or were released into the wild. They now inhabit many of the more inaccessible areas, where they do great damage to the vegetation. Feral goats quickly overpopulate an area and destroy the vegetation by eating seedlings and ringbarking adult plants. This destruction quickly leads to accelerated erosion of the affected area.

Like goats, pigs were originally domestic but some escaped or were released. Feral pigs cause great destruction in wetlands and the moister forests. Their habit of digging for roots leads to large areas of disturbed soil, which is either eroded or invaded by weeds. They are also a hazard to livestock, as they will actively hunt for lambs.

With Australia's relatively poor, thin soils and irregular rainfall in many areas, the effects of feral animals are not only felt by farmers, but also by our unique native fauna and flora. Overgrazing has led to a reduction in plant cover, and selective feeding by ferals puts increased pressure on some species of plants. Native herbivores and the herbivorous ferals compete directly for food and carnivorous ferals threaten native fauna. The problem of feral animals is massive and difficult to address, and each year vast amounts of money are spent to reduce their numbers.

Case study: Domestic and feral cats

It has been speculated that domestic cats were first introduced into Australia's environment during the seventeenth century, although some scientists think the introduction was much earlier. It is probable that they escaped during shipwrecks of the Dutch ships that carried them on board to control vermin. When Australia was colonised, settlers brought cats with them as pets and to control pests. Many cats managed to escape captivity and by the 1850s colonies of feral cats were established in the wild. People were happy to have many cats because they controlled the rats, mice and rabbits that endangered crop yields. In fact, in the late 1800s, cats were purposely introduced into Australia's wilds.

The numbers of cats in Australia has grown so rapidly that they now represent a danger to the environment. The average female cat becomes sexually mature after one year and will have two litters a year (but may have up to four), each of which averages four kittens. These mating patterns, coupled with the availability of vast amounts of food, have led the feline population of Australia to balloon at an almost uncontrollable rate. It is estimated that there are over 12 million feral cats in Australia.

These cats are voracious hunters. Most hunt even when they are not hungry and many kill more food than they could possibly eat. Domestic cats, despite being well fed, also hunt and kill wild animals. A feral cat may kill as many as 1000 animals per year, while a domestic cat may kill around twenty-five animals per year.

Australia's feral cats prey on 'pests' (rats, mice and rabbits) for the most part. However, in absence of those animals, cats will attack any animal that is roughly their equivalent in size and weight, as well as smaller creatures. They feed on small mammals, birds, reptiles and insects. In Australia, they tend to prey on mammals weighing up to 2 kg and birds weighing up to 3.5 kg. However, most of their prey is comprised of smaller species that weigh less than 220 g. Among the species that they prey on are several endangered species, including wallabies, finches and bandicoots.

The main problem is that there are too many cats. Since the other animals in the ecosystem did not evolve with the capacity to defend themselves against cats, cats are disproportionately successful hunters. This phenomenon is quite common wherever humans have introduced a species into an environment to which the species is not indigenous. Attempts have been made to slow the growth of these populations (see Figure 4.2.4), but feral cat colonies seem to be too well established to be reduced.

ADDRESSING THE PROBLEM

There have been a series of proposals on how to limit the environmental damage caused by cats, most of which concern reducing their numbers. Among these ideas have been poisoning, neutering, capturing or killing them, and introducing a deadly disease to curb their numbers. The problem is that if cats are eliminated from one area alone, it is likely that they will migrate from another area to take the place of the cats that have been eliminated. Therefore, it is necessary to eliminate them all over the continent. Unfortunately, reducing their numbers would only be a temporary solution since a single pair of breeding cats can exponentially produce 420 000 offspring over a seven-year period.

Furthermore, it is likely that any of the plans mentioned above would have drastic side effects on the ecosystems of which the cats are now part. Poisoning the cats may adversely affect the scavengers that feed on dead cats. A disease would only temporarily decrease the number of cats, but would eventually bring back a group of much stronger cats. It may also infect other animals, perhaps even the ones it is designed to protect. The other options do not seem feasible because they can only eliminate small numbers of cats at a time. Migrating cats or new kittens would quickly increase these numbers.

Figure 4.2.4 The end of one of Australia's most effective killers—the feral cat.

ACTIVITIES

1
Recount the ways in which introduced species threaten Australia's native flora and fauna.

2
Outline the impacts of introducing new species into the environment.

ACTIVITY

3
Investigate whether any introduced species are known to have caused the extinction of any specific native species.

SUMMARY

An environment is the product of the interaction of numerous factors, all of which can be divided into biotic (biological) factors and abiotic (physical) factors.

Biotic factors are living features, while abiotic features are non-living features.

The physical aspects of Australian environments that are vulnerable to some introduced species includes its soils, climate, bushfires and aquatic environments.

Fire has long been part of the Australian environment and has played an important role in shaping the flora and fauna.

Australia's geographical isolation has made our flora and fauna vulnerable to introduced species.

Introduced species threaten the survival of many native species because they grow faster, survive better and outcompete natives for shelter, food and nesting.

PRACTICAL EXERCISE
Looking at the local environment

In this exercise you will carry out a detailed examination of a local area in order to note the presence of introduced species.

Procedure
Choose a local environment, such as a river cutting, the verge of a main road, an old abandoned garden or an abandoned industrial site. Then complete the following activities.

ACTIVITIES

1
Using field guides and your local State of the Environment Report, identify the introduced plant species in the area.

2
Draw a plan of the area. On the plan, mark in the areas that are covered by the introduced species.

3
Calculate what percentage of the total area is covered in introduced species.

4
In the areas where the introduced species are growing, describe what has happened to the indigenous plants in those areas.

5
Research the mode of dispersal for each plant and relate it to the area in which the introduced species is growing.

6
Compare the chosen area with a natural area. Compare the:
a types of introduced species
b number of introduced species
c coverage by the introduced species.
Give reasons why there may be differences.

Caution: It is important that you are careful in these environments. Do not enter old buildings, take care of old workings and be aware of traffic. Take care at all times. If necessary, wear protective clothing.

OUTCOMES

At the end of this chapter you should be able to:

assess the relative contributions of the following conditions to two named introduced plants and two named introduced animals becoming pests
- suitable habitat
- suitable climatic conditions
- range of food resources
- relative lack of natural predators/grazers
- high reproductive capacity
- well-developed dispersal mechanisms

summarise for each of the named introduced plants and animals
- the history of introduction
- the environmental conditions leading to the organism becoming a pest
- the impact on the physical environment
- dispersal techniques
- reproductive capacity
- control strategies

examine and critically analyse the environmental impacts of the named plants and animals.

WHY ARE INTRODUCED SPECIES SUCCESSFUL?

Several factors affect how well an introduced species will do in its new environment. The degree of influence of each factor varies between species and is directly influenced by the habitat in which the species is attempting to establish itself. If the introduced species is to survive and flourish it must have no, or few, predators or grazers. It must be able to outcompete the indigenous species in the habitat it has colonised for food, oxygen, light, shelter or living space. The introduced species must be able to reproduce rapidly and preferably in sufficient numbers to colonise an area rapidly and sufficient numbers to colonise new areas. The new climate to which the introduced species has to adapt must be favourable, or else the organism will use its energy and survival strategies to survive instead of outcompeting indigenous species.

If the introduced species succeeds with all the above factors, the species is likely to survive and flourish. If it has a high success rate in establishing itself it can become a weed or pest.

Many plant and animal species have been introduced into Australia, and a number of these have become major pests. We will examine four of these.

RABBITS

The wild European rabbit *(Oryctologus cuniculus)* is Australia's most serious animal pest. (See Figure 4.3.1.) Economic damage by wild rabbits in Australia is estimated at around $600 million annually. Additionally, they cause environmental damage that is often irreparable. The loss of vegetation from rabbit grazing threatens the survival of native birds, mammals and insects that rely on plants for food and shelter. Rabbits have contributed to the extinction of many native plant and animal species.

Figure 4.3.1 Under certain conditions, rabbit populations can reach plague proportions.

They compete with livestock for available pasture and kill young trees and shrubs. The consequent failure of natural regeneration and gradual loss of mature trees is a serious problem in many areas. In addition, rabbits contribute to soil erosion by burrowing, removing vegetation and disturbing soil.

The rabbit was introduced into England during medieval times. It had originally come from the hot grasslands of Spain. Therefore, when it arrived in Australia it was already adapted to the general climatic conditions it encountered here.

The first fleet in 1788 brought rabbits to Sydney as part of their food stocks. By 1827, rabbits posed a major problem in parts of Tasmania. Yet, the introduction that is considered to have led to the establishment of the mainland rabbit population occurred in 1859. In that year, Thomas Austin brought twenty-four rabbits to his property in Victoria and set them free, with the intention of hunting them. The European rabbit spread during the following sixty years across the southern half of the continent to occupy an area of $4\,000\,000$ km^2, and it continues to increase its range today.

Rabbit populations are distributed in characteristic patterns within each major type of habitat. For instance, in arid habitats they live close to watercourses and swamps, whereas in subalpine areas they live in grassy valleys. In agricultural areas, the location of the rabbit's living area is dependent on the relationship between food, predation and shelter.

The rabbit is a grazer, feeding on grass and other plants. It eats the more nutritious parts of the plants, such as buds and roots. Its feeding habits have led to changes in the composition of plant communities. Under dry conditions, the rabbit will strip the bark off shrubs and eat the shrub's roots and seeds in order to obtain the moisture it requires. This may help to kill the shrub. This causes a change in the species composition within a habitat, resulting in a change from woodland to grassland. The rabbit is thought to have been responsible for the extinction of at least several native herbaceous plants (such as peppercress) and for the eradication of several species of small wallabies and bandicoots as they have taken their food and grazed down their cover. Overgrazing by rabbits is thought to be a major factor in the desertification of the north-east of South Australia, and a number of cases of soil and sand dune erosion.

The rabbit becomes sexually mature when it has reached the age of three to four months. It produces a litter size of four to five rabbits after a gestation period of twenty-one days. In arid conditions, an adult may have one to two litters per year, but in favourable conditions an adult may have five or more litters. The mean number of young produced annually by a rabbit may be between eleven and twenty-five.

The average longevity is very short. About 80% of the population die before the age of three months. Once adulthood has been achieved, the mortality rate drops. The principle causes of death in the rabbit population are **myxomatosis** and predation by foxes and cats. Predation has not been able to keep up with the rabbit's reproductive rate, and this has resulted in an increasing number of rabbits. The history of the interaction of rabbits and the myxoma virus has become the most completely documented example of the interaction of a host animal species and a disease organism.

definition

mxyomatosis
a highly infectious disease of rabbits

Controlling rabbits

There are a wide range of control methods used for rabbits. Some of them are more effective than others. They include:

• *Poisoning.* This involves the used of poisoned baits that contain pesticides such as sodium monofluoroacetate (commonly referred to as '1080').
• *Fumigation of burrows.* Burrows are fumigated using phosphine (World War II mustard gas) or carbon monoxide.

- *Ripping.* This is the destruction of the rabbit's warren system and is regarded as one of the most effective means of reducing rabbit numbers.
- *Trapping.* There are two types of traps: jaw style traps and cage traps. Jaw style traps are now illegal in most states and territories and are not regarded as an effective rabbit control. Cage traps are used for research purposes.
- *Shooting.* The vast numbers of rabbits makes this an ineffective method.
- *Explosives.* This creates lots of noise and dust, but is an ineffective rabbit control. It is of some use in rocky or inaccessible country.
- *Predators.* The predators of rabbits (such as cats, ferrets and foxes) can be released.
- *Biological control.* Examples are the myxoma virus, calicivirus and rabbit fleas.
- *Rabbit-proof fencing.* This fencing is designed to prevent the rabbits moving from the bush into pastoral land.

In late 1950 and early 1951, myxomatosis swept through the Murray-Darling River system. Heavy rains at the time meant that mosquitoes bred and then carried the virus from infected to uninfected rabbits. Within three years, the disease had been carried to all parts of Australia and rabbit numbers were drastically reduced. By 1953, scientists studying the virus and rabbits noticed that the virulence, or strength, of the virus had changed from being 99.9% effective to 95% effective; a small but significant drop. The less-virulent virus took three to four weeks to kill a rabbit instead of six to ten days. The milder strain was therefore more successful in infecting rabbits, and it spread rapidly. Through this selection the virus evolved to a less-virulent form.

Evolutionary selection processes were also working in the rabbit population. If a rabbit survived it would have offspring and the offspring's chance of survival would be greatly increased by lack of competition from other rabbits. They would also inherit their parents' resistance. Within a short time the proportion of resistant rabbits would increase. Today the myxoma virus may kill only 50% of the rabbit population during an epidemic.

The rabbit haemorrhagic disease is a viral disease caused by the calicivirus. It affects only European rabbits and results in a high mortality. The virus has spread throughout most of Australia where rabbits occur. This has mainly been by natural spread since its accidental escape in 1995 from Wardang Island where tests were being carried out to measure its effectiveness. To date, its impact has generally been greatest in drier areas, with reductions in rabbit numbers of 65% to over 90%. Results from wetter areas have produced reductions of less than 65%.

REVIEW
ACTIVITIES

EXTENSION
ACTIVITIES

1
Outline the conditions that must exist for an introduced species to be successful in colonising a new habitat.

2
Explain why the rabbit was well suited to the Australian environment when it was first introduced.

3
Outline how the rabbit may alter the biodiversity of an area.

4
Outline the methods of controlling rabbits.

5
Evaluate whether the calicivirus has been an effective control agent.

6
Compare the biological controls used on rabbits.

CANE TOADS

The cane toad, *Bufo marinus*, (see Figure 4.3.2) was introduced to Australia from South America by the sugarcane industry to control two pests of sugarcane: the grey-backed cane beetle and the frenchie beetle. In 1935, the government sanctioned the introduction of 101 toads into about eleven sugar-growing locations in Northern and coastal Central Queensland. After this date, introductions were non-official. Unseasonal breeding occurred almost immediately, and within six months over 60 000 young toads had been bred in captivity and then released.

The cane toad adapted well to the Australian climate and has found sufficient habitats to colonise. Australia has no natural predators to limit the cane toad's population size, and so the toad has been able to spread rapidly throughout coastal Queensland. The rate of spread was accelerated by humans transporting them, both accidentally and intentionally, ahead of their natural rate of spread.

Since 1935 they have expanded their range to include about half of Queensland. By the early 1980s, the northern front line had crossed into the Northern Territory and there are fears that they may soon colonise areas of Kakadu National Park. They have also spread to coastal Northern New South Wales to just north of Lismore, and occasionally individuals have shown up as far south as Sydney. It appears that these have been carried south in plants, for example. There is no evidence of a breeding population as far south as Sydney, as only individual cane toads have been discovered. The natural rate of spread of *B. marinus* is now 30–50 km/year in the Northern Territory and about 5 km/year in Northern New South Wales.

Figure 4.3.2 The cane toad.

The most distinctive features of cane toads are bony ridges over each eye and on each shoulder a pair of enlarged poison glands containing venom. The call of the male cane toad is a high-pitched 'brrrr', which sounds like a telephone dial tone.

Like all frogs, cane toads are insect feeders. However, they will attack anything that moves and is small enough to fit in their mouths. Their diet includes small lizards, frogs, mice and snakes and even younger cane toads. They have also been known to steal food from dog and cat bowls. As cane toads have a variable diet, little energy is used in the animal's search for food, at times preferring the food to come to them. In urban areas, they are often seen gathered around street lamps eating insects attracted by the light.

Cane toads are very adaptable. They are much more tolerant of variations in water salt content than most amphibians, and can survive and breed in brackish water. They need only a small pool of water for breeding. A female toad can produce vast quantities of eggs: up to 30000 a month. The males fertilise the eggs once they have been laid in long strands. Within three days the eggs hatch into small (3 cm) jet black tadpoles, which are unlike those of any native frog. These tadpoles become toadlets unusually early, so they are out of the water and hopping around faster than most other frogs.

One of the most important factors in the success of the cane toad is that they are highly poisonous to eat, at every stage of their life cycle. All frogs and toads have enlarged chemical-secreting glands at particular points on their bodies or small glands spread over the whole skin. The chemicals they produce are varied and, in some cases, may be highly toxic. A cane toad's reaction to a threat is to turn side-on to its attacker so that the venom glands face them. Cane toad venom is also found all over its skin. Animals touching a cane toad can receive a dose of venom, which may cause death within fifteen minutes. The glands on the cane toad's shoulders are also capable of oozing venom or even squirting it over a distance of up to 2 m if the toad is particularly roughly treated.

Research is being carried out to study what effects cane toads have on native wildlife. It is known that countless birds, mammals and reptiles have been killed by coming into contact with cane toads. However, there are some reasons for optimism. In the areas where cane toads have lived for the longest time, their populations have declined after the initial population explosion. Although cane toads have no known predators, frog-eating organisms will try to eat them. Some native animals are learning to avoid eating them, but others have shown they can eat the toad. The keel back snake can detoxify the venom, while water rats, ibis, crows and some other birds turn the toads over and eat only the non-poisonous internal organs.

As yet, cane toads still present a major threat to the biodiversity of the areas into which they have spread. Scientists have been unable to control either their numbers or their spread.

REVIEW ACTIVITIES

1
Explain why cane toads were introduced.

2
Recount the natural rate of spread of cane toads.

3
Describe what cane toads eat.

4
Why are cane toads dangerous to potential predators?

5
How do some predators cope with eating cane toads?

EXTENSION ACTIVITY

6
Ecologists are very concerned that the cane toad is expected to enter Kakadu National Park. Explain why scientists and the National Parks and Wildlife Service are so concerned.

SALVINIA

The waterfern *Salvinia molesta* was introduced to Australia from Brazil as a pond ornamental in the late 1950s and early 1960s. It escaped from garden fish ponds and entered waterways. Salvinia has been successful in Australia because it is well suited to the climate, has few competitors or grazers and is well adapted to the aquatic habitats in which it flourishes.

Salvinia is classified as a noxious weed and affects waterways by:
• trapping and causing the build-up of sediment
• altering water flow patterns
• disrupting recreational activities
• causing stagnation, which may lead to fish deaths
• excluding native plants, causing loss of habitat for birds and other animals
• reducing light penetration.

Salvinia is considered to be one of the world's worst aquatic pests. It is an aggressive, competitive species that can have impacts on aquatic environments, local economies and human health.

This plant is a rapidly growing, free-floating aquatic fern. (See Figure 4.3.3.) Its leaves range from a few millimetres to 4 cm in size. Salvinia forms a thick mat of fleshy leaves (see Figure 4.3.4), which blocks sunlight and causes underwater plants to die. The decomposing plants cause oxygen levels to drop, killing aquatic wildlife. Excessive growth of salvinia results in complete coverage of water surfaces, which degrades natural habitats in several ways. Heavy growth of salvinia competes with and shades desirable native vegetation. Mats of floating plants prevent atmospheric oxygen from entering the water while decaying salvinia drops to the bottom, greatly consuming dissolved oxygen needed by fish and other aquatic life. The animal

Figure 4.3.3 The leaves of salvinia range from a few millimetres to 4 cm in size.

habitat is most noticeably altered by the obliteration of open water. Migrating birds may not recognise or stop at water bodies covered with salvinia. Fishermen have found it impossible to cast into smothered lakes and are abandoning infested fishing spots. Salvinia clogs water intakes, which interferes with agricultural irrigation and electrical generation. The floating mats also provide an excellent habitat for disease-carrying mosquitoes.

Salvinia can reproduce extremely rapidly—it can double its numbers in as little as two to ten days—and thus completely dominate waterways. Salvinia reproduces asexually and is easily spread by dumping, animals, water movement, wind, boats and flooding. Salvinia is sterile and produces no seeds. Instead, it reproduces from fragments that form daughter plants. No matter how much it is broken up, this water plant continues to grow, with large fragments simply drifting away to start their own colonies. Salvinia is so aggressive that it has spread to all mainland states and territories, including the Northern Territory where it has invaded the wetlands of Kakadu National Park.

Controlling salvinia

Salvinia reproduces so rapidly that infestations quickly become impossible to eradicate. The mats may be up to 1 m thick, which hinders management by chemical control. The most straightforward management technique is to prevent introduction.

Salvinia plants found in aquariums should be dried and burned or buried so they cannot enter the stormwater system. Environmental management alternatives for salvinia include the three traditional weed management methods: mechanical, chemical and biological control.

Mechanical shredders for control of salvinia are not effective because the plant reproduces vegetatively and any plant part with a bud can form new plants. Physical removal has been effective in small-scale locations, but is very expensive.

Salvinia can be controlled locally with herbicides, such as paraquat. The control is difficult because the salvinia's leaves are hairy, which hampers the herbicide's absorption. Herbicide use has limitations:

- it has impacts on non-target plants
- some salvinia infestations are not accessible or easily located
- it is costly and time consuming.

Controlling salvinia with mechanical or chemical methods is unsustainable over a region of any size. Researchers quickly realised that biological control was the only practical regional approach to solving the problems posed by this weed.

In 1980, the CSIRO Division of Entomology in Brisbane released a weevil called *Cyrtobagous salviniae* (see Figure 4.3.5), the fern's natural predator in Brazil. The weevil was released at Lake Moondarra, which provides the Queensland mining town of Mt Isa with its drinking water. The results were spectacular. Within eighteen months, the weevil had eaten its way through the salvinia, and what it did not eat, simply sank into oblivion. The weevil has also been effective in Kakadu. However, trials have shown that it can take a year or two of activity to deal with a salvinia colony because during the treatment period the fern can regrow. Patience is required and it is necessary to resist the temptation to use herbicides. This short-term chemical solution causes water pollution and also kills the weevil.

Use of the weevil is a good application of biological control, but it is not always successful. The further south the weevil goes, the less effective it is. Being cold-blooded, its activity drops with temperature. So around Sydney it may be inactive for large parts of the day as well as nights: not eating, not laying eggs, and often being killed off by the cold temperatures. However, weevil-based control of salvinia has been successful near Liverpool in the Botany Wetlands.

Figure 4.3.4 Salvinia covering open water.

Figure 4.3.5 *Cyrtobagous salviniae*, the natural predator of salvinia.

1

Where did salvinia originate from?

2

Explain why salvinia is regarded as a noxious weed.

3

Describe how salvinia colonises a new area.

4

Outline how salvinia can be controlled.

5

Evaluate the effectiveness of the methods available for controlling salvinia.

Figure 4.3.6 *Lantana camara.*

Figure 4.3.7 A dense, tangled lantana bush on disturbed soil by the road.

LANTANA

Lantana is an invasive plant that originated in Central and South America, although a few species of lantana are native to Africa and Asia. *Lantana camara* (see Figure 4.3.6) was imported from Brazil as a garden plant and was first recorded in Australia at Adelaide's Botanic Gardens in 1841. Lantana has some uses other than as an ornamental plant. The plant has been used to treat ailments, such as the common cold, chickenpox, fever, fresh cuts, sores, toothaches and inflammation.

Two main species of lantana are known to occur in Australia: *L. camara* and *L. montevidensis*. *L. camara* easily forms hybrids. In Australia alone, twenty-nine forms are recognised. Lantana grows well in dry areas because it prefers well-drained soil and can tolerate drought. A prolific seed producer, lantana has quickly spread by birds carrying the seeds in their droppings. Lantana is believed to infest 4 million ha of coastal land in Australia, from North Queensland to south of Sydney. Lantana has a relative lack of natural grazers, which allows the plant to expend its energy in outcompeting native plants instead of in defence mechanisms or repair.

Lantana is particularly damaging to Australia because it is a menace to both farmers and the natural environment. Lantana outcompetes other plants, releasing a chemical that inhibits the growth of native species. It then can quickly colonise and dominate a site where the overstorey has been disturbed. This reduces biodiversity and the ability of native systems to re-establish themselves.

The weed grows in thickets in tropical and subtropical zones around the world. This is a fast-growing shrub with brittle climbing and scrambling branches. It can grow as a vine and in compact clumps and dense thickets. (See Figure 4.3.7.) The leaves are bright green above and paler beneath, with rounded-toothed edges. Flower colours vary from pale cream to yellow, white, pink, orange, red, lilac and purple. The glossy, rounded fruits are fleshy and purplish-black when ripe.

It often escapes as a weed and inhibits the growth of other species of plants. It uses established plants for support and ends up suffocating them as they are starved of light and nutrients. Dispersal is by birds, dumping and branches rerooting at ground level.

Lantana's leaves and seeds are toxic to many animals and can cause gastrointestinal disturbances and death to sheep and calves. Poisoning in stock depends on the relative toxicity of the particular kind of lantana and the amount eaten.

Controlling lantana

L. camara is regarded as one of the ten worst weedy species invading pasture lands and natural communities in Australia. Control measures have included the use of chemical sprays and the deployment of biological control agents. Despite a massive effort involving many countries and the release of more than thirty insect species, control has been spotty and unpredictable, ranging from none to significant in some countries. The main reason for this overall lack of success has been ascribed to the extreme variability of the plant, coupled with its colonising ability.

Disturbances to a forest's canopy lead to increases in the amount of light reaching the forest floor. This was the most significant cause of lantana's establishment. Removal of the understorey and shrub layer also increases the establishment rate. Controlled burning off, which reduces the risk of catastrophic fire and the destruction of the forest canopy, may lead to lantana infestation. Lantana increases rapidly after high-intensity fires. In terms of management, it is necessary to protect the overstorey of the forest by avoiding these wildfires. But in trying to avoid wildfires by burning off fuel loads, it is necessary to make sure the shrubs in the understorey are not burnt out. There is a very fine line to tread in preventing wildfires by using low-intensity fires.

The use of fire as part of a management program is recommended for the control of dense infestations. A suggested control program is:
- exclude stock to establish a fuel load and then burn
- sow improved pastures
- continue to exclude stock until pasture has established and seeded
- burn again in the hot, dry months before rain
- spot spray regrowth when it is vigorously growing between 50 cm and 1.5 m tall
- carry out follow-up controls after each burn for the next few years.

Treatment of large lantana infestations with herbicides is not economically feasible. Fire, bulldozing and slashing, or cutting, can reduce dense infestations and make spot treatments with chemicals more economically effective.

In Australia, *L. camara* has been the target of many biological control attempts. Since 1914, twenty-eight insect species have been introduced into Australia in an attempt to control lantana. Of the seventeen species that have become established, four are important because they have reduced the vigour and competitiveness of lantana in some areas. The four most important biological control agents are the sap-sucking bug (*Teleonemia scrupulosa*), the leaf-mining beetles (*Uroplata girardi* and *Octotoma scabripennis*) and the seed-feeding fly (*Ophiomyia lantanae*).

To improve control, a number of new insect species and a pathogen from Central and South America are currently being assessed by Queensland's Department of Natural Resources.

REVIEW ACTIVITIES

EXTENSION ACTIVITY

1
Recount why lantana was introduced to Australia.

2
Recount how lantana can colonise an area quickly.

3
Describe the growth habit of lantana.

4
Why are bushfires beneficial to lantana?

5
Outline how lantana can be physically controlled.

6
Outline the different biological controls that are used on lantana.

7
Explain why it is necessary to use a number of different biological controls to treat lantana.

SUMMARY

If an introduced species is to survive and flourish it must have no or few predators or grazers, and be able to reproduce rapidly and outcompete the indigenous species.

The wild European rabbit is Australia's most serious animal pest. In 1859 Thomas Austin brought twenty-four rabbits to his property in Victoria with the intention of hunting them. Within sixty years they occupied an area of 4 000 000 km².

Myxomatosis was released in Australia in 1950. Within three years it had spread throughout Australia, initially resulting in a drastic reduction in rabbit numbers.

Rabbit haemorrhagic disease is a viral disease that affects only European rabbits.

Cane toads were introduced to Australia to control two pests of sugarcane.

The cane toad adapted well to the Australian environment and spread throughout coastal Queensland. The rate of spread was accelerated by humans transporting them ahead of their natural rate of spread. The natural rate of spread of *B. marinus* is now 30–50 km/year in the Northern Territory and about 5 km/year in Northern New South Wales.

Salvinia is a noxious weed that can affect waterways by causing the build-up of sediment, altering water flow patterns, disrupting recreational activities, causing stagnation, causing loss of habitat for birds and other animals and reducing available light.

Lantana is an invasive plant that originated in Central and South America, and was imported to Australia as a garden plant.

Lantana outcompetes other plants. It can quickly colonise and dominate a site where the overstorey has been disturbed.

PRACTICAL EXERCISE
Controlling introduced species

In this exercise you will carry out research that allows you to analyse different control methods for rabbits, cane toads, salvinia and lantana.

ACTIVITIES

1
Draw a table with the following headings:
• introduced species
• mechanical control
• chemical control
• biological control.
Complete the table using your answers to the following activities.

2
Identify the control methods that have a local or widespread affect on each organism, that is, rabbits, cane toads, salvinia and lantana.

3
For each control method, give one advantage and one disadvantage of its use.

4
For each organism, choose the best control method. Give a reason for your choice.

5
For each biological control, identify what sort of organism the biological control is, for example an insect, virus, mammal or bird. For each organism describe how it controls the introduced species.

OUTCOMES

At the end of this chapter you should be able to:

explain what is meant by biological control

describe the following types of biological control and give examples of the use of each
– predator–prey
– bacterial/viral parasites
– release of sterilised males

outline the criteria used to determine the conditions under which an organism can be used for biological control

describe the history of control of prickly pear as an example of successful biological control

recount the role of the Bradley sisters in establishing the bushland regeneration program known as the 'Bradley Method'

identify broadscale environmental imports of one or more introduced species on a local ecosystem

examine and critically evaluate the strategies being used to rehabilitate this ecosystem or to minimise threatening processes

recommend ways in which the strategies could be refined

extrapolate current level of effectiveness of the identified strategies to the future in terms of
– costs
– sustainability of the ecosystem
– monitoring
– management of the program.

BIOLOGICAL CONTROL

As was seen in Chapter 4.3, many of the major introduced species that have become pests have at least one **biological control**, or biocontrol, organism that is a predator or grazer of the pest. As society has become more aware of the adverse effects of using chemicals in the environment, scientists have increased the use of biological control organisms. This increased use of biological controls has required careful checking prior to their release in order to prevent a repetition of the problem presented by the cane toad.

An advantage of using biological control organisms instead of chemical means to destroy pests is that there is no bioaccumulation or biomagnification within the environment. Nor is there an increase in resistance. There are a number of other advantages:

• a biological control organism is specific for a particular pest and can be targeted at a specific stage of the pest's life cycle
• they are generally more economical than other means
• there is a reduced risk to both the environment and to water quality within the area being targeted.

definition

biological control
the use of a pest's natural enemy to control and contain the population of that pest

Biological control methods are not the complete panacea to the problem of pest species. Use of biological controls has a number of disadvantages:

- releasing a biocontrol organism takes more intensive management and considerably more planning than other means
- planning and implementation of their use can take a considerable length of time, as can the wait for results
- it requires a detailed understanding of the relationship of the biocontrol organism, the pest and the environment.

It is successful against only a limited number of weeds, has not been tried with marine pests and has never been 100% successful against vertebrates. In his book *Feral Future*, biologist Tim Low states that only 6% of the biological controls released in Australia succeed completely, while 18% do some good and 76% fail. These figures are somewhat pessimistic because they relate to the effectiveness of a single biological control. As Low states, it is commonplace to release a number of agents against one pest.

A number of biological control organisms have been released with somewhat disastrous consequences. This is exemplified by the cane toad. Due to the problems caused by this animal and other future biological control organisms, a number of guidelines were put forward by the International Union for the Conservation of Nature:

- the release of non-indigenous species should occur only if there are advantages to both humans and the environment
- non-indigenous species should be used only if an indigenous species is unsuitable
- no non-indigenous species should be released in a natural area unless there are exceptional reasons
- releases should only occur after considerable testing and risk assessments have been carried out and, once released, close monitoring must be undertaken
- non-indigenous species should be eradicated once the problem has been negated.

The use of biological controls has occurred for over 100 years. In 1878, scientists experimented with producing and releasing a fungus that had been seen killing beetles in cereals. Australia has been a world leader in biological control for nearly ninety years and has an impressive record with regard to safety and the effective management of major invasive pest species.

The use of **natural enemies** as biological control agents was set back by the growing use of synthetic organic chemical pesticides, such as DDT, in the 1950s. Once the problems of using these chemical pesticides were identified, priority was given to biological agents that could be produced and used like chemicals. The growth of interest in integrated pest management (see Chapter 3.3) is currently creating a demand for biocontrol technologies worldwide.

Biological control methods

There are three main methods of biological control: classical biological control, conservation and augmentation.

Classical biological control involves importation of a natural enemy. It is long lasting and—other than the initial costs of collection, importation and rearing—little expense is incurred. When a natural enemy is identified it must fulfil certain requirements in order to be a good biological control. It must have host (that is, pest) specificity, be **synchronous** with its host's life cycle, be adaptable to different environmental conditions and have a high reproductive rate. The latter is important so that populations of the natural enemy can rapidly increase when hosts are available. The natural enemy must also be effective at searching for its host and it should be searching for only one or a few host species.

definitions

natural enemies
organisms that may attack and destroy the pest organism in its native environment

classical biological control
the importation of the pest species' natural enemies from the country of origin

synchronous
occurring at the same time or coinciding

Successfully established, it rarely requires additional input and it continues to kill the pest with no direct help from humans and at no cost. Unfortunately, classical biological control does not always work. It is usually most effective against exotic pests and less so against native insect pests. The reasons for failure are often not known, but may include the release of too few individuals, poor adaptation of the natural enemy to environmental conditions at the release location, and lack of synchrony between the life cycle of the natural enemy and host pest.

The **conservation** of biological controls involves the identification of any factors that may reduce their effectiveness in attacking the pest species. These factors may include reproductive rate, climatic needs and food requirements. Once these factors have been identified, steps may be taken to eradicate these problems.

Augmentation is a method of increasing the population of a natural enemy that attacks a pest. This can be done by mass producing a pest in a laboratory and releasing it into the field at the proper time. Another method of augmentation is breeding a better natural enemy that can attack or find its prey more effectively. Mass rearings can be released at special times when the pest is most susceptible and natural enemies are not yet present, or they can be released in such large numbers that few pests go untouched by their enemies. The augmentation method relies upon continual human management and does not provide a permanent solution, unlike the importation or conservation approaches.

conservation
identifying and negating factors that limit the effectiveness of natural enemies

augmentation
increasing the population of an existing natural enemy

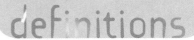

Types of biological control

PREDATOR-PREY

Certain species are extremely useful natural enemies of insect pests. Predators of insects and mites include beetles, true bugs, lacewings, flies, midges, spiders, wasps and predatory mites. Insect predators can be found throughout plants, including the parts below ground, as well as in nearby shrubs and trees. Predators may be specialised with respect to their choice of prey, or they may be more generalist. Unfortunately, a number of predators prey on beneficial insects as well as pests. Some species may provide good late season control, but appear too late to suppress the early season pest population. Many beneficial species may have only a minor impact by themselves, but contribute to overall pest mortality.

A second type of predator–prey relationship is the parasite–host relationship. Insect predators immediately kill or disable their prey, whereas pests attacked by parasites die more slowly. Some hosts are paralysed, while others may continue to feed or even lay eggs before succumbing to the attack. Parasites can be the dominant and most effective natural enemies of some pests. Different parasitic species can attack different life stages of the host. The major characteristics of parasites include the following:

• they are specialised in their choice of host
• they are smaller than their host
• they usually lay their eggs or larvae in, on or near the host
• immature parasites remain on or in the host, while adults are free-living and mobile and may be predaceous.

A third type of predator–prey relationship is the herbivore–plant relationship. Insects can control weeds by feeding on the roots of plants or by transmitting plant pathogens, which will infect plants. Other natural enemies of weeds include plant pathogens, nematodes and vertebrates (fish).

In considering species for introduction to control pest weeds, researchers first determine that the species feeds exclusively on the pest species. Other considerations include the effectiveness of the control, favourable host-plant synchronisation with the natural enemy's life cycle, and the natural enemy's ability to produce young. Researchers have also determined that a successful introduction is more likely if the new locale is similar in ecology and climate to the area of origin of the predator, and if trials are made where population densities of the pest plant are similar to those in the new location. The reasons for success or failure are not yet well understood.

Using biological agents for weed control is beneficial because there is less overall expense compared to herbicidal sprays. Once a population of biocontrol agents is established, minimal effort is required to conserve it. Also, the use of natural enemies does not require a high level of technology. Another major benefit is that natural enemies are less disruptive ecologically and so natural biodiversity is maintained.

Some desirable characteristics of weed-feeding natural enemies are as follows:
- they are specific to one plant species
- they have a negative impact on plant individuals and the population dynamics of the target weed
- they are prolific
- they thrive and become widespread in all habitats and climates that the pest weed occupies
- they are good colonisers.

BACTERIAL AND VIRAL PARASITES

Most bacterial and viral parasites act as pathogens or disease-causing organisms. They kill or debilitate their host and are relatively specific to host species. The effectiveness of a bacterial or viral parasite depends on two factors: its capacity to kill pests, and its capacity to reproduce on or in pests and therefore compound its killing action. Under some conditions, such as high humidity or high pest abundance, these naturally occurring organisms may multiply to cause disease outbreaks that can decimate a pest population. Diseases can be important natural controls of some pests.

Some pathogens have been mass-produced and are available in commercial formulations for use in standard spray equipment. These products are frequently referred to as microbial insecticides, or bioinsecticides. Some of these microbial insecticides are still experimental, while others have been available for many years. The bacteria *Bacillus thuringiensis* (Bt) is a good example of a bacterial bioinsecticide. It can reproduce in insects and acts like a pesticide. Unlike chemical insecticides, microbial insecticides can take longer to kill or debilitate the target pest. To be effective, most microbial insecticides must be applied to the correct life stage of the pest, and so some understanding of the target pest's life cycle is required.

Microbial insecticides are compatible with the use of predators and parasites, which may help to spread some pathogens through the pest population. Beneficial insects are not usually affected directly because of the specificity of a microbial product.

Major characteristics of pathogens are as follows:
- they kill, reduce reproduction, slow growth or shorten the life of pests
- they usually are specific to target species or to specific life stages of the target species
- their effectiveness may depend on environmental conditions or host abundance.

RELEASE OF STERILISED MALES

The biological control of a pest species by genetic manipulation requires the release of genetically altered individuals. The genetically altered individuals generally either have chromosomes removed or added or the alteration may occur at the gene level. Genetic manipulation is directed at a pest's ability to reproduce and generally reduces the numbers of individuals.

This sterilising of a sex has been successfully carried out in populations of the Mediterranean fruit fly in some of the fruit-growing areas of South Australia, New South Wales and Queensland. Large numbers of male Mediterranean fruit flies are grown and then irradiated with gamma radiation. This radiation sterilises the fruit flies, which are then released in the target area, approximately 100 000 per square kilometre. They then mate with fertile females, who lay unfertilised eggs. If this is successfully repeated over a number of years, the population will die out. Even if it is only partly successful, it will result in a fall in the population size. This method has also been used on other pests, such as the screw worm and mosquito.

Recently proposals have been put forward to use a daughterless gene as a control method for the European carp. This would necessitate the introduction of a gene into the wild population that would prevent the production of female carp in the population. This would cause the number of females to decline and the reproductive levels to drop, resulting in a population decline over a number of years.

REVIEW ACTIVITIES

EXTENSION ACTIVITIES

1
Define the term 'biological control'.

2
Explain the advantages of using biological controls.

3
Detail the guidelines that are followed when considering the use of biological control.

4
Differentiate between the three methods of biological control.

5
Describe what sort of organisms may cause disease.

6
Explain how biological control is achieved by genetic manipulation.

7
Evaluate whether biological control is a more effective and economical method of control than pesticides or mechanical means.

CASE STUDY: PRICKLY PEAR AND CACTOBLASTIS

It is not known how the prickly pear (*Opuntia inermis*) came to be introduced to Australia. It may have been brought in as a botanical curiosity: a potplant, in fact. We do know that it existed as early as 1839 in New South Wales. By 1863, the pear was established in Queensland, sometimes as hedges. (See Figure 4.4.1.)

Each plant, even individual leaves or parts of leaves, can readily take root and grow. It can easily survive long periods of desiccation before taking root. By 1900, some 40 400 km² of the country were affected.

Following on from the great drought in 1902, the spread became more rapid. This occurred because natural grasses were scarce, and so starving cattle and sheep were being fed on the pear. Cattle, in particular, thrived on the plant. Additionally, native animals, notably crows and emus, fed on the pear's fruit and spread the seeds in their droppings. By 1920, an estimated 234 000 km² were infested. The problem peaked in 1925, by which time 264 000 km² were affected by the pest. This led to many large properties being abandoned.

The common method attempted for control was poisoning. However, any reasonable control by this method sometimes cost twenty times the value of the land. As early as 1899, it was realised that some natural means of control was required. Of course, it was unlikely that such a device would occur naturally in Australia.

Figure 4.4.1 The prickly pear dominating a field.

In 1920, the Federal Government set up an organisation to investigate potent insect predators in the countries where various pears originated. Ultimately, the insect cactoblastis (*Cactoblastis cactorum*) was introduced from South America, and the results were spectacular. Eggs from the insect were placed on the pear plants throughout the country. The caterpillars hatched and hungrily fed on the pear, boring deep into the leaves. The results were immediately apparent. The initial release of the insect was in 1926, and a widespread release occurred during 1928–30. Between 1930 and 1932, a tremendous cactoblastis population explosion occurred, resulting in the literal collapse of hectares of pear, the insects having reduced the plants to decayed pulp. By 1933, the last great primary stands of pear were gone from Queensland. Complete control was achieved by 1940. Some pear does, of course, still exist, as do some numbers of the cactoblastis. The remaining pear is not considered to be a risk.

REVIEW EXTENSION
ACTIVITIES   ACTIVITY

1
Explain how the prickly pear was introduced to Australia.

2
What happens to the prickly pear if part of the plant is cut off?

3
Prior to the introduction of cactoblastis, what was the normal control method for the prickly pear?

4
Explain how cactoblastis controls prickly pear.

5
Research how many insect predators were tested by the Federal Government prior to the selection of cactoblastis. Explain why some of the predators were not selected.

THE BRADLEY METHOD

The names Eileen and Joan Bradley are linked to bush regeneration. They established the bushland regeneration program known as the Bradley Method. These pioneers came from a family of scientists and they themselves followed that tradition. They were employed by the National Trust in 1976 to train workers in their techniques.

The Bradley Method's basic principles are:
• work from the least weed infested areas to the most densely infested areas
• minimise soil disturbance
• allow native plant regeneration to dictate the rate of weed removal.

In the least infested areas there are abundant native plants and seeds to colonise the area from which weeds have been removed. In the dense weed infestations, the number of weed **propagules** far outnumber native propagules and so weed growth, rather than native plant growth, is favoured.

Soil disturbance can be minimised by replacing topsoil in its correct position so that the stored seed is not buried too deeply. It is also important to keep the soil deeply mulched.

Native plants that regenerate must be allowed to form a dense and healthy group before the adjoining weeds are removed. These natives can then successfully colonise the newly weeded area. By tipping the balance of power towards the natives, weeds will be inhibited and finally eliminated so that very little attention needs to be given to the area—possibly a follow-up once every year or two. If too large an area is cleared before native plants are capable of colonising it (that is, if the area is overcleared), weeds will successfully compete with native plants.

propagules
plants growing from seed

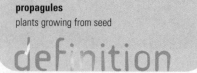
definition

1
Outline the basic principles of the Bradley Method.

2
Describe how soil disturbance can be minimised.

3
List the advantages and disadvantages of the Bradley Method.

SUMMARY

The advantages of using biological control organisms instead of chemical means to destroy pests is that there is no bioaccumulation or biomagnification within the environment and no increase in resistance.

Only 6% of the biological controls released in Australia succeed completely, while 18% do some good and 76% fail.

There are strict guidelines for the release of biological controls.

There are three main methods of biological control: classical biological control, or importation; conservation; and augmentation.

The three types of biological control are:
- predator–prey relationships
- bacterial and viral control
- genetic manipulation.

The Bradley Method of bushland regeneration has three basic principles: work from the least weed infested areas to the most densely infested areas; minimise soil disturbance; and allow native plant regeneration to dictate the rate of weed removal.

PRACTICAL EXERCISE
Bush regeneration

In this exercise you will evaluate the effectiveness of a bush regeneration program in your local area.

Procedure

Choose a piece of bushland in your local area that is undergoing bush regeneration. Then complete the following activities.

ACTIVITIES

1
Describe how the area is being rehabilitated.

2
Evaluate how effective the regeneration method has been.

3
Describe the steps being taken to minimise any threatening processes, such as weed invasion and soil erosion.

4
Identify the ways in which the strategies being carried out in the area could be improved or modified.

5
Research how the rehabilitation program is managed and monitored.

6
Research the costing of the rehabilitation program.

4.5 Modern quarantine methods

OUTCOMES

At the end of this chapter you should be able to:

- outline the quarantine procedures in place in Australia to prevent introduction of new species

- use the example of the introduction of new species through off-loading of ballast water as an example of accidental introduction

- assess the effectiveness of procedures in place to prevent the spread of new species.

REASONS FOR QUARANTINE PROCEDURES IN AUSTRALIA

As Australia was isolated from other landmasses for most of its history, today its biodiversity is very different from that of other countries. European-style farming brought new animals and plants to the continent. When these organisms were introduced, only a few diseases that affected them existed here. Most pests and diseases were left back in their countries of origin. Therefore, today Australia is relatively free from most of the world's serious pests and diseases and the Australian Quarantine and Inspection Service (AQIS) helps keep it that way.

The increase in global trade and travel, while bringing great benefits, puts Australia's agricultural industry and unique natural environment at risk of devastation from a range of serious pests and diseases. If new pests and diseases were allowed to enter Australia it would destroy our agricultural industry and seriously damage, if not devastate, our economy. There could also be major threats to the health of the Australian people. **Quarantine** procedures aim to protect Australia and Australians against the entry of unwanted exotic pests. A modern Australian quarantine station is shown in Figure 4.5.1.

quarantine
strict isolation to prevent the spread of disease

definition

QUARANTINE PROCEDURES AND THE ROLE OF AQIS

AQIS is responsible for the administration of the Commonwealth *Quarantine Act 1908* and its related legislation: the Quarantine Proclamation and the Quarantine Regulation. The Act provides powers for quarantine officers to deal with quarantine matters, sets out the legal basis for controlling the importation of goods, animals and plants, and determines the offences for breaches of the Act.

Certain goods, animals and plants may be prohibited imports, which means they are forbidden from import unless a permit has been obtained. Others may require inspection or treatment before being imported. Many items, such as pottery and wooden furniture, pose little or no risk and will not be a quarantine concern. Often the problem may not lie with the goods being imported but with the packaging.

Failure to comply with legislative requirements may, in serious cases, result in heavy fines or terms of imprisonment. All serious breaches of quarantine laws are investigated by specialist investigators, and prosecution can be expected where evidence of a breach of the quarantine laws is established. Minor breaches may result in a warning or the issue of a quarantine infringement notice and fine. AQIS investigators may also use the provisions of other Commonwealth Acts where relevant.

AQIS takes the position that quarantine is a shared responsibility between government, industry and the Australian public and everyone has a part to play in protecting Australia from incursions by foreign pests and diseases.

Each year, approximately 8 million passengers and 20 million tonnes of cargo pass through Australia's airports and shipping ports. Also, about 160 million items of mail enter the country. All these movements into and out of the country pose a threat and so it vital that they are closely monitored.

Figure 4.5.1 A modern Australian quarantine station.

As a regulatory agency, AQIS provides screening services for goods and passengers at airports, seaports and mail centres. Quarantine officers have wide-ranging powers to search, seize and treat goods suspected of being a quarantine risk. Detection of these goods is aided by the use of X-ray equipment and detector dogs. In addition, passengers and goods may be subject to random inspection. This makes importation of undeclared goods very risky for those who may try to do so. It may result in charges, fines and/or imprisonment. A side effect of the September 11 terrorist attack is that it has led to increased baggage checks at Australian airports and, as a result, the interception rates for **contraband** goods has increased dramatically.

A number of pests have entered Australia by means of freight containers shipped in from overseas. They include the fire ant, the giant African land snail, the Asian longhorn beetle and the Asian gypsy moth. AQIS is unable to check every single container and so it must rely on importers providing correct paperwork. AQIS does, however, carry out a huge number of container inspections each year as these containers pose a high level of threat. The containers chosen for inspection are based on risk-assessment procedures. The level of risk is based on the country of origin and what is thought to be in the container. Recently, a heavy duty X-ray unit has been installed at Botany Bay container terminal in Sydney. This device is able to X-ray an unopened container, thus revealing its contents, and has increased the inspection rate from ten containers per day to 100 containers per day. AQIS intends to install more X-ray units at container terminals around Australia.

AQIS is a small organisation and, considering the vast length of Australia's coastline and the enormity of the Australian continent, it has a limited budget with which to enforce the quarantine regulations. AQIS cannot be everywhere at once and so many breaches of the regulations occur. AQIS realises it cannot keep every pest out of Australia and so it relies heavily on risk analysis. It looks at the levels of risk posed by different imports and targets the worst. Limited checks are placed on high-volume, low-risk imports and therefore pests are likely to enter the country this way.

AQIS is not only responsible for preventing the importation of unwanted organisms from overseas, it also has internal restrictions. Certain fruit-growing areas are free from fruit flies and other pests. It is AQIS's role to prevent the introduction of those pests into vulnerable areas, whether they be the continent itself, specific agricultural areas or bushland.

1

Explain the possible consequences for AQIS of global trade and travel.

2

Recall the Act that AQIS is responsible for administrating.

3

Recall how many items of mail enter Australia per year.

4

Recall the problems that are faced by AQIS in controlling the movement of goods and people into and out of Australia.

5

List the screening methods carried out by AQIS.

6

Explain why certain imported living things have to undergo a period of quarantine while other species do not have to pass through a quarantine period.

CASE STUDY: AN INTERNAL QUARANTINE ISSUE

Fruit flies, such as the Mediterranean fruit fly, damage many types of ripe fruit and vegetables by using the fruit or vegetables as a breeding site. The fruit fly female lays eggs and deposits bacteria in the fruit. The eggs hatch and the developing larvae feed on the rotting fruit. Fruit flies attack a wide variety of soft, fleshy fruit and vegetables.

If an area of Australia has fruit fly, it is placed under strict quarantine. Farmers in the quarantined area cannot sell their fruit until it has been treated and has successfully passed the tests conducted by AQIS. Treatments for fruit flies are very expensive.

Four farming areas in Australia have been declared by AQIS as free from fruit fly. (See Figure 4.5.2.) Other countries will accept fruit and vegetables from farms in these areas. Australia can use the fruit fly free status of these areas to open up new markets for Australian produce.

ROLE OF BALLAST WATER IN SPECIES INTRODUCTION

The introduction of invasive marine species into new marine environments by ships' ballast water has been identified as one of the greatest threats facing the worlds' oceans. It poses serious ecological, economic and health threats.

Approximately 80% of the world's commodities are transported by ships. Ships travelling to and from ports carry 10 to 12 billion tonnes of ballast water across the world each year and 60 million tonnes in Australian waters. Ballast is required by partially unladen or fully unladen ships to provide balance and stability. Ballast is often water, but can be any material used to weight or balance an object. Ballast water is taken on board at the port before the voyage begins and is pumped into special ballast tanks. The water that is used is untreated harbour water. When it is pumped on board, marine organisms are sucked up into the tanks. During the voyage, temperature changes in the tanks and lack of food and light kill most, but not all, of the organisms. On arrival at the port, the ship loads the cargo and pumps out its ballast tanks with the surviving organisms into the harbour. (See Figure 4.5.3.) Some of the surviving organisms may survive the new conditions to establish populations in the new environment.

Over 100 species of marine organisms are known to have been introduced into Australia via ballast water. They include species of plankton, shellfish and starfish. It

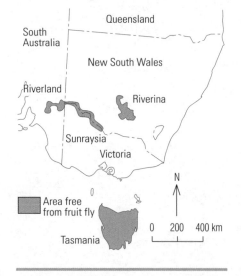

Figure 4.5.2 Areas of South-eastern Australia that are free from fruit fly.

has been discovered that ballast water can also carry disease-causing pathogens, such as cholera. Some species introduced to Australia via ballast water are benign and therefore harmless. Others, such as the Northern Pacific sea star, have become pests, threatening biodiversity, fisheries and aquaculture. Some introduced species severely deplete native populations or deprive them of food. Others form colonies that can smother existing fauna. Some introduced micro-organisms, such as some of the dinoflagellates, cause red tides and algal blooms and are toxic to shellfish, fish, seabirds and humans.

It is not only water from the ballast water that poses a problem. Many ships use solid ballast material, and bitou bush is one of the major introduced pest species that has arrived on Australian shores by this means.

Until July 2001, Australia had a voluntary system operating where ships exchanged their ballast water for ocean water. The ocean water may contain organisms but they are unlikely to survive in shallow waters. Unfortunately, this system was voluntary and difficult to monitor. Also, there was a risk of instability, and therefore capsize, while the ship was exchanging its ballast water.

In July 2001, Australia's new ballast water management requirements came into force. They require all ships that intend to enter Australian territorial waters or dock at an Australian port to carry out a ballast water risk assessment.

Under the new requirements, AQIS requires each ship to provide information about the amount of water in each ballast tank, the location where the uptake took place, and where the intended discharge is to take place for each ballast tank. This information must be supplied prior to entry into Australian waters. Based on this information, each ballast tank is given a risk rating of either low or high risk. If the ballast tank is in the high-risk category, an AQIS approved management option has to be carried out prior to arrival in Australian waters. These options include:
• the non-discharge of high-risk ballast water in Australian waters
• a full ballast water exchange in deep water away from the Australian coast
• other treatment methods acceptable to AQIS.

Ballast water contamination is a major problem to international shipping companies. Many countries are carrying out research into how to treat ballast water to prevent the introduction of foreign species. Treatment methods include irradiation, bubbling ozone through the water or adding pesticides to the water. At present, no one method has proved totally successful. Therefore, governments are devising strategies to prevent the dumping of ballast water into shallow coastal waters.

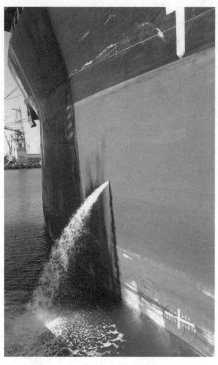

Figure 4.5.3 A ship discharging ballast water.

REVIEW
ACTIVITIES

EXTENSION
ACTIVITY

1

Recall why ballast water poses one of the greatest threats to the world's oceans.

2

Explain what ballast water is used for.

3

Recall what sort of pathogens can be carried in ballast water.

4

Solid ballast has been responsible for introducing which major noxious weed?

5

Outline the new Australian ballast water management requirements.

6

Using a list of Australian introduced species, identify which species have been introduced via ballast water. (A list of introduced species can be found in Tim Low's book *Feral Future*, published by Penguin Books.)

SUMMARY

- Quarantine procedures aim to protect Australia and Australians against the entry of unwanted exotic pests.

- If new pests and diseases were allowed to enter Australia it would destroy our agricultural industry, seriously damage, if not devastate, our economy and potentially threaten the health of the Australian people.

- AQIS is responsible for the administration of the Commonwealth Quarantine Act 1908 and its related legislation: the Quarantine Proclamation and the Quarantine Regulation. This legislation is aimed at preventing the introduction of diseases and pests.

- AQIS provides screening services for goods and passengers at airports, seaports and mail centres.

- AQIS is also responsible for imposing internal restrictions aimed at preventing the introduction of pests into certain fruit-growing areas. If fruit flies are allowed to spread to certain fruit-growing areas, overseas trade may be affected.

- Ballast is required by partially unladen or fully unladen ships to provide balance and stability.

- A number of invasive marine species have been introduced into new marine environments via ship's ballast. Ballast water can also carry disease-causing pathogens, such as cholera.

- Over 100 species of marine organisms are known to have been introduced into Australia via ballast water. They include species of plankton, shellfish and starfish.

- In July 2001, Australia's new ballast water management requirements came into force. AQIS is responsible for ensuring these requirements are met.

PRACTICAL EXERCISE
The Australian Quarantine and Inspection Service

In this exercise you will gather, process and present information to summarise the methods used by AQIS to prevent or control the entry into Australia of new species and analyse the effectiveness of these methods.

ACTIVITIES

1
Research how and why AQIS is preventing or controlling the entry into Australia of the following:
a foot and mouth disease
b the giant African land snail
c foodstuffs in luggage
d straw and wooden materials
e soil on machinery
f insects on commercial planes
g aquarium fish.

2
Prepare a report on the results of your research.

3
In your report, outline the procedures used by AQIS in detecting each of the imports you researched and analyse the effectiveness of these procedures.

Organic geology: A non-renewable resource

5

Organic resources (such as coal, oil and gas) play an extremely important role in our society. We rely on them for our wellbeing, but they are not resources that will last forever. In this section we will examine the nature of fossil fuels and the role of science in finding and processing them. In addition, we will look at the environmental impacts of fossil fuel use and the alternatives that are being developed to conserve fossil fuels and reduce their impacts on the environment.

CONTENTS

OUTCOMES

At the end of this chapter you should be able to:

- distinguish between the natures of renewable and non-renewable resources

- assess estimates of known reserves of non-renewable resources in light of technological innovation

- define fossil fuels as 'useful organic-matter-derived Earth materials'

- describe the changes in coal with increasing rank in terms of
  - physical properties
  - composition
  - grade
  - energy yield

- describe properties of liquid petroleum in terms of composition and energy yield

- describe properties of gaseous fossil fuels in terms of composition and compare the energy yields of coal-derived gas and petroleum-derived gas.

FOSSIL FUELS ARE NON-RENEWABLE RESOURCES

We rely on energy in every aspect of our lives. Energy is the ability to do work; the thing required for change to occur. In our homes, energy is used for cooking, lighting, heating and cooling. The things we use are created using energy. It is vital for agricultural and industrial production and for transportation.

Today, we obtain energy predominantly from fossil fuels, such as coal, oil and natural gas. These three fuels account for almost 90% of global commercial energy needs. Nuclear power and hydro-electric power supply smaller amounts of energy. There are also a growing number of technologies that can supply some of our energy requirements. Our society is, however, built around the use of fossil fuels and we will continue to depend on them for some time to come.

A fossil fuel is a **combustible** material, of biological origin, that is found as altered carbon compounds within geological deposits. Fossil fuels are not only useful as fuels. More than 85% of the **crude oil** produced each year is used to generate heat, but the rest is used as raw materials for manufacturing industries. Oil, coal and natural gas are all derived from the remains of ancient plants and animals that have been altered within the Earth. As materials that have been deposited, buried and modified, fossil fuels are sedimentary deposits. Coal is a sedimentary rock and petroleum (oil and gas) are sourced from, and generally found in, sedimentary rocks.

A brief history of fossil fuel use

Fossil fuels have not always played the important role they do today. Our dependence on fossil fuels has evolved over time. In Europe during the fifteenth century—and for perhaps 80% of the Earth's human population today—wood, vegetable waste and animal dung provided fuels for heating. Farming and manufacturing relied on the muscle power of people and livestock. The widespread use of metals was also limited by the amounts of heat that could be produced from the low-grade fuels available.

combustible
able to burn and release heat energy

crude oil
the liquid component of petroleum

definitions

Wind and water were used in some places as sources of energy, but they were, and are, dependent on particular environmental conditions.

The adoption of fossil fuels changed the way societies functioned and the quality of people's lives. Coal was the first fossil fuel to be widely used. By the sixteenth century, in England coal was not only being used for fuel in homes but played an important role in industries such as metallurgy, brewing, dying and the manufacture of glass, bricks and tiles. The industrial revolution was literally fuelled by coal. It powered the steam engines that ran industry and played a vital role in the relatively inexpensive smelting of high-quality iron and steel as well as other metals. It was the availability of cheap metals that transformed the nature of industry and transportation.

In the nineteenth century the adoption of oil and natural gas saw these fuels begin to displace coal. From the late 1800s to the early twentieth century, oil was principally used to make kerosene for lighting and heating. Today, oil is used predominantly for transportation but it is also used to make a wide variety of products. These products include chemical **feed stocks**, which are used to make plastics, drugs and other chemicals. Other products derived from oil and natural gas include coke for metal smelting, fertilisers for agriculture, and lubricants for use in industry.

The problem we face today is our reliance on such fossil fuels. In Australia we live in an affluent society that is classified as a developed country. Although 76% of the world's human population lives in what are termed less developed countries, it is the population of countries such as ours that accounts for over 70% of the world's annual energy resource consumption. To put this in personal terms, you, as an affluent Australian, will on average use the equivalent of 200 tonnes of coal and 170 cubic metres of crude oil during your lifetime.

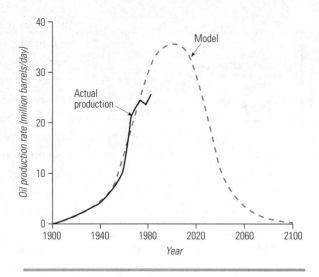

Figure 5.1.1 A model of resource use applied to world oil production.

feed stock
the raw materials used in a process

resource
the total amount of a valuable material that exists

finite
limited

definitions

Fossil fuels as a resource

The total amount of fossil fuels in the world—the fossil fuel **resource**—is effectively **finite**. The processes that lead to the formation of new coal and petroleum take hundreds of thousands, or millions, of years to occur. The slow rate of renewal does not keep pace with the rate of consumption. Resources that have long replacement times are referred to as non-renewable resources. Metal ores, fossil fuels and naturally occurring fertilisers are all non-renewable. Renewable resources are those things that are replaceable within a short time. Timber is a renewable resource because trees can be grown and harvested in a cycle measured in tens of years. Food is renewable on an annual scale.

Because fossil fuels are non-renewable there must come a time when the resource is fully utilised, assuming we continue to use it at the current rate. Figure 5.1.1 shows a model of resource use applied to world oil production. Initially production rates are small because new technologies do not produce the resource efficiently, the process is costly and the market for the new resource is small. Production increases as the efficiency of production improves and the utilisation of the resource becomes more diverse. The increasing production rate of a resource will ultimately slow as the resource's scarcity and growing cost make it less affordable. This may, in turn, lead to the production rate declining, not only because the material is scarce, but also because new substitutes are found.

There is always a strong economic dimension to the amount of a reserve we can access. Two important terms used in describing energy and mineral resources are 'proven reserves' and 'recoverable reserves'. Proven reserves are those that are known in some detail and can be mined using current technologies. Recoverable reserves are

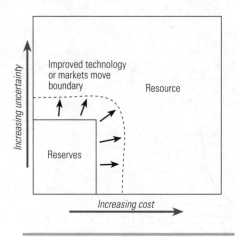

Figure 5.1.2 The relationship between resource, reserves, cost and uncertainty.

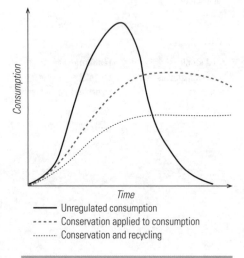

— Unregulated consumption
---- Conservation applied to consumption
········ Conservation and recycling

Figure 5.1.3 How conservation and recycling modify the rate of resource consumption.

those that are known well enough to estimate that they may be recoverable in the near future. Other resources are undiscovered, in the sense that science suggests they exist but exploration has not found them yet. In Australia we have about 36 billion tonnes of proven coal reserves and an estimated 400 billion tonnes of additional recoverable reserves.

Figure 5.1.2 is an adapted McKelvey diagram that shows the relationships between a resource, cost, technological uncertainty and reserves. The reserve is the total amount of an economic material that is known and predicted to exist. The reserves include both proven and recoverable reserves. The proportion of a resource that are reserves can increase in a number of ways, two of which are implied in the diagram. If the price of the material rises, the acceptable cost of acquiring the material will also rise. This leads to parts of the known resource becoming reserves. As new technologies develop, uneconomic deposits may become reserves because the cost of recovery falls. New technologies may also lead to the discovery of more of a resource, increasing the reserves.

Rapid consumption can be modified in two ways: by conservation and through **recycling**. (See Figure 5.1.3.) Both approaches depend on scientific, economic, political and social factors. (See Figure 5.1.4.) All these factors are interrelated and a change in one area can affect others. For example, social perceptions of need affect the cost of something and also affect scientific priorities in research. Oil is a good example of where demand and technology affect, and are affected by, social and political change. Figure 5.1.5 shows how oil costs have varied during much of the nineteenth and twentieth centuries. Increasing demand and production during the period from the 1860s to 1890 saw the oil price fall. The First World War created interruption to supply but greater demand for oil, causing the price to increase. Political events in the Middle East during the 1970s saw a rapid increase in the cost of oil during that decade. This gave rise to more attention being directed towards alternatives and the importance of more efficiency in the way petroleum is used.

Conservation occurs when a resource is utilised in an efficient manner. Political regulation may slow the rate of consumption and indirectly increase the value of the resource. Regulation of tree felling, for example, has not only preserved old trees but also encouraged the development of new technologies, such as metal house frames. Science and technology may produce more efficient recovery and utilisation of a resource. Goldmines once needed to yield at least 5% gold to be economic. Technological advances have reduced the percentage to 0.5%, allowing the mining of

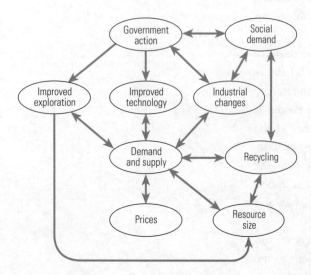

Figure 5.1.4 The interaction of factors that affect commodity prices.

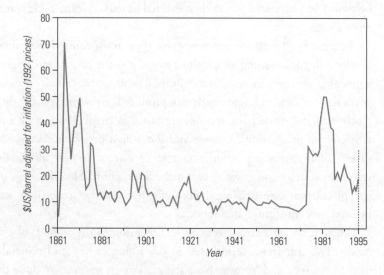

Figure 5.1.5 How oil price has varied from 1861 to 1995.

gold reserves that were once considered uneconomic. Education also increases social awareness of conservation issues.

Recycling lengthens the period during which a resource is utilised. Recovery and reuse reduces the demand for new mines. Copper is a good example of a resource that has benefited from recycling. The metal is vital in the electronics industry and in situations where a slow-rusting metal is an advantage. The cost of reprocessing the copper is relatively small compared with mining and refining the metal from ores.

It must be recognised, however, that conservation and recycling reduce the rate at which a resource is used. They do not make non-renewable resources renewable. In the end, new resources are needed to replace other resources as they become depleted. It is interesting to consider the fact that it is the energy sources of Europe in the fifteenth century (such as vegetable waste, wind and water) that are increasingly seen as alternatives to fossil fuels.

REVIEW ACTIVITIES

1
Define the term 'fossil fuel'.

2
Outline how the use of energy resources has changed since the fifteenth century.

3
Describe the property of a material that determines whether it is renewable or non-renewable.

4
Explain how economic considerations can affect the amount of a resource available for people to use.

5
Outline the ways by which the rate of use of a resource can be modified. Give examples in your answer.

EXTENSION ACTIVITIES

6
The HSC Syllabus defines a fossil fuel as 'useful organic-matter-derived Earth materials'. Contrast this definition with the one used in this section of the text.

7
The world's fisheries are an example of a resource that has been rapidly consumed. Describe how science and technology have led to more of the ocean's fish resource becoming fish reserves.

8
Analyse how the amount of fossil fuel resources is affected by the rate of development of technological innovation. (Treat this activity as a long-response activity.)

THE PROPERTIES OF COAL

Coal is an organic sedimentary rock. It contains the remains of plant material that has been altered by pressure and temperature resulting from burial. Different types of coal are recognised by their physical and chemical properties. These properties include such things as the carbon content and **ash** content of the coal, the amount of energy it produces, the amount of water present and the **density** and hardness of the coal. These properties are a result of the amount of change the coal has experienced.

Ash and sulfur are important economic characteristics of a coal. Ash is composed of inorganic material incorporated into the coal. This material may be clay or sand deposited at the same time as the organic matter that formed the coal. High ash contents are undesirable because they build up in combustion chambers and must be removed and disposed of periodically. The ash may also escape from a furnace as fine particles called fly ash. Filters or electrostatic precipitators are needed to stop fly ash escaping and to control the air quality around the chimney of the furnace. Sulfur in the coal reacts to form sulfur dioxide and then sulfuric acid. Sulfur compounds are significant air pollutants and measures must be taken to reduce the levels of such gases emitted from a furnace. This is sometimes done using lime slurries, which react with and bind the sulfur before it escapes to the atmosphere.

The amount of moisture and inorganic impurities in a coal determines the **grade** of the coal. Coals that contain high amounts of sulfur and ash are classified as low-grade coals and are less attractive than high-grade coals, which give more energy per weight and cleaner combustion products. The grade of a coal depends on its origin. Swamps near the sea contain sulfates from sea water and often produce coals with higher sulfur contents than areas of freshwater deposition. Coals deposited near the mouth of deltas may also contain more sediment than coals deposited in deeper, quieter areas of deposition.

The degree to which the original organic matter has been altered in a coal is described as the rank of the coal. The progressive change in **rank** that the coal experiences is referred to as the rank advance. During rank advance the physical and chemical properties of the coal change. Five ranks are recognised in coals: brown coal (lignite), sub-bituminous coal, bituminous coal, semianthracite and anthracite. Each rank is described below.

Peat

Peat is not a coal, but the material from which coal forms. It is a soft, spongy mass that contains recognisable plant fragments. It also contains a lot of water and so, in places where peat is used as a fuel, it has to be dried before it can be used.

Brown coal

Brown coal, or **lignite,** is a dull and earthy material. While coal is a stronger material than peat, it is reasonably soft and easily broken. Plant remains, particularly woody fragments, are recognisable in the coal. Brown coal contains less water than peat but still has a relatively high water content. In order to use it as a fuel, the coal is compressed and dried.

Brown coals contain high amounts of volatiles, which are relatively small molecules that are driven off as vapour when a coal is heated. Many volatile gases are flammable and the amount of volatiles indicates how easy a coal is to ignite. Brown coals are particularly rich in volatiles and the coals can spontaneously combust if they are not stockpiled properly.

Sub-bituminous coal

Sub-bituminous coal is a relatively hard, tough, black rock. It sometimes appears blotchy in appearance and has a slight lustre. It has a moisture content of 10–25%. Some plant fragments are still recognisable under a microscope, but otherwise the coal shows little obvious evidence of its plant origin. Sub-bituminous coals are mined commercially in Australia, including at Collie in Western Australia and Leigh Creek in South Australia.

Bituminous coal

Bituminous coals are hard, tough, lustrous and black. They are often strongly banded. Bituminous coals differ from sub-bituminous coals in having less moisture (usually less than 10%). They may sometimes be richer in volatiles than some sub-bituminous coals, but generally contain fewer volatiles.

Semianthracite

Semianthracite is hard and dense. Vitrinite, dark shiny bands within coal, reflect more light than bituminous coals but less than anthracite. Semianthracites have fewer volatiles and less moisture than coals with a lower rank.

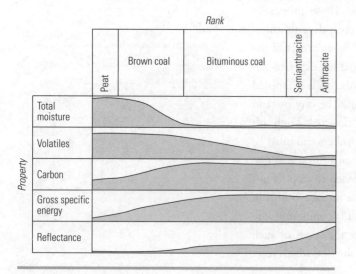

Figure 5.1.6 The changes in properties of coal with increasing rank.

Anthracite

Anthracite is very shiny and hard. It breaks with a conchoidal fracture, similar to that found in quartz or glass. The anthracite is very rich in carbon and produces a large amount of heat for a given mass of coal. It is poorer in hydrogen than bituminous coals and contains almost no water. Coals of this rank are rare in Australia. However, some coals near Mittagong in New South Wales approach this rank due to heating by nearby igneous intrusions.

Table 5.1.1 shows a comparison of four Australian coals, and Figure 5.1.6 indicates how the properties of coal change with rank. Note that volatiles and moisture decrease as the rank increases. Carbon content and, consequently, the energy produced per kilogram increases with increasing rank. The mineral content increases too because it is unreactive and its proportion of the coal becomes larger as other materials are lost from the coal.

Table 5.1.1   Characteristic properties of four Australian coal samples

Characteristic	Coal type and origin			
	Brown Yallourn, Vic.	Sub-bituminous Callide, Qld	Bituminous Lithgow, NSW	Semianthracite Yarrabee, Qld
Total moisture (%)	66.5	15.5	8.0	9.0
Inherent moisture (% ad)	45.0	9.6	2.9	2.0
Ash (% ad)	1.7	16.4	14.3	10.0
Volatiles (% ad)	50.3	24.6	32.3	9.5
Fixed carbon (% ad)	48.0	49.4	53.4	78.5
Gross specific energy (MJ/kg ad)	26.1	29.38	34.47	35.45
Carbon (% daf)	67.5	76.8	82.6	91.7
Hydrogen (% daf)	4.8	3.81	5.33	3.6
Nitrogen (% daf)	0.57	1.18	1.8	1.8
Sulfur (% daf)	0.24	0.36	0.82	0.8
Oxygen (% daf)	26.9	17.9	10.1	2.1
Reflectance (R_v max)	–	0.53	0.78	2.59

Note: ad = ash dry; daf =dry ash free

REVIEW ACTIVITIES

1
Define what the terms 'rank' and 'grade' mean when applied to coal.

2
Summarise four characteristics of a coal that affect its quality.

3
Outline the characteristics of peat, lignite, sub-bituminous coal and anthracite.

4
Identify the characteristics in Table 5.1.1 that show a progressive change as the rank of the coals increases.

EXTENSION ACTIVITIES

5
Describe the relationship between rank, fixed carbon and the amount of specific energy in coals.

6
Graph the changes in carbon, oxygen and sulfur with increasing rank of coal shown in Table 5.1.1.

THE PROPERTIES OF LIQUID PETROLEUM

Petroleum is a term that describes a complex mixture of organic compounds. Depending on the types of compounds present, petroleum can occur as a liquid (called crude oil) or as a gas (commonly called natural gas). Petroleum, like coal, is the result of a continuous process that alters organic material. In the case of petroleum, the process is a continuous series of chemical reactions that progressively change large, complex organic molecules into smaller, simpler ones.

Liquid petroleum, or crude oil, is made up of three important classes of **hydrocarbon** molecules. Hydrocarbons are molecules containing hydrogen and carbon that are covalently bonded together. The three groups of hydrocarbons making up crude oil are alkanes, naphthenes and aromatic hydrocarbons. (See Figure 5.1.7.)

Alkanes, or paraffins, are straight chain hydrocarbons. (See Figure 5.1.7a). The carbons form a long, unbranched, or branching, chain. Except in methane (CH_4), each carbon atom has bonded to it one to three hydrogen atoms. The general formula for alkanes is $C_nH_{(2n + 2)}$, where 'n' represents the number of carbon atoms in the molecule. In crude oil the number of carbons in alkanes may vary from five to forty. Above forty carbons, molecules are solids at normal surface temperatures. Crude oils composed almost entirely of alkanes are both valuable and rare. They make up 2% of crude oils and can be used to synthesise a host of other chemicals.

Naphthenes are similar to alkanes except that the carbon chains form rings rather than straight molecules. (See Figure 5.1.7b.) The naphthenes have a formula C_nH_{2n} and, like alkanes, they are quite stable compounds. Cyclopentane (n = 5) and cyclohexane (n = 6) are the most common naphthenes in crude oil.

Aromatic hydrocarbons contain structures called benzene. (See Figure 5.1.7c.) A benzene molecule consists of six carbon atoms in a ring. Unlike cyclohexane, the carbon atoms in benzene share a double bond with at least one adjacent carbon atom. As a result, the ratio of hydrogen to carbon is quite low in aromatic hydrocarbons.

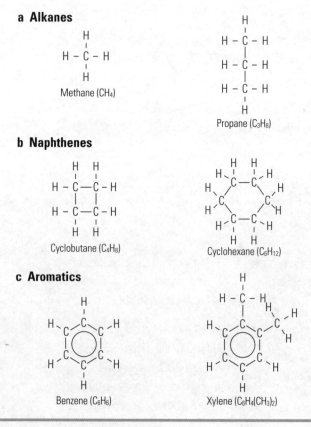

a Alkanes

Methane (CH_4)

Propane (C_3H_8)

b Naphthenes

Cyclobutane (C_4H_8)

Cyclohexane (C_6H_{12})

c Aromatics

Benzene (C_6H_6)

Xylene ($C_6H_4(CH_3)_2$)

Figure 5.1.7 Important groups of hydrocarbons.

Crude oil may also contain small amounts of other organic compounds that contain oxygen, sulfur and nitrogen. Heavy metals, such as nickel and vanadium, also occur. Sulfur in crude oil presents similar problems to sulfur in coal, and better prices are paid for crude oils with a low sulfur content. While crude oils always contain a certain amount of sulfur, some contain sulfur as hydrogen sulfide, which gives the oil a noticeable rotten-egg smell. Such oils are called sour crudes and low-sulfur crude oils are called sweet crudes.

The composition of a crude oil affects certain physical properties of the oil. The physical properties that are measured include refractive index, density and viscosity. The refractive index describes how an oil bends, or refracts, light. Light oils refract light less than heavier oils and refractive indexes vary from 1.42 to 1.48. Water, for comparison, has a refractive index of 1.33 and glass has a refractive index of about 1.50.

The density of an oil is most commonly described using the American Petroleum Institute (API) gravity scale. The density (p) of the oil is used to calculate the scale using the following formula:

$$\text{API scale} = \left(\frac{141.5}{p}\right) - 131.5$$

Water has a density of 1.00 g/cm³ at 25 °C and a value of 10° API. Because the components of oil have densities less than water, crude oils have API scale values greater than 10°. Light crude oils are crude oils with API values greater than 30°. Heavy crude oils have API values less than 22° and intermediate crude oils are those with API values that lie between 22° and 30°.

Viscosity is a measure of how easily a substance flows. A measure called the pour point can be used to describe viscosity. It describes the minimum temperature at which a crude oil will flow under standard conditions. Some crude oils have pour points below –30 °C, while those that are rich in paraffin waxes may have pour points above 40 °C.

REVIEW ACTIVITIES

1
Define the term 'petroleum'.

2
Summarise the three major classes of organic compounds that make up liquid petroleum.

3
Outline three physical properties that are used to describe crude oils.

EXTENSION ACTIVITIES

4
Explain the relationship between the size of molecules in a petroleum sample and the probable boiling point of the sample.

5
A crude oil has an API value of 20°.
a Identify whether it is a heavy, intermediate or light crude.
b Use the equation for the API scale to determine the oil's density.

THE PROPERTIES OF GASES DERIVED FROM COAL AND PETROLEUM

Natural gas is composed of relatively small hydrocarbon molecules. These include methane, ethane, propane and butane. Natural gas is usually processed to produce two gaseous products: dry natural gas and liquefied petroleum gas (LPG). At low temperatures larger molecules, such as propane and butane, condense. Natural gas that contains these components is called wet natural gas and their removal produces dry natural gas. Dry natural gas consists almost entirely of methane and propane and will produce about 37 kilojoules of energy when a cubic metre of gas is burnt at 15 °C.

In recent years, gaseous fuels have been produced from coal by a process called coal gasification. One of the products of such processes is called **syngas**, which is a mixture of carbon monoxide and hydrogen. Syngas generates about 9 kilojoules of energy per cubic metre of gas burnt. Another product of coal gasification is synthetic natural gas, which contains mostly methane. This gas produces approximately 30 kilojoules of energy per cubic metre of gas burnt.

LPG produces a lot of energy per molecule and more heat per unit volume than dry natural gas. This fuel produces about 27 000 kilojoules per cubic metre of gas burnt, which is similar to the energy output of a cubic metre of crude oil at the same temperature. A cubic metre of oil is equal to 6.293 standard barrels of oil.

OTHER USES OF FOSSIL FUELS

While petroleum and oil are used widely for heat production we should not forget that there is a wide range of other uses for fossil fuels. Table 5.1.2 shows a selection of these. The development of alternative fuels (such as ethanol to fuel cars) and alternative energy sources (such as wind power and solar electrical technologies) will no doubt conserve fossil fuels.

Table 5.1.2 Some uses of coal and petroleum
Some uses of coal
Electrical generation
Coke production for metal production and other uses
Production of lime, plaster, cement and other building products
Engineering and metal fabrication
Chemical synthesis of medicines and other products
Food production: brewing, milk and meat processing, and sugar refining
Woodworking: wood curing and wood treatments
Manufacture of soaps, oils, inks, paints and other products
Paper and printing processes
Laundry and dry-cleaning processes
Production of textiles and clothing
Production of gases and synthesis of chemical feed stocks
Some uses of petroleum
Petrol, distillate and other automobile, ship and aircraft fuels
Feed stocks for the chemical industry, producing research grade chemicals, drugs, fertilisers, solvents, plastics, rubbers and other products
Production of coke for metallurgy
Production of kerosene
Lubricants
Heating for many of the industries listed under coal uses

1
Describe the composition of natural gas.

2
Explain how a wet natural gas is different from a dry natural gas.

3
Describe the properties of two gaseous fuels derived from coal.

4
Compare the heat generated by natural gas and coal-derived fuels.

5
Summarise the differences between dry natural gas and LPG.

6
Summarise the properties of Earth materials derived from organic materials.

7
Analyse the differences to your life if fossil fuels were unavailable for the production of electricity, steel and cement.

8
Analyse how the amount of fossil fuel resources is affected by the rate of development of technological innovation. (Treat this activity as a long-response activity.)

SUMMARY

- Fossil fuels are useful materials composed of organic materials derived from once-living organisms. They include oil, natural gas and coal.

- Fossil fuels are non-renewable resources because they take a long time to form.

- Conservation and recycling reduce the rate at which resources are used, but the need for resources will continue.

- The size of fossil fuel reserves increases as technology allows us to recover material that was once unrecoverable.

- Fossil fuels are used for energy production and for raw materials for the chemical industry.

- Coals increase in rank over geological time.

- Coals are described in terms of their carbon (energy), ash and moisture content and their composition in terms of organic particles.

- Liquid petroleum is a complex mixture of organic compounds.

- Coal-derived and petroleum-derived gases may differ in their energy yields, but have similar uses in chemical manufacturing.

PRACTICAL EXERCISE
The properties of fossil fuels

In this exercise you will process information to classify some fossil fuels according to their properties and compositions.

Fossil fuels include solid, liquid and gaseous materials. In this exercise you will examine fossil fuels using their state and other properties. Seven materials are described in Table 5.1.3.

ACTIVITIES

1
Use the information in Chapter 5.1 to identify the coals in Table 5.1.3. List the properties that allow you to identify the rank of each sample.

2
The remaining samples consist of gaseous fuels and liquid fuels. Identify each one. If you have access to a chemical data book you may be able to identify some of the hydrocarbons present using their melting and boiling points.

3
Can you identify the fuels that are renewable? Name them and give reasons for your choices.

4
Summarise the properties that were most useful in identifying the samples.

Table 5.1.3 Properties of some fossil fuels

Material	Appearance	Melting point (°C)	Boiling point (°C)	Energy produced by combustion (joules/kg)	Density (g/cm³)	Components	Name
A	A brown, easily broken material	Not applicable	Not applicable	25 600 000	1.19	Moisture 60.5% Ash 7.3% Total sulfur 1.73%	
B	A colourless gas	Begins at −183	Begins at −161	37 000 000	0.46	Two alkanes with low molecular mass	
C	A black, heavily banded rock with a glassy lustre and curved fracture surfaces	Not applicable	Not applicable	35 500 000	1.55	Bands of vitrinite Low moisture and volatiles Mainly carbon	
D	A colourless gas	Begins at −126	Begins at 82	50 000 000	0.54	Two alkane hydrocarbons	
E	A volatile liquid	Begins at −26	Begins at 36	48 000 000	0.74	Carbon, hydrogen and oxygen	
F	A colourless liquid	−114	78.3	29 700 000	0.785	Carbon, hydrogen and oxygen	
G	A colourless gas	−182	161.5	55 625 000	0.448	Carbon and hydrogen	

OUTCOMES

At the end of this chapter you should be able to:

- outline the characteristics of coal-forming environments

- discuss the process of coalification—transferring vegetable matter into peat and coal

- describe the characteristics of petroleum-forming environments

- outline the maturation of petroleum—diagenesis, catagenesis, metagenesis

- outline the process of oil and gas migration

- describe the features of source rocks, reservoir rocks and cap rocks

- analyse the conditions under which petroleum accumulates in structural and stratigraphic traps.

FOSSIL FUELS EVOLVE

Fossil fuels are the products of processes that occur over time. Temperature and pressure alter the organic material in a variety of ways. Water is removed and chemical changes alter the composition of the raw material. Some of the important changes that coal undergoes were described in Chapter 5.1, and are summarised in Figure 5.2.1 (page 232). Changes in organic composition occur as buried organic material is converted to various types of oil and natural gas. (See Figure 5.2.2, page 232.) Heat and pressure are important, but so are the activities of bacteria. In this chapter we will examine the environments in which the raw materials of coal and petroleum are generated and how those raw materials are transformed into the products that we use.

FORMATION OF COAL

In order for coal deposits to form, the accumulation and preservation of plant material must occur at a faster rate than its destruction by oxidation and decay. Oxygen reacts with carbon compounds in ways that break down the molecules. Organisms such as bacteria and fungi are also capable of breaking down the organic material so that it is of no use as a fossil fuel precursor. The net result of breakdown by micro-organisms is carbon dioxide and water.

Coal-forming environments

Coals form in environments characterised by particular climates, plant communities and sedimentation. Suitable conditions for plant growth are required together with active burial of the resulting peat and a lack of oxygen where the material is buried. The quality of the coal produced depends on there being a relatively low amount of

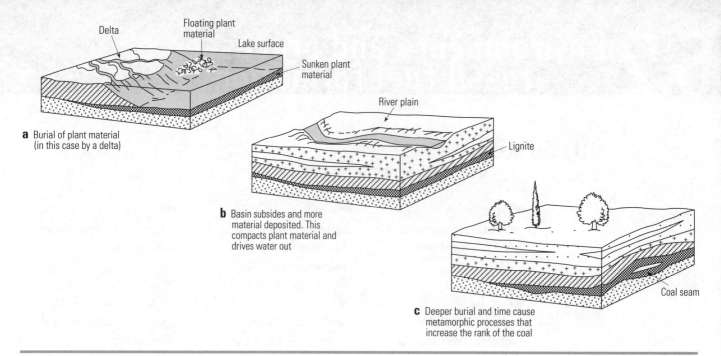

a Burial of plant material (in this case by a delta)

b Basin subsides and more material deposited. This compacts plant material and drives water out

c Deeper burial and time cause metamorphic processes that increase the rank of the coal

Figure 5.2.1 The process of coal formation.

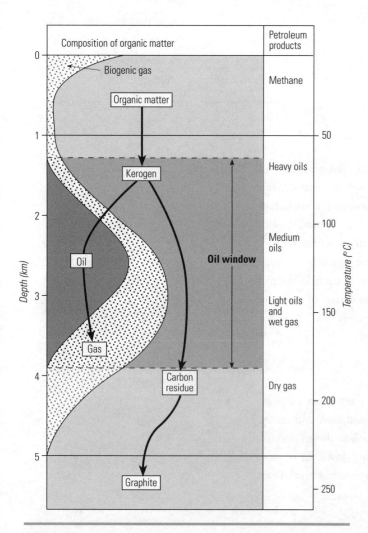

Figure 5.2.2 Changes in the composition of petroleum with depth.

inorganic sediment, such as sand or clay, being deposited with the organic material. Coal buyers impose penalties on companies if the amount of inorganic material in a coal is too high.

CLIMATE

Climate plays an important role in forming the raw materials of coal. Warm conditions mean faster growth rates for plants and regular rainfall. Water is particularly important for coal deposition. It not only enables plant growth but also prevents oxidation. If peat is exposed to the air the oxygen reacts with the organic material, causing it to decompose. If rainfall is regular, the water level in a lake or swamp remains fairly constant and there is less chance of the water body evaporating and the organic material being exposed to the air.

PLANT COMMUNITIES

The evolution of plants, together with tectonic and climatic conditions, has affected the availability of plant material for coal formation. Land plants evolved during the Late Silurian but it was not until the Devonian that significant accumulations of plant material gave rise to coal deposits. (See Figure 5.2.3.) At the end of the Permian the nature of plant communities from which coal formed began to change. Coals formed during the Carboniferous are dominated by plants such as lycopods, which do not carry much foliage. After the Permian extinction event, the dominance of conifers and other seed plants meant that much more foliage was incorporated into coal.

There have been three significant periods during which significant amounts of coal were deposited. The Carboniferous (350–280 Ma) was a significant period of peat accumulation in what is now the Northern Hemisphere. The sea level rose at the beginning of the Carboniferous causing warm, shallow seas to spread across much of modern Europe and the USA.

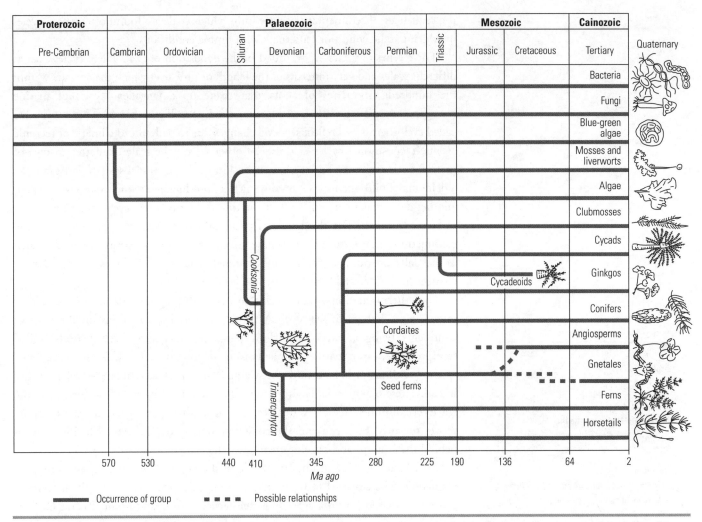

Proterozoic	Palaeozoic								Mesozoic			Cainozoic		
Pre-Cambrian	Cambrian	Ordovician	Silurian	Devonian	Carboniferous	Permian	Triassic		Jurassic	Cretaceous	Tertiary	Quaternary		Bacteria

(Figure columns: Pre-Cambrian, Cambrian, Ordovician, Silurian, Devonian, Carboniferous, Permian, Triassic, Jurassic, Cretaceous, Tertiary, Quaternary. Plant groups listed right: Bacteria, Fungi, Blue-green algae, Mosses and liverworts, Algae, Clubmosses, Cycads, Ginkgos, Cycadeoids, Conifers, Angiosperms, Cordaites, Gnetales, Seed ferns, Ferns, Horsetails. Fossil labels: Cooksonia, Trimerophyton. Time axis: 570, 530, 440, 410, 345, 280, 225, 190, 136, 64, 2 Ma ago.)

Occurrence of group ▬▬▬   Possible relationships ▪ ▪ ▪ ▪

Figure 5.2.3 Evolution of plants.

These areas were part of a large landmass called Euramerica, which lay close to the equator.

Significant coal reserves of Carboniferous to Permian age (350–225 Ma) exist in Australia and other areas that were once part of Gondwana. During this time, Gondwana collided with a continent composed of what is today Europe and North America to form the supercontinent Pangaea. The coal deposited in Australia, India and Antarctica at this time differed from the plants found in the Carboniferous coals of North America in being cold climate adapted plants. Cold environments are highly seasonal and the different conditions between summer and winter produce distinctive tree rings. The rate of growth in tropical areas, by comparison, is fairly constant.

The Mesozoic to Early Tertiary (250–15 Ma) was an important period of coal formation in Australia and other parts of the world. Triassic and Jurassic coals occur in Queensland and South Australia. Widespread coal deposits in Canada and the Western USA are of Cretaceous age. As the warm Mid-Cretaceous climate gave way to periods of cooling and Australia broke away from Antarctica, the Tertiary saw peat accumulation in Germany, the USA and Southern Australia.

SEDIMENTARY ENVIRONMENTS

The active burial of sediment occurs only in particular environments. In general terms, environments of sediment deposition can be divided into two types: limnic and paralic. A limnic environment is one in which the deposition occurs where the water is fresh. A paralic environment is one where the water is marine for at least part of the time. The most common environments in which coal is formed are paralic, but the sequences of sedimentary rocks within which coals are found often show cycles in which paralic conditions alternate with limnic conditions.

Three types of cycles are recorded in the deposition of coal. These cycles exist at different scales and vary in terms of the length of time over which they occur. Within the hundreds of metres of **strata** that make up a **formation**, cycles of marine transgression and regression can be recognised. (See Figure 5.2.4.) These cycles are called cyclotherms and reflect sea level changes due to widespread climatic or tectonic events. Depositional cycles can also be observed on the scale of metres to tens of metres. Such cycles are due to changes within the depositional system. For example, a delta may build out across a muddy surface, producing a sequence of layers. Later, another delta may be deposited over the top of the first one, repeating the sequence of layers within it. The third style of cyclic development is the smallest, involving centimetre to metre thick units. These cycles may be due to seasonal changes, such as leaf falls providing more organic material than at other times of the year, or changes in the watertable that occur in a yearly cycle.

Cyclotherms are important cycles that may result in large coal deposits and they reflect large changes in sea level. A rising sea level leads to areas on the edge of continents being covered by shallow parts of the ocean. Such periods, called transgressions, cause marine sediments to be deposited over non-marine sediments. As the coastline moves landward, the swamps where peat accumulates develop further inland. These depositional environments may be buried by marine sediments as the sea level continues to rise. When sea level falls, during a regression, the coastline moves towards the ocean and the coastal swamps move seaward. Moving seaward behind the swamps and burying them are deltas.

Plate tectonics play an important role in shaping suitable environments for coal formation. The sea level changes that cause cyclotherms are the result of tectonic processes, such as sea-floor spreading, **subsidence**, rifting and uplift. Passive margins formed by plate divergence, together with **intracratonic basins** that form within continents, are major environments of coal formation. The subsidence of such areas as sediment builds up allows further sedimentation to occur and ensures that the watertable protects material laid down during deposition.

strata
layer of rock in the Earth

formation
a fundamental rock unit that contains distinctive rock strata and well-defined upper and lower boundaries

subsidence
an event in which a broad area of the crust sinks without major deformation

intracratonic basin
a sedimentary basin formed on, and within, a continental shield

definitions

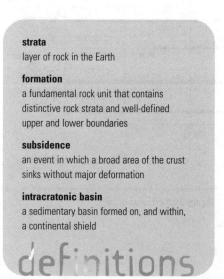
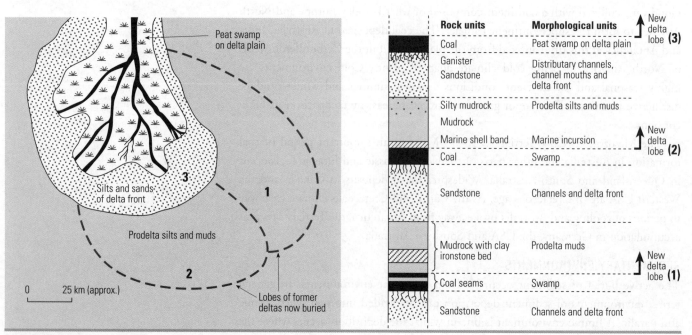

Figure 5.2.4 Cycles of sedimentation reflecting changes in sea level.

Four modern depositional environments that may lead to **coal seams** are alluvial fans; rivers and upper delta plains; lower delta plains; and back barrier lagoons. Figure 5.2.5 shows the types of structures that develop within each environment. Note that the environments may grade into each other or replace each other as sea level changes or as tectonic processes alter adjacent areas.

Alluvial fans are lobe-shaped wedges of coarse material that build up at the base of mountains. The upper parts of these fans are sites of rapid sedimentation and oxidation. The lower parts of the fans can host swamps where peat accumulates. As the fan grows it moves out over the swamp, burying the peat. Coals within the Sydney Basin formed in such environments as a result of actively rising mountains to the east of the current coast. Two modern environments where alluvial fans are burying future coal deposits, and which operate in very different climates, are Southern Papua New Guinea and South-eastern Alaska.

Rivers and upper delta plains are sites of coal deposition because back swamps and billabongs are sites of peat accumulation. The build-up of peat is better in these environments because the water level within these areas remains fairly high. Out on the rest of the floodplain, plant growth may be vigorous but the plant material is unlikely to be buried without being oxidised or broken down by micro-organisms. As the river channels change their course they may cut across the peat, burying it under sands and muds. The resulting coal seams may show splits due to the levee and overbank sediments deposited by the moving channel.

Along coasts, the types of coal-forming environments present are determined by the relative strength of processes derived from the land and the ocean. A lower delta plain may develop as the sediment carried by the river builds a delta out into the ocean. This process occurs if the flow of the river is greater than the strength of the currents and wave action from the sea. In places where wind and waves dominate river processes, or where rivers are small, lagoons and barrier islands may form parallel to the coast.

When a delta forms, sediment is deposited along the edges of channels and along the front edge of the delta. The sediment is carried along a number of interconnected channels, called distributary channels, which form and reform as sediment blocks the older channels. As the channels build out into the ocean, marshes and swamps develop on their flanks. The thin blankets of peat that form in the wetlands are buried as the basin containing the delta subsides. As subsidence occurs, younger channels of the delta deposit sediment above the peat. A classic example of such an environment is the Mississippi Delta in the Gulf of Mexico. Alternatively, a marine transgression may cause the peat to be buried by marine sediments as the sea moves inland. The marine sediments cause the buried peat to acquire sulfur and the quality of the coal is not as high as in the coals formed during a marine regression.

definition

coal seam
a stratum of coal

Alluvial fans

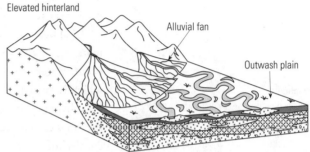

River and upper delta plain

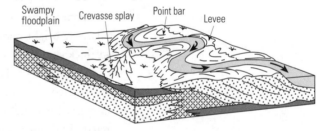

Lower delta plain

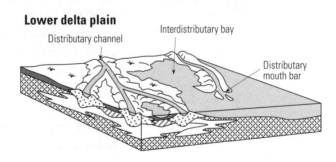

Back barrier lagoon

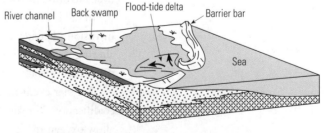

Figure 5.2.5 Environments of coal accumulation.

Along coasts where barrier islands of sand form, peat may accumulate in back barrier lagoons found between the mainland and the barrier islands. At the edges of the lagoons, deltas may form when rivers and creeks bring sediment from the mainland side of the lagoon. On the barrier island side, sand may move out into the lagoon with large tides and storms. These processes cause the resulting coal seams to have interbedded muds and sands between relatively thin seams. The presence of salt water also leads most coals from this type of environment to have a relatively high sulfur content.

Coalification: The making of coal

The process that produces coal, coalification, begins with peat and ends with anthracite. The peat is initially converted to brown coal or lignite. As the material becomes harder and more mature it passes from being lignite to sub-bituminous coal and then bituminous coal. Finally, the coal may become anthracite. The rank increases as a result of low temperatures over a long time, or higher temperatures and pressures over a shorter time.

Coalification is due to the heat and pressure that result from the burial of organic material. As the amount of sediment overlying peat increases, a number of changes occur. Water is driven off and gases, called volatiles, also escape. The initial material, which is easily broken, becomes harder and more durable.

Obvious plant remains become altered to the point of being unrecognisable and the physical characteristics of the materials in the coal changes. Initially the peat consists of discrete fragments and organic compounds called humic acids. Humic acids are produced by small amounts of alteration in the atmosphere before burial. By the time a coal's rank has advanced to being **bituminous** a number of new components can be seen in hand specimens. Vitrain and clarain are bands of relatively bright material. Vitrain appears glassy and forms bands from 3 to 10 mm in thickness. It is brittle and has a characteristic pattern of closely spaced fractures that cut across the layer. Clarain has a finely laminated structure. Fine vitrain layers are situated in a dark, fine-grained matrix. The overall appearance of clarain is a dark, laminated material with a silky lustre. Fusain and durain are coal materials that are relatively dull. Fusain occurs as bands of soft powder, which resembles powdered graphite. Durain is a harder and more homogeneous material.

At a microscopic level, coals can be studied in terms of their organic particles, called macerals. Macerals are the remains of plant parts or the products of decomposition before burial and they are recognised by their properties, such as structure, hardness and light reflection. The macerals that are plant remains include parts of woody tissues, spores, leaves and algae. Macerals derived from decomposed material include such things as organic gels, fungal remains and oxidised woods. Low-rank coals contain a variety of macerals that are recognisable as plant remains, but as the rank of the coal increases the characteristics of the macerals change until they are difficult to distinguish from each other.

In addition to coalification, post-depositional changes can affect coal. The intrusion of magma into coal seams can lead to coal turning into graphite. Besides altering the quality of the coal, the intrusions present problems for miners because the intrusive rock is much harder than the coal and associated sedimentary rocks.

Another event that will affect the size and nature of a coal seam is erosion before coalification. Stream channels that cut accross peat beds remove material and may allow oxygen to alter some of the organic matter. The alteration subsequently affects the quality of the coal.

bituminous
naturally occurring tar-like material; a type of coal

definition

1

Describe the conditions necessary for coal deposits to form.

2

Explain how climate affects the accumulation of peat in coal-forming environments.

3

Outline how coal-forming environments have changed over time.

4

Summarise the characteristics and causes of the cycles found in sedimentary rocks surrounding coal measures.

5

Outline the characteristics of four modern environments in which future coals may be accumulating.

6

List the stages through which a coal passes as it moves from peat to anthracite.

7

Summarise the changes that occur as a coal's rank increases.

8

Discuss the process of coalification and the changes that occur during the process.

9

Contrast coal-forming environments in equatorial locations with those at high latitudes. In your answer, outline the characteristics of the environments, and the process of coalification.

FORMATION OF PETROLEUM

Petroleum-forming environments

Petroleum, like coal, is formed by the alteration of organic materials deposited in sedimentary environments. Figure 5.2.6 (page 238) shows the worldwide distribution of sedimentary basins that do, or may, contain petroleum reserves. The tectonic settings in which petroleum reservoirs form are quite varied. They include foreland and forearc basins, basins associated with rifting and wrenching, accretionary prisms and passive margins. Like coal, petroleum requires heat and pressure to alter the original organic material.

Petroleum differs from coal in a number of ways. Petroleum is produced from organic materials that are disseminated throughout sedimentary rocks. Coal is concentrated in strata and, to be economic, needs to be relatively free of inorganic sediment, such as sand or mud. Petroleum forms within one area and migrates to sites of accumulation. Coal forms where the raw materials are deposited. A third difference is that while coal is a solid material, petroleum may vary from being nearly solid (bitumen) to gas.

Three important factors control the generation of petroleum hydrocarbons. They are:

• the nature of the organic raw materials
• the abundance of the raw materials
• the heating history of the material once it is buried.

The nature of the raw materials determines the types of petroleum products that are formed. The abundance of the raw materials is important because not all of the material is converted to petroleum or escapes to the site of accumulation. The heating history is important because it determines the relative amounts of oil and gas formed. Each of these factors is described in more detail on the following pages.

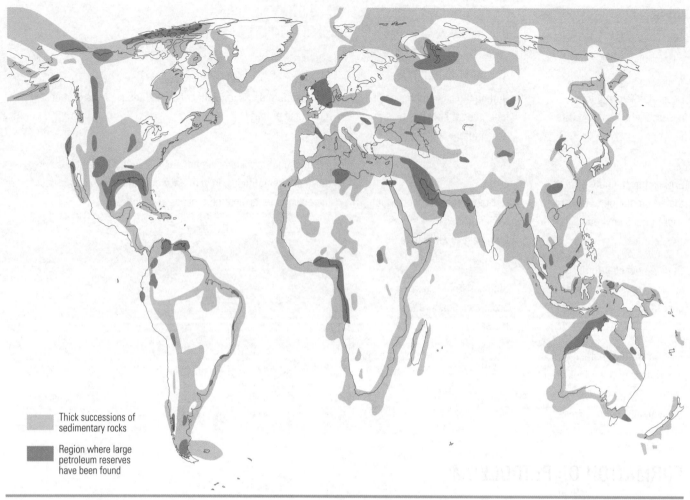

Thick successions of
sedimentary rocks

Region where large
petroleum reserves
have been found

Figure 5.2.6 Sedimentary basins.

ORGANIC MATTER AND HOST ROCKS

The majority of Earth scientists believe that the organic material from which petroleum forms begins as the remains of microscopic organisms. Some scientists have argued that some of the Earth's fossil fuels are derived from deep within the Earth, but this view has not achieved much support. The chemical fingerprint of most petroleum deposits contains evidence of living organisms that live in large water bodies, such as lakes and oceans. The compounds also show temperature effects caused by relatively shallow burial.

Three types of organic material produce kerogen, the raw material from which petroleum forms. Freshwater lakes in warm climates produce large volumes of algae. These organisms form sapropelic kerogen, which is relatively rich in hydrogen and low in oxygen. This type of kerogen gives rise to light, high-quality oils rich in waxes.

Waxes are large organic molecules with very high melting points. Where upwelling nutrient-rich waters occur in warm areas of the oceans, huge quantities of plankton grow. On deposition these organisms give rise to mixed planktonic kerogens. These kerogens are relatively rich in hydrogen and produce medium to heavy oils, low in waxes. The third type of kerogen, humic kerogen, is derived from terrestrial plant remains that are deposited in swamps and shallow marine environments. Being relatively low in hydrogen and rich in oxygen, this type of kerogen gives rise to some waxy oils and relatively large amounts of gas. Figure 5.2.7 shows cross-sections of three environments in which kerogens are laid down.

Mixed planktonic kerogens give rise to most of the world's petroleum supplies. This is because of the large volumes of organic material that are deposited. In order

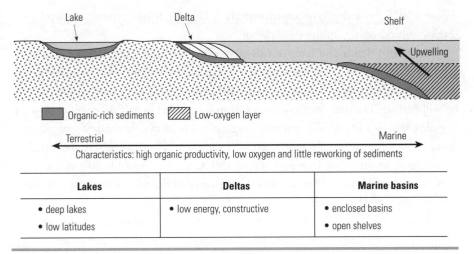

Lakes	Deltas	Marine basins
• deep lakes	• low energy, constructive	• enclosed basins
• low latitudes		• open shelves

Figure 5.2.7 Some environments in which the raw materials of petroleum are deposited.

for a rock to act as a **source** for petroleum it needs to contain at least 0.5% kerogen. (The average amount of organic carbon that is generated in the world's oceans each year and is preserved as carbon in sediments is about 0.4%.) An outstanding source rock may contain as much as 10% kerogen, but not all of it will escape from the rock when petroleum forms.

Like coal, the organic material that forms petroleum must escape decay and breakdown. The best conditions for this are the low-oxygen conditions that can occur in both shallow and deep water bodies. These conditions are referred to as **anoxic**. Rapid burial of the organic material occurs when sediment settles on it. The sediment is fine grained, like the organic remains, and the rocks that are formed as a result are mudstones, shales and fine-grained limestones. The source rocks formed from the sediments are often laminated. Pale bands of inorganic sediment alternate with organic-rich layers in the shales. The colours of the rocks range from brown to greens and black.

Petroleum maturation: Chemical and physical changes

The first stage in the production of petroleum is the formation of kerogen from organic remains contained in sediments. The kerogen is produced by anaerobic bacteria within the sediment. They convert the organic material within the sediment into kerogen until the temperature reaches 50 °C. This occurs in the sediment at a depth of up to 1000 m. The decomposition of the organic material produces methane as a by-product. This gas usually escapes, but may be trapped in very cold conditions.

As the depth of the sediment hosting the kerogen increases, the kerogen breaks down to form smaller hydrocarbon molecules. This occurs at depths ranging from 1000 m to 3500 m where the temperature ranges from 50 °C to 145 °C. As part of the reactions that occur, oxygen is incorporated into water and carbon dioxide. The first petroleum products produced are heavy oils, which are generated at temperatures ranging from 70 °C to 90 °C. As the temperatures increase, further reactions break the hydrocarbon molecules into smaller molecules. The nature of the petroleum changes and light oils and gases form.

The maximum production of liquid petroleum takes place within a relatively narrow range of temperatures and pressures. The breakdown of molecules to form liquid petroleum takes place principally within a temperature range of 100 °C to 150 °C. Above 200 °C only dry gas is produced. This corresponds to a depth of about 4500 m. The nature of the gas also changes with temperature. As the temperature increases, smaller gas molecules are formed until only methane remains.

source
the place where petroleum is formed

anoxic
without oxygen; environments containing little oxygen

definitions

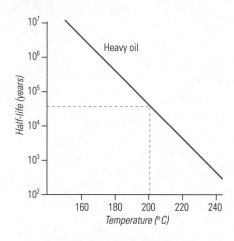

Figure 5.2.8 The half-life of oil deposits. At 200 °C a body of oil will decrease by half in about 50 000 years. The oil is converted to gas.

migration
the movement of petroleum from its site of formation to the site of accumulation

porosity
the capacity of a rock to hold water in the spaces, or pores, between the mineral grains

Above 230 °C, at a depth of approximately 5 km, the hydrocarbons are destroyed, leaving behind only graphite-pure carbon.

While both depth and temperature play roles in forming petroleum, time is an important factor too. Imagine a reservoir of oil that is being held at a high, but constant, temperature. Over time, the oil will be converted to gas and the amount of oil will decline. The time for an oil reservoir to be reduced to half the original amount is called the half-life of the reservoir. Figure 5.2.8 shows the half-life of oil as a function of temperature. At 160 °C an oil reservoir will have a half-life measured in tens of millions of years. However, above 180 °C the oil's half-life falls to less than 1 million years. This would mean that the existence of the oil would be very unlikely in many geological environments.

Three terms used to describe the maturation of petroleum are diagenesis, catagenesis and metagenesis. These are three phases that follow each other during petroleum formation. Diagenesis refers to the physical and chemical changes that a sediment undergoes after deposition. It is usually used to describe how a sediment becomes a rock, but in relation to petroleum formation it refers to the stages that precede petroleum generation. Catagenesis refers to the generation of oil and gas together with the subsequent formation of gas from previously generated oil. Metagenesis describes the production of dry gases from kerogens that have not previously formed oil or gas.

Petroleum accumulation: Stratigraphy and structures

In order to recover petroleum, scientists and engineers look for sites where petroleum is concentrated. The rocks that contain the kerogen from which the petroleum is formed are called source rocks. The source rocks are basin sediments, which are usually older than the rocks in which the oil accumulates.

The movement of the petroleum from the source rocks to the site of accumulation is called **migration** and it occurs in two stages. (See Figure 5.2.9.) The first stage, called primary migration, involves the petroleum moving from the host rock into nearby porous rocks. The second stage is referred to as secondary migration and this involves the petroleum moving through porous rock until it reaches the area of accumulation.

Primary migration is not well understood and mechanisms are debated within the scientific community that studies petroleum generation. It is thought that pressure build-up within the source rock causes the formation of minute fractures, through which the petroleum escapes. Many reasons for the build up of pressure have been proposed. Some of them involve the increasing temperature of the source rock as it is buried deeper within the basin. One hypothesis is that the pressure increases occur when fluids expand with increasing temperature. Another hypothesis is that the pressure increases due to changes in clays as the temperature increases. At a certain point the clays give up water bound within them and the released water helps to build up pressure within the rock. Primary migration appears to be an ongoing process with cycles of pressure build-up, microfracture and petroleum release. The migration of oil from the source rock may be upwards or downwards.

Secondary migration depends on the oil moving through porous and permeable rocks to the site of accumulation. The main driving force for this process is the buoyancy of the petroleum. The droplets of petroleum are surrounded by water, and the lower density of the petroleum causes it to move upwards. It will move up the dip of strata until it is stopped. The rate at which the petroleum moves depends on both the **porosity** and permeability of the rock. Permeability reflects how well the pores are connected and how easily materials flow through the rock. Pores in sedimentary rocks are usually lined by thin films of water, which can impede the movement of the petroleum. The presence of clays within the rock can also reduce the permeability of a rock to the flow of petroleum. Sedimentary rocks consisting of grains that are well sorted (of similar size) and coarse have an increased porosity and flow. Wind-deposited sandstones and limestones, which contain joints and porous material, have high porosity and permeability.

Secondary migration will move the petroleum to the surface unless it flows into a suitable reservoir and trap. The reservoir is a rock that is both porous and permeable. A trap is a structure where the reservoir is overlain by an impervious surface called a cap, which prevents the petroleum moving higher in the formation. **Cap rocks** may be impervious rocks, such as shale, or the ground-up rocks along faults, called fault gouge. In places where traps do not occur, the petroleum may reach the surface and form tar sands and oil seeps. Smaller molecules within the petroleum evaporate, leaving a tarry residue. Over time, bacteria will break down the material.

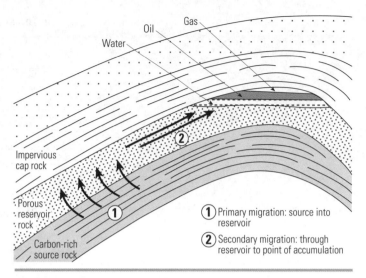

Figure 5.2.9 Stages in petroleum migration.

The geometry of source rocks, reservoirs and cap rocks shows a range of forms and sizes. Size is important because the amount of oil in a trap needs to be large enough to make its recovery economic. It is important to realise that the maturation of the petroleum continues once it is in a trap. As the gas increases above the liquid petroleum, the liquid may be pushed downwards and escape around the sides of the trap.

Classification of traps allows scientists to estimate the size of reservoirs and the risks involved in exploration. Classification also allows petroleum scientists to identify the particular features to be sought in particular tectonic environments.

Forces that shape sedimentary basins produce structural traps. Such processes include tectonic, gravitational, **diapiric** and compactional processes. Stratigraphic traps arise from sedimentary processes within a basin during its evolution. The geometry of a stratigraphic trap arises from the original deposition of materials within the basin and the transformation of the sediment into rock. Combinations of structural and stratigraphic traps comprise about 9% of the known sources of petroleum. Such structures arise when sedimentation within the basin and forces forming the basin interact with each other.

cap rock
the rock that seals an oil trap

diapir
a vertical intrusion that is formed of less-dense material rising through surrounding rock

definitions

STRUCTURAL TRAPS

Structural traps have generated the majority of the world's petroleum production. In this section we will briefly examine a range of structural traps: those caused by compressional forces and those formed within extensional environments. Table 5.2.1 classifies the structural traps described.

Anticlinal traps

Anticlines are often large-scale structures and can hold large volumes of petroleum. They are the structures from which the greatest amount of petroleum has been recovered. (See Figure 5.2.10, page 242.) Surface mapping and geophysical methods can map anticlines relatively easily. The limbs of an anticline allow easy migration of petroleum and the hinge zone acts as a good trap, assuming that a suitable cap rock is present. Anticlines often have fractures in the hinge area. Shales are particularly good cap rocks as they are ductile and flow to seal the trap.

In fold belts, anticlines and other folds are associated with low-angle thrust faults. These faults can both generate anticlinal structures and seal them. (See Figure 5.2.10.)

Table 5.2.1 A classification of structural petroleum traps

Structural
Extension structures
Compression structures
Salt or mud movement
Drape structures
Stratigraphic
Depositional, such as reefs, channels and pinch-outs
Unconformities

a Thrust structure

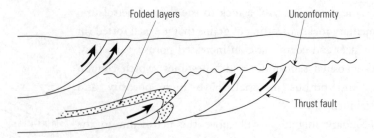

b Reverse fault trap

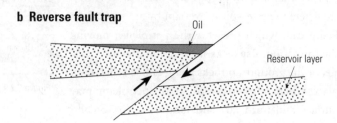

c Anticline

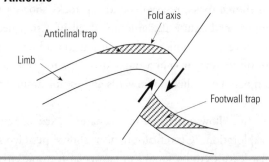

Figure 5.2.10 Thrust belt related structures in which petroleum may occur.

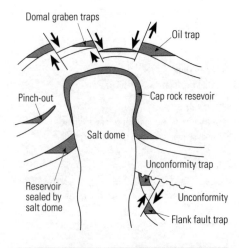

Figure 5.2.11 Traps associated with salt domes.

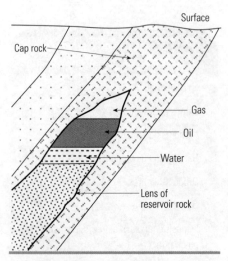

Figure 5.2.12 A pinch-out stratigraphic trap.

Fault traps

Faults create traps when they cut across inclined strata. (See Figure 5.2.10.) Both compressional and tensional faults can produce traps. The ideal situation is one where movement along the fault is relatively large and where a good cap rock lies above the reservoir. The fault itself can help to seal the trap. Crushed rock along the fault can act as a seal, making it impermeable to petroleum. Faults in tensional tectonic settings can also give rise to horst and graben structures, which become part of composite traps.

Salt domes

Salt domes are particularly successful traps. A salt dome forms when variations in salt layers, or layers above the salt, allow the salt to rise. At depths greater than 600 m, salt is less dense than the rocks above it and so it flows. It rises due to its buoyancy, forming a plug that penetrates the sedimentary layers above it. Figure 5.2.11 shows some of the traps that are produced on the sides of and above salt domes.

Drape anticlines

Horsts and tilted fault blocks are structures that form by tensional forces. They create topographic high points and, over time, sediment may bury the structures. Subsequent compaction of the sediment forms drape anticlines that are excellent petroleum traps.

STRATIGRAPHIC TRAPS

Stratigraphic traps are created by variations in the composition and geometry of sedimentary rocks as they are deposited. Such traps account for about 13% of the large oil accumulations identified to date. They are harder to locate than the structural traps and it is likely that future discoveries of petroleum reserves will involve such structures.

Pinch-outs

Pinch-outs are created when a stratum of reservoir rock tapers to an edge. The strata in the area are often slightly inclined and the cap rock is the material surrounding the reservoir. (See Figure 5.2.12.)

Depositional traps

Depositional traps (see Figure 5.2.13) are formed when porous sediments that become reservoirs are deposited within sediments that act as source and cap rocks. Three examples of such traps are **aeolian** sands, fluvial channels and reefs. Aeolian sands are sands deposited by wind. Fluvial channels are the channels created by meandering rivers in which coarse sands are deposited. Reefs create porous, high-relief structures. All three features become traps when they are surrounded by sediments that act as source (trap) strata.

Unconformities

Unconformities are important trap-forming structures. Erosion of tilted strata is followed by deposition of relatively flat-lying sediments. The sediments above suitable reservoir layers act as cap rocks, allowing the accumulation of petroleum. (See Figure 5.2.14.)

Australian sites of petroleum accumulation

The Australian sedimentary basins that do, or may, contain petroleum are shown in Figure 5.2.15. Intracratonic basins are those that form on the craton itself in shallow seas. Australia's Amadeus, Cooper and Surat Basins are examples of intracratonic basins. **Rift basins** are those basins formed when divergent boundaries either succeed or fail. These basins contain some of the largest petroleum reserves in Australia and include the North-west Shelf Basin, the Gippsland Basin, the Browse Basin and the Canning Basin. Barrow Island in Western Australia is an example of a marginal shelf basin. Organic-rich marine sediments have been deposited over and between fault blocks, creating traps and oxygen-poor conditions.

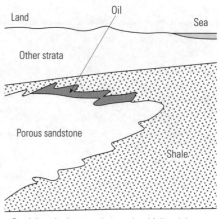

Sand deposited seaward as sea level fell and then retreated as sea level rose

Figure 5.2.13 Depositional traps.

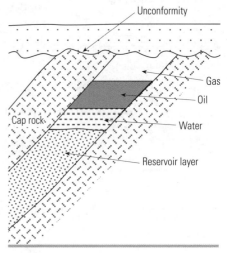

Figure 5.2.14 Unconformity traps.

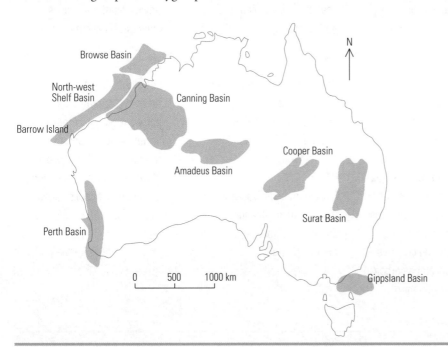

Figure 5.2.15 Major Australian sedimentary basins that do, or may, contain petroleum.

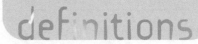

aeolian
refers to sediments that are carried or deposited by wind

rift basin
a depositional basin formed by block faulting and resulting from tensional forces that pull apart the crust

definitions

1

Describe how petroleum is different from, and similar to, coal.

2

Outline the three important factors that control petroleum generation.

3

Describe the three origins of kerogen and how kerogen is produced.

4

Define the terms 'diagenesis', 'catagenesis' and 'metagenesis'.

5

Describe the sequence of events in petroleum production, starting with kerogen production and finishing with petroleum accumulation in a trap.

6

Outline the features of source, reservoir and cap rocks. For each type, describe the environments in which the sediments giving rise to such rocks may be deposited.

7

Compare the accumulation of petroleum in a stratigraphic trap with the accumulation of petroleum in a structural trap.

8

Sketch a cross-section of strata that contain five petroleum traps: three structural and two stratigraphic. Show clearly where the petroleum will accumulate.

9

Explain why petroleum geologists pay particular attention to the time that petroleum has existed in a reservoir.

10

Analyse the similarities that exist between environments of coal generation and petroleum generation.

SUMMARY

Coals form in environments that are characterised by particular climates, plant communities and sedimentary environments.

For coal to be formed, relatively large amounts of peat need to be rapidly buried.

Coalification involves changes due to the heat and pressure that follow burial. The process involves water and volatiles being driven off and changes in the appearance and carbon, energy and ash content of the coal.

Coalification is reflected in the increasing ranks of coal: peat, lignite, sub-bituminous coal, bituminous coal and anthracite.

Petroleum, like coal, is formed by the alteration of organic materials deposited in sedimentary environments.

Petroleum is derived from the remains of small aquatic organisms. Most remains are marine in origin.

As a petroleum deposit is heated it undergoes a series of changes caused by the processes of diagenesis, catagenesis and metagenesis.

Petroleum forms in source rocks and migrates through porous rocks into areas of accumulation in porous reservoir rock. The movement of the petroleum is stopped by impervious rocks called cap rocks.

Comparing fossil fuel localities

In this exercise you will use information from Chapter 5.2 and other sources to analyse similarities and relationships between localities where coal is produced and those where petroleum is produced.

ACTIVITIES

1
Use the resources available to you to describe the environment of Eastern Australia during the Permian. In what sort of environment were the coals of the Sydney Basin deposited?

2
Earlier in this chapter you read about three types of kerogen from which petroleum is derived. Identify the types of kerogen that might have been produced in the Sydney Basin during the Permian.

3
Methane is produced as part of petroleum maturation and by coal deposits. Describe how the methane is generated in each case.

4
Review the environments of deposition described in the text for both coal and petroleum. Compare the role of deltas in the deposition of coal and petroleum.

5
In which climatic conditions would you find high rates of coal and kerogen accumulation?

6
Describe the similarities and differences between modern environments in which coal and oil are mined. You may approach this task by considering tectonic environments, climates or political situations.

7
Summarise the similarities between localities where coal is produced and those where petroleum is produced in terms of:
a the original environments of deposition
b modern environments of production.

5.3 Exploration methods for coal and oil

LOCATING AND DEVELOPING ECONOMIC DEPOSITS

The location and subsequent development of an economic deposit is an expensive process that takes time. At each stage of the process a company must make judgments about continuing the development or cutting their losses and concentrating on other prospects. Less than one in 1000 prospects may end as a profitable venture.

Two principles that guide resource **exploration** are working from the broad scale to the small scale and using less expensive methods before more expensive methods. An outline of the process is shown in Figure 5.3.1. The early, broad-scale work may involve searching records and using satellite or photographic images. This may cost tens of thousands of dollars, but the subsequent work to bring an identified deposit to the point of mining may cost many millions of dollars.

Miners strive for efficiency, and forward planning helps them to achieve this. At each stage, future stages have to be planned for. During the exploration phase of a mine, information necessary for later rehabilitation may be collected. Early samples gathered during exploration may be used to gather initial information about appropriate uses and processing of the material to be mined. Moving from one phase to the next is critically dependent on the quality of the information gathered at any stage and on the correct interpretation of that information.

The seven phases that make up the development of a mining prospect are:

1 identification of a likely area
2 obtaining legal title to explore the area
3 gathering information through exploration
4 evaluation of a deposit that is located
5 development of the mine or drilling platform
6 extraction of the resource
7 closure of the mine or drill site.

Identifying an area involves developing knowledge of the broad-scale geology of the area and how the material that is the target occurs. The mining history of an area can also provide useful information, and government and social conditions in an area need to be understood if the mining company aims to develop a prospect overseas. State and federal geological surveys, such as the NSW Department of Mineral Resources or the Australian Geological Survey Organisation, play an important role in providing basic information about areas. By mapping large areas, **surveys** encourage exploration that may otherwise be too expensive to conduct.

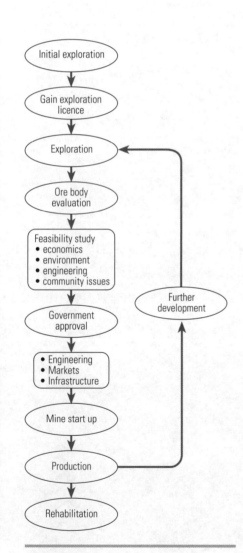

Figure 5.3.1 Stages in the development of an economic deposit.

Once a possible area has been identified as prospective for an economic deposit, access to the area has to be obtained. In Australia, a licence to explore has to be obtained before exploration can begin. This involves checking to ensure that other people do not already hold exploration rights to the area. An exploration licence also imposes certain conditions on how exploration is to occur and be reported. Evidence of progress needs to be provided as part of the process. It is not in the government's interest, nor in the interest of other companies, for a company to obtain an exploration licence and then do nothing.

Exploration involves a range of activities and scientific methods. (See Figure 5.3.2.) Preliminary surveys of an area using satellite images or aerial photographs are used to develop detailed maps of the area. The maps are subsequently used for gathering and recording information about the area. Some of the information gathered enables the location, size and quality of a deposit to be determined. Some of the information establishes pre-existing environmental conditions and is later used in the development and rehabilitation of a site.

Besides gathering information, the exploration phase often involves other processes. These include negotiating access to land from landowners and sometimes buying an area to ensure access. Preliminary drilling and access for other reasons may require the building of roads. Also, planning for closure has to occur in case the prospect is judged uneconomical. The total time invested by the end of the exploration phase is usually several years, but it may be longer in some petroleum areas.

Once the initial exploration evidence is compiled and examined a further period of assessment occurs if the identified deposit is judged worth developing. This phase may take as much as five years to complete and involves a range of detailed and more expensive investigations. Drilling, trenching and underground exploration are used to obtain bulk samples for **assaying**. During assaying, the quality and value of the deposit are assessed and the information may inform work on refining or processing the material. As the estimates of the amount of metal ore, petroleum or coal are developed and refined the feasibility of developing the mine also develops. Towards the end of this phase, attention begins to focus on the ways the material will be extracted and processed.

After the assessment phase a relatively short period of development occurs. During this phase the infrastructure of the mining site is developed. Infrastructure describes the facilities needed to service the mine or drilling platform. Towns, or

exploration
the process of determining the characteristics of an area

survey
to make measurements of and map an area

assay
to determine the quality of something

definitions

Remote sensing methods

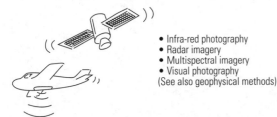

- Infra-red photography
- Radar imagery
- Multispectral imagery
- Visual photography
(See also geophysical methods)

Geochemical methods

Pb Gu
Au As
Na

- Stream sampling
- Weathered outcrop testing
- Ground water analysis
- Sediment and soil analyses

Down-hole logging and drilling

- Rock, fossil and chemical analyses
- Density, resistance, electromagnetic and radiometric logging

Geophysical methods

- Magnetic methods
- Electrical methods
- Gravity methods
- Seismic methods
- Radiometric surveys

Geobotanical methods

- Vegetation sampling and analysis
- Vegetation distribution mapping

Figure 5.3.2 Methods used in exploration.

accommodation, may need to be developed in remote locations. Transport facilities for moving people and material will be needed. These may include airports (for moving personnel to and from oil rigs), railways, ports and roads. At the mine or drilling rig, facilities for storing or refining products are built. The construction of an oil rig or mine also involves infrastructure costs.

Production from the mine or oil well may be reasonably short or long depending on the size and quality of the material being extracted. Ongoing processes during production may involve extraction, milling and refining of the coal, ore or petroleum. As part of the production process, further exploration may identify new economic deposits and allow mines to expand and increase the period of production for the facility. An ongoing part of production in modern mines is environmental monitoring, and rehabilitation may occur at the same time as production.

At the end of a mine's, or an oil rig's, productive life the facilities must be closed. Closure may occur several times. Sometimes a closure caused by low economic prices may be reversed by economic changes or the development of better technologies. But when a mine or oil rig has exhausted a deposit, the site must be rehabilitated and, in many cases, a program of monitoring must be implemented to ensure there is no subsequent environmental impact. Infrastructure is removed, either being moved to another site or sold. Well holes need to be plugged and wastes need to be dealt with. The rehabilitation of the mine or rig draws on information gathered at an early stage of the process and is monitored both by the company and by government agencies.

Mining is a long and involved process. It is risky in that few prospective areas give rise to profitable mines, but the rewards of success can be substantial. In some cases, a decade of development may be paid for in two or three years of production. In large mines that exist for fifty years, it is often the early years of production that are the most profitable, but the mine also benefits those who work there and the wider community through taxes and foreign exchange.

In the next section we will look in more detail at some of the techniques used to identify the location and characteristics of an economic deposit.

REVIEW ACTIVITIES

1
Outline the steps involved in developing a fossil fuel deposit.

2
Why must a company exploring for a fossil fuel deposit obtain an exploration licence?

3
Explain the importance of planning in the exploration and development of a mining venture.

EXTENSION ACTIVITY

4
Construct a flow chart to show the steps involved in developing a mine.

THE ROLE OF GEOLOGY, GEOPHYSICS AND GEOCHEMISTRY

Geology is the study of the rocks and structures that make up the Earth. We have no way of directly seeing into the Earth and so we rely on the careful and detailed mapping of the Earth's surface to determine the structures that lie beneath it. Field geologists map the surface using rock exposures at the surface. **Palaeontologists** use information gathered by field geologists to date the age of sedimentary layers. Palaeontologists also play an important role in oil exploration in that they can use microfossils obtained during drilling to date strata and provide information about the likely maturation of any possible petroleum.

Geophysics is the study of the physical properties of the Earth. Geophysicists use careful measurements of gravity, magnetic fields, and electrical and wave behaviour to work out the Earth's structure. By using measuring devices down drill holes, geophysicists can provide information about rock layers that are not easily determined from the surface. Geophysics is not restricted to the ground. Many instruments are carried on aircraft and ships and large areas can be surveyed in a short period of time.

Geochemistry is the study of the chemistry of the Earth. Samples of soil on land may contain hydrocarbons that have escaped from traps and made it to the surface. The geochemistry of carbon compounds in rocks obtained by drilling may allow geochemists to determine the past history of heating and pressure in sedimentary basins. In coal exploration, geochemists can assess the grade of coals and help predict trends in quality during mining and exploration.

COAL EXPLORATION METHODS

Coal has been mined in Australia since its discovery in 1791. While the sedimentary basins in which coal occurs are reasonably well understood, the detail needed to provide efficient recovery requires ongoing measurement and exploration. Geological surveys have produced detailed information about many areas and remote sensing information may be used to identify regional structures and assist in the development of detailed maps.

Coal exploration methods occur at the surface and within the ground. Working coalmines have teams planning future mining at the same time as mining is proceeding. Forward measurement and planning ensure that the mine manager knows the location, quality and potential problems that may occur when mining moves into new areas. Mine geologists may use drilling data from the surface, but surveying and geological mapping information obtained in the mine itself is important for planning.

Detailed field mapping and geophysics are important in determining structures present in a coal area. Base maps produced from aerial photographs are used to plot information gathered on the surface of the Earth. Maps and cross-sections of the area are developed from the data, and depths and interpretations of depositional environments can be made.

Geophysical methods

Geophysical methods used in coal exploration include magnetic and seismic methods. Both methods involve geophysicists taking systematic measurements along lines, or transects, over the area of interest. Both seismic and magnetic methods can be carried out on the ground, but magnetic methods can also be carried out using instruments installed in planes.

Magnetic methods use the magnetic behaviour of certain rocks within the Earth. Some rocks under and within a sedimentary basin are more magnetic than others. Such rocks alter the magnetic field of the Earth in subtle ways and geophysicists can identify these rocks by carefully mapping the magnetic field, using an instrument called a magnetometer (see Figure 5.3.3), and processing the information using computers. By removing the part of the field due to the Earth as a whole, geophysicists can determine the depth and location of rocks that locally contribute to the magnetic field.

Figure 5.3.3 Obtaining information using a magnetometer.

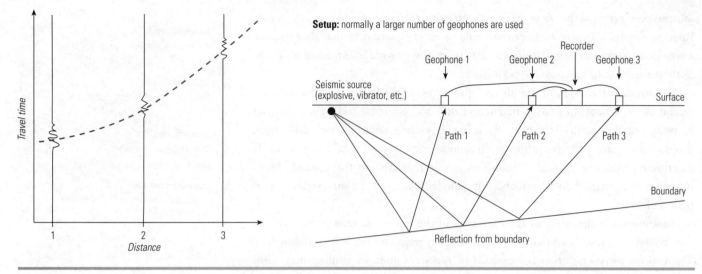

Figure 5.3.4 How seismic reflection works.

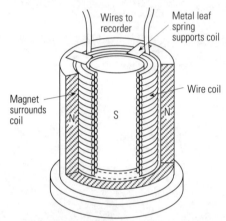

Figure 5.3.5 caption labels: Wires to recorder; Metal leaf spring supports coil; Wire coil; Magnet surrounds coil; N; S; N

A vibration causes the magnet to move (the coil remains relatively still) and the moving magnetic field produces an electrical current in the coil

Figure 5.3.5 A geophone.

Magnetic profiles can be used to work out the structure of the basin bottom and igneous intrusions within the basin. Sedimentary structures within the basin can be related to structures in the basement and igneous rocks, which alter coal quality and present problems for mining.

Seismic methods involve creating shock waves that travel into the Earth, and recording them when they return to the surface. The shock waves are usually called seismic waves and are produced by explosives, or by vibrating plates placed on the ground. The waves travel out in all directions from the site at which they are produced. When a wave hits a boundary some of the wave passes through but some of it is reflected back towards the surface. (See Figure 5.3.4.)

The reflected waves are detected by a line of instruments called **geophones** (see Figure 5.3.5) at the surface. A geophone records the vibration as the wave reaches the surface, and the record is combined with others to produce a cross-section of the area being surveyed. The processing of such information is done by computers and involves the use of mathematical methods to produce clear, accurate descriptions of the area studied.

In coal exploration, geophones have to be very responsive to the waves they detect. The depths being investigated are usually no more than 1 km, and layers reflecting waves to the surface may occur on an average of every metre. Because the seismic waves are travelling at speeds of 1.5–6 km/s a great deal of information can reach a geophone in a very short time.

A type of seismic method used in underground coalmines is the in-seam seismic (ISS) technique. The coal seam acts as a guide for the seismic wave. This means that the material around the seam causes the wave to travel along the seam. Breaks and thinning in the seam are detected by changes in the seismic transmission behaviour. Both the source and receiver may be located at either a coalface within a tunnel or within a borehole.

Drilling and well logging

The most detailed information about the geology of a coal deposit comes from drilling and well logging. Drilling provides rock samples as a cylindrical core or as broken rock chips. It provides valuable information about the depth, thickness and quality of a coal seam. Importantly, such information can also supply information about the strength of overlying strata. This is important in designing underground mines and in determining the best way to remove **overburden** in **open-cut** mines.

Drilling is carried out in a variety of ways. (See Figure 5.3.6.) Diamond drilling involves a drilling bit that has industrial diamonds attached to rotating teeth. These teeth grind down through hard rock as they rotate and create a continuous core of rock that is encased in the hollow drilling rod. Because coal and oil exist in relatively soft sedimentary rocks, a method called mud-rotary drilling is often used. This process involves a drilling head with three cone-shaped bits. The bits have hardened teeth and as they rotate they grind a circular hole in the rock. Mud, a special mixture of clay and water, is pumped down into the hole through the centre of the drilling rod. The mud exerts pressure on the walls of the hole and stops it collapsing. At the same time, the mud lubricates the drill bit and carries chips of rock from the drill head back towards the surface.

Cores from drilling allow geologists to log and sample rocks very accurately. Rock chip studies are also useful but do not supply the same amount of information as a whole core. Cores are usually 4.5–8.5 cm in diameter, but larger cores up to 20 cm in diameter are sometimes used. Larger cores deal better with soft, poorly consolidated materials and provide useful amounts of material for analysis. A drill hole can also be used for the geophysical method called well logging. While the information provided by well logging is not as detailed as that derived from a core, it can be useful.

Well logging involves lowering instruments into drill holes and assessing he physical properties of the rock as the instrument descends. Such instruments have been used since the late 1920s and they use electrical, radioactive and seismic methods to measure the rocks they pass through. We will briefly examine four measurements that can be made in well logging to illustrate the way such instruments work and their uses. They are resistivity, gamma-ray logging, density logging and source logging. What is covered here also applies to oil exploration. There are many other well logging techniques, and each technique is widely used.

Resistivity measures how well a rock conducts electricity. Wet, porous rocks contain salts in the fluids and have high conductivities. On the other hand, coals and dry quartz sandstones do not conduct electricity well at all. The instrument used in this method works by simply placing two electrodes in contact with the rock in the hole wall and measuring the conductivity between the two electrodes.

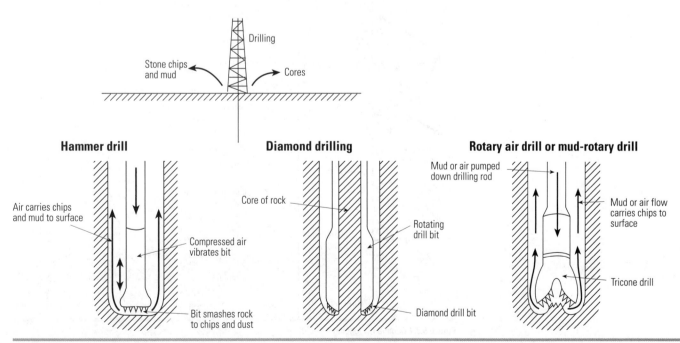

Figure 5.3.6 Drilling methods.

Some potassium atoms are radioactive and as they break down they give out radiation in the form of **gamma rays,** which can be measured by gamma-ray logging. Potassium is a common metal in micas and clays, but is rare in quartz, sands and coal. By measuring the natural gamma rays emitted by rock in the hole wall, the gamma-ray log can indicate changes in rock type according to the clay content and the gamma rays emitted. (See Figure 5.3.7.)

Sometimes, gamma rays are produced by well logging instruments to measure the density of rocks. The instrument is held up hard against the wall of the hole and gamma rays are emitted into the rock. Above the emitter, and shielded from it, is a detector. It measures the amount of gamma rays reflected within the rock. Low-density, or porous, rocks produce higher counts than less porous rocks. The density log can provide good information about shale within coal seams and identify boundaries to a high degree of accuracy.

Within a drill hole it is possible to use sonic logging to measure the speed of seismic waves in a layer. The information can be used to increase the accuracy of seismic investigations, but it also gives evidence of porosity. Porous rocks contain more water than non-porous rocks and therefore porous rocks have much slower seismic velocities.

Figure 5.3.8 shows a model set of well logging results from a borehole through a coal-bearing sequence. Note how the coal stands out relative to the sandstones and shales. Note also how the upper part of the coal seam has more variation in density than the lower part of the seam. This may be due to more shales and impurities within the upper seam.

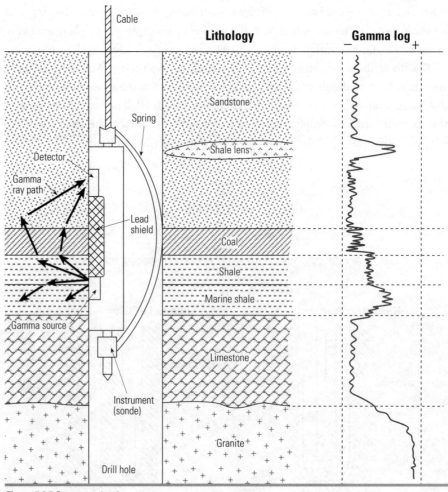

Figure 5.3.7 Gamma-ray logging.

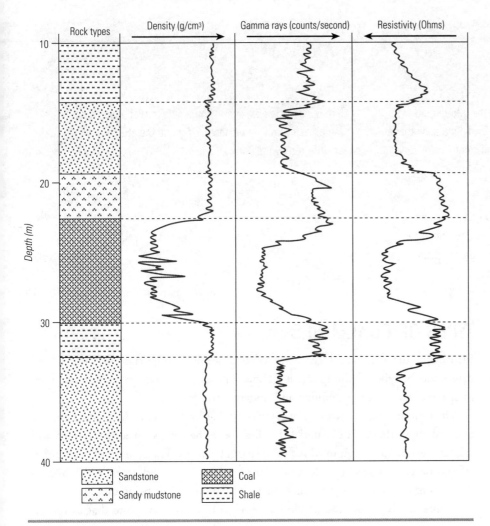

Figure 5.3.8 Rock types and well logging results.

Rock types — Density (g/cm³) — Gamma rays (counts/second) — Resistivity (Ohms)

Depth (m)

Legend:
- Sandstone
- Sandy mudstone
- Coal
- Shale

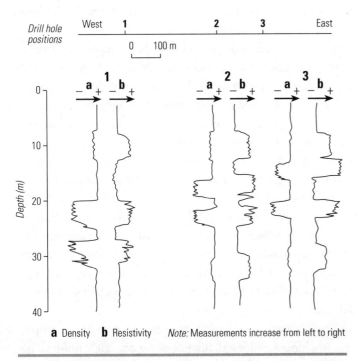

Drill hole positions

West   1   2   3   East

0   100 m

Depth (m)

a 1 b   a 2 b   a 3 b
− + − +   − + − +   − + − +

a Density   **b** Resistivity   *Note:* Measurements increase from left to right

Figure 5.3.9 Three sets of well logging data (for use in extension activity 6, page 254).

1

Outline the techniques used by scientists in coal exploration.

2

Describe how seismic reflection works.

3

What is the role of field mapping in coal exploration?

4

Compare the information obtained by magnetic methods and seismic methods in coal exploration.

5

Describe two well logging methods used to obtain information about a coal seam.

6

Figure 5.3.9 (page 253) shows three sets of well logging data recorded across a coal prospect. Describe the structure of the coal seam inferred by the data.

---

drag-line
a method of extracting coal using a specialised crane and bucket scoop

definition

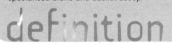

a

Overburden

Coal

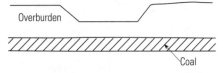

b

First bench (working areas and storage)

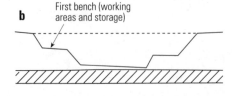

c

Benches reduce slope of pit sides

Original surface

Second bench

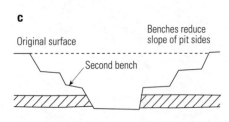

Figure 5.3.10 Development of an open-cut pit.

TYPES OF COALMINES

Identifying the depth and geometry of coal seams is important because it will determine the type of mining that is feasible. There are two basic types of mines used in coal recovery: open-pit mining and underground mining.

Open-pit mining is used for flat-lying coals that have a relatively small amount of material above them. The material that lies above the coal is called overburden and it is removed to form a pit in which the coal can be mined The amount of overburden affects the cost of reaching the coal. An extra metre of overburden adds an extra 2 tonnes of material for every square metre of the pit surface being dug up.

Figure 5.3.10 shows the development of an open-cut pit. Note that as the pit increases in depth it also increases in width. The levels within the pit are called benches. They provide areas for work and for the construction of roads for machines to travel along. The benches also reduce the overall slope of the pit walls and help prevent wall collapse.

The overburden may be weakened by explosives and then removed by trucks, shovels and **drag-lines**. In some cases machines may simply rip the surface before shovels remove the material. As the depth of the coal and the size of the pit increases, the volume of overburden to be removed increases too. Ultimately there comes a point where the value of the coal is not equal to the cost of removing it and then the option of underground mining may become attractive.

A particular type of open-pit mining that can be seen operating in coal areas in New South Wales is strip mining. (See Figure 5.3.11.) A long, elongated pit is produced by blasting and mechanical excavation. First, the surface is carefully surveyed and then the topsoil is removed and stored. As the subsoil and overburden is removed it is also stockpiled. When the coal is exposed it is broken up and transported from the pit in huge trucks. Then the pit is widened to expose the next strip of coal. On the area that has been mined the overburden is replaced and contoured. Next, the topsoil is replaced and steps are taken to revegetate it. All these phases may occur simultaneously.

If you have the opportunity to visit such a mine you may see a drag-line in action. Drag-lines are enormous machines weighing thousands of tonnes. They move along the mine by raising, moving and lowering broad legs that support the weight of the machine. It is important that the machines do not travel too near the edge of the bench because their weight can cause the bench edge to collapse. Drag-lines move rock and coal in a bucket that carries tens of tonnes of rock at a time. Hundreds of tonnes of material are moved every hour by such machines and they operate continuously except for periods of maintenance.

Underground coalmines are found in areas towards the centre of sedimentary basins or below water bodies, such as lakes and the sea. In such areas, the amount of overburden is too great to be removed economically. Underground mines allow economic recovery of the coal. Underground coalmines in New South Wales are found where the depth of the coal is greater than about 150 m below the surface. Generally open-pit mines are economic for overburden to coal ratios less than 10:1, but at this ratio the coal needs to be thicker than about 2.5 m.

Figure 5.3.12 shows the features of an underground coalmine. This is an example of a mine using a room-and-pillar system. In such mines, machines called continuous miners cut

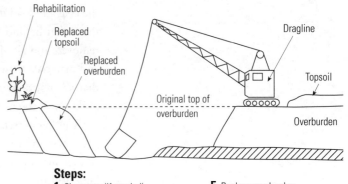

Steps:

1 Clear area (if needed)
2 Remove topsoil and stockpile
3 Remove overburden and stockpile
4 Remove coal

5 Replace overburden
6 Contour surface
7 Replace topsoil
8 Revegetate area

Figure 5.3.11 How strip mining occurs.

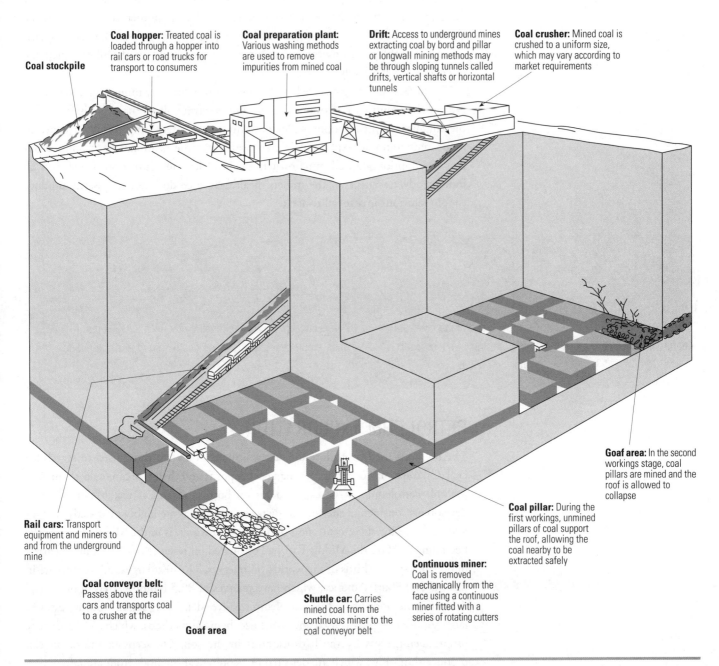

Coal stockpile

Coal hopper: Treated coal is loaded through a hopper into rail cars or road trucks for transport to consumers

Coal preparation plant: Various washing methods are used to remove impurities from mined coal

Drift: Access to underground mines extracting coal by bord and pillar or longwall mining methods may be through sloping tunnels called drifts, vertical shafts or horizontal tunnels

Coal crusher: Mined coal is crushed to a uniform size, which may vary according to market requirements

Goaf area: In the second workings stage, coal pillars are mined and the roof is allowed to collapse

Coal pillar: During the first workings, unmined pillars of coal support the roof, allowing the coal nearby to be extracted safely

Rail cars: Transport equipment and miners to and from the underground mine

Coal conveyor belt: Passes above the rail cars and transports coal to a crusher at the

Goaf area

Shuttle car: Carries mined coal from the continuous miner to the coal conveyor belt

Continuous miner: Coal is removed mechanically from the face using a continuous miner fitted with a series of rotating cutters

Figure 5.3.12 An underground coalmine.

the coal and load it onto a conveyor belt system for removal from the mine. The coal that is left behind forms pillars that support the roof. Although some pillars may be removed, not all of the coal is recoverable. Note the inclined drift that carries miners and materials between the surface and the seam. Two important parts of such mines that are not shown in Figure 5.3.12 are the pumps, which ensure water does not build up in the mine, and the ventilation system, which provides clean, fresh air.

Another system, called long-wall mining is more efficient than the room-and-pillar system. Two tunnels are driven through a coal seam up to 200 m apart. A connecting tunnel is created between the two parallel tunnels and a machine called a long-wall miner is installed. The coal is cut by a shearer, which consists of rotating cutters that move along the length of the long wall. The coal falls onto a conveyor belt and is carried along the long wall to other conveyors at the ends. Once a cut has been made along the working face, the long-wall machine moves forward. Hydraulic ramps that hold up the roof of the area also move forward and the roof is allowed to collapse in a controlled manner.

Open-pit mining has a number of economic and safety advantages over underground mining. More coal can be removed and thinner coal seams can be utilised. Also the rate of extraction is usually higher. In terms of safety, breathing dust is less of a problem, ceiling collapse is not an issue and build up of explosive methane or coal dust is less likely.

Explosions and lung disease have been major problems for miners since before the industrial revolution. Explosions due to candles setting off methane gas in mines led the famous chemist Humphrey Davies to invent the safety lamp. While it helped to reduce underground explosions if used properly, some historians have pointed out that its adoption allowed mine owners to spend less on ventilation, which had an indirect adverse effect on the miners' health. Science does not necessarily produce simple solutions in a social context.

REVIEW
ACTIVITIES

EXTENSION
ACTIVITY

1
Contrast open-pit and underground coalmines.

2
Draw a diagram to illustrate the features of a long-wall mining system.

3
Use secondary sources to research the technologies that make underground mining less hazardous than it was 200 years ago.

PETROLEUM EXPLORATION

The methods used to explore for petroleum are similar to those used to search for coal. Oil traps may occur deeper than coal deposits and their size can make them hard to find. Petroleum can be explored for on both land and at sea. Major oil fields develop on continental shelves and deltas so these areas are highly prospective for gas and oil. The largest concentration of oil fields, however, is in the Middle East. At the beginning of 2000 the Middle East had estimated oil reserves of 675.7 billion barrels of oil: almost two-thirds of the world's oil reserves. The next largest oil reserves are in Central and South America, which have reserves of 89.5 billion barrels of oil.

Initial methods of finding petroleum involve identifying regions of interest. On land, surface mapping is used and so are geochemical methods, which aim to identify oil-producing areas by the hydrocarbons in the soil. Oil seeps in the ocean can produce thin films, which are detected by geophysical methods from aircraft.

Exploration licences

Once an area of interest is identified, an exploration licence is needed. In Australia, onshore licences are granted by the relevant state government but offshore licences are granted by the Federal Government. To obtain a licence, the company or companies seeking the licence must present maps showing the precise areas to be explored, a plan for the works program to be conducted, an assessment of the petroleum potential in the area and an environmental impact statement related to the area. A government is most likely to award the licence to a company with sound technical, financial, environmental and safety credentials.

The exploration program

Large-scale surveys of basins involve the use of remote sensing, gravity and magnetic methods. Remote sensing can identify regional structures and smaller structures, such as salt domes. The other methods help to identify the structure of basins and whether heat and pressure conditions suitable for petroleum generation exist.

The most useful tool for petroleum exploration is seismic imaging. It works in the manner described in the section on coal exploration, but the scale of the projects is many times larger. Satellites are used for positioning surveys and the surveys are conducted both on land and at sea. At sea, air guns fire pressurised air into the water, forming rapidly expanding bubbles. The bubbles create seismic waves, which travel through the water, sediments and rocks, and are reflected back to geophones trailed in a line behind the boat. On land, the seismic signals are generated by vibrators mounted on survey vehicles. The vibrator lifts and drops a heavy steel plate onto the ground in a regular way and produces a series of useful seismic waves. The array of geophones for petroleum exploration is often spaced more widely than the arrays used for coal. The travel times for the waves are also larger.

Sometimes **seismic reflection** may detect the presence of oil, water or gas within the Earth. The presence of fluids slows the speed of the seismic waves, and when the records are processed the gas areas may appear as characteristic bright spots in the images. Merging surveys can generate three-dimensional models of areas, but the complexity of the methods used in processing the data still makes finding suitable targets a challenging task.

seismic reflection
a seismic method that involves the study of reflected vibrations from rock layers

definition

Drilling

Once a target has been identified by geophysical methods or mapping it is drilled. Around Australia at present about 100 wells are drilled each year. Some are drilled to adjust production rates but many are drilled in the hope of locating petroleum. The drilling methods are the same as those described for coal exploration but the hole is cased with steel to stop it collapsing. On land, multiple drilling rigs may work in an area, but in the ocean a single drilling platform may drill multiple holes from the same position. This is possible because drilling teams can carry out what is called deviational drilling. This method causes the drill hole to form a curve, making it possible to drill areas that lie other than directly below the drilling rig. Low-angle drill holes also mean that more petroleum can be recovered if it is found and increases the likelihood of intersecting a trap.

Three common types of drilling rigs are used for exploration in the ocean. Jackups are drilling rigs that are towed into location. Legs are lowered to the sea floor and the hull of the rig is raised above the sea surface. There is a limit to the size of the legs on jackups and they are restricted to use in water depths less than 160 m.

Drill ships look like ordinary ships with a drill rig mounted on them. The drill passes through a hole, called a moonpool, in the centre of the ship, and the ship maintains its position while the drill works on the sea floor. The technology used to

semisubmersible

something that can be submerged or floated by adding water to tanks or pumping them out

definition

maintain position involves satellite positioning systems and computer-controlled propellers to adjust the ship's position. The advantages of drill ships are that they are self-propelled, can move from site to site relatively quickly and have no limitation to the depth in which they can drill. The deep-sea drilling program that has produced so much information about the structure of the world's oceans has utilised ships like these.

Semisubmersibles are also mobile drilling platforms. They have large ballast tanks, which are attached to the platform by long columns. When the semisubmersible is in position, water enters the ballast tanks until the ballast tanks are sitting below the depth of wave action. Here, the ocean is still and the tanks keep the platform very stable. These are quite remarkable structures and they allow drilling to occur up to depths of 300 m.

Once a hole has been drilled, well logging is used to describe the strata through which it passes. The instruments described earlier are used and additional electric sensors may be used to generate more information. An additional sensor that may be used is a dip meter, which measures the inclination of the individual layers through which the hole passes. The information derived from this form of well logging allows geophysicists to identify structural and stratigraphic features, such as anticlines, faults, cross-bedding and delta sequences.

If the geologists drilling the hole find significant amounts of petroleum they may choose to test the well. This process involves isolating an area of interest and then shattering the rock around the drill hole to release any hydrocarbons present. If there is petroleum present it will flow into the drill hole and rise to the surface. High flow will change the well from being an exploration well to a production one. Once a productive well is in operation further wells may be drilled around it to determine the size of the field.

On land, drilling rigs tend to be smaller and more portable. Onshore rigs can be moved by trucks, and sometimes helicopters. Once erected they operate continuously until the well is declared dry, that is, unproductive, or until something is found.

REVIEW ACTIVITIES

1
Describe the types of exploration methods used in the search for petroleum.

2
Describe the drilling platforms used in petroleum exploration.

EXTENSION ACTIVITIES

3
Explain why drilling is left to a later stage of exploration when it generates such large amounts of detail about the geology in an area.

4
Contrast the exploration methods used in coal and petroleum exploration.

5
Draw a diagram to show a drilling ship drilling into an anticline below the sea floor. Label your diagram to show features of the process and of the ship.

SUMMARY

- Exploration involves a range of activities and scientific methods, including the use of satellite images, field mapping, geophysical surveys and drilling programs.

- Exploration methods used for petroleum are similar to those used for coal. Petroleum exploration differs in having targets that are deep and are frequently offshore.

- Petroleum localities in Australia include offshore areas, such as Bass Strait and the North-west Shelf, together with land areas in Queensland.

- Coal reserves in Australia occur in sedimentary basins of a number of ages.

PRACTICAL EXERCISE
The technologies of modern fossil fuel exploration

In this exercise you will gather information about methods and technologies used to explore for fossil fuels. A number of technologies are described in this chapter and further information can be obtained from organisations involved in coal and petroleum industries.

Part A: Seismic reflection
Seismic reflection is a method used in both coal and petroleum exploration. In this part of the exercise you will use resources provided by your teacher, or those you locate, to outline the stages involved in seismic reflection surveys. Use your resources to complete the following activities.

ACTIVITIES

1
What are the sources of energy used in seismic reflection?

2
Outline how a seismic wave carries information about a layer within the Earth.

3
Describe the structure of a geophone.

4
How are seismic reflection surveys conducted at sea?

5
Draw a diagram summarising how seismic waves are produced, reflected and detected using the reflection method.

6
See if you can locate an example of a seismic section produced by reflection.

Part B: Another exploration technology
Select another exploration technology. Prepare a short (five-minute) presentation suitable for explaining the technology to others in your class. Prepare a graphic to illustrate an important aspect of the technology you have chosen. Ensure that you cite the references you have used for your information.

OUTCOMES

At the end of this chapter you should be able to:

describe the refining of coal by washing

describe the refining of petroleum, including distillation and catalytic cracking

describe and evaluate the uses of coal and oil as fuels and raw materials for industry.

THE REFINING OF COAL

When coal is initially mined it is a mixture of particles with different sizes and compositions. The individual pieces of coal may range in size from fine dust to pieces that are too heavy for an average person to lift. The particles also vary in composition. Some are composed of higher-grade coal than others. The original organic matter in one part of a seam may be different from the material in another part and the resulting coal may vary in composition and quality. There will also be non-organic materials mixed with the coal. They may be from shale lenses that were within the coal seam or other sedimentary material from the roof or floor bordering the seam.

Coal preparation improves the quality of the coal in three ways. The energy value of the coal is improved by removing mineral particles that, if burnt, would make up ash. The range of particle sizes is standardised. Fine dust is removed and large lumps are crushed to a size that is easier to handle. The coal from one area may be blended with coal from another area to meet specific requirements set by the buyers. Blending allows sulfur content and ash content to be adjusted.

Four steps may occur in coal preparation: crushing and sizing; cleaning; dewatering; and blending. (See Figure 5.4.1.) These processes occur in a coal treatment plant, which is generally situated close to the mine from which the coal is recovered. Preparation close to the mine ensures that wastes from the preparation are contained in the mine site and it ensures that the coal is in a suitable state for transportation and storage.

Crushing and sizing

The crushing and sizing stage involves reducing the size of the mined lumps of coal and sorting the product into particular size ranges. At an open-pit mine it is not unusual to see pieces of coal a metre in diameter being loaded onto a transport truck. Pieces of this size present difficulties in handling and transportation. Conveyor belts, for example, are engineered to carry certain weights and extremely large pieces would present problems.

At the processing plant, mechanical crushers reduce the coal so that the largest particles are no more than 15 cm in diameter. The crushers may be rotating drums in which the coal is picked up and dropped, breaking up the larger lumps. When the pieces of coal are of an appropriate size they are removed from the drum. Jaw crushers, as the name suggests, consist of two pieces that close and open. Coal is placed in the top of the jaws and as it is broken it moves deeper into the crusher.

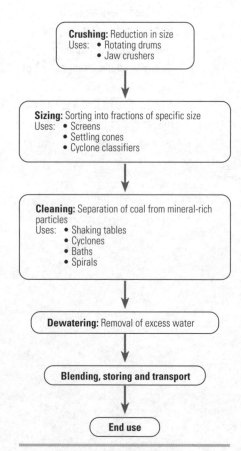

Crushing: Reduction in size
Uses: • Rotating drums
• Jaw crushers

Sizing: Sorting into fractions of specific size
Uses: • Screens
• Settling cones
• Cyclone classifiers

Cleaning: Separation of coal from mineral-rich particles
Uses: • Shaking tables
• Cyclones
• Baths
• Spirals

Dewatering: Removal of excess water

Blending, storing and transport

End use

Figure 5.4.1 The stages in coal preparation.

Because the jaws are arranged in a V shape, the size of the coal pieces is gradually reduced as they move through the crusher. Pieces that are the right size fall through the crusher relatively quickly.

Crushing may involve the coal passing through more than one machine. It is important to prevent the coal from being broken up too much because particles less than 0.5 mm in diameter (called fines) can clog up equipment that is used later in the preparation process.

After crushing, the coal is **sized**. One way this is achieved is by passing the coal over a series of screens. The screens are made of metal and contain holes of a given diameter. Coal of an appropriate size can fall through the screen, but material that is too large is retained above the screen. A series of screens is arranged in layers. As the coal passes down through successive layers the size of the coal pieces becomes smaller and smaller.

In some cases water is used to sort the coal. Coal pieces that are large settle in a water faster than smaller pieces. In a machine called a cyclone classifier, water and coal swirl through the machine and at certain points differences in the water speed cause coal pieces of a particular size to settle out so that they can be collected.

Cleaning

Cleaning follows crushing and sizing. During cleaning, inorganic lumps consisting of shale and sandstone are removed. The processes make use of differences in the densities or surface properties of the coal and non-coal material. The processes are made easier because the crushing and sorting separate the coal from other materials. Water is used during cleaning and the process is also known as **washing**.

Density differences mean that in a moving body of water the less dense coal floats or sinks more slowly than the non-coal contaminants. Machines that clean the coal using the density differences include shaking tables, spirals, cyclones and dense medium baths. A dense medium bath has chemicals added to the water so that the coal floats more effectively than it would in water alone. The way in which a shaking table and dense medium bath work is shown in Figure 5.4.2, page 262.

Froth flotation (see Figure 5.4.3, page 262) uses the surface properties of the coal. Coal is different from shale in that it does not have charges on its surface. In shales, the charged surface is caused by the presence of charged atoms on the surface of the clay particles making up the rock. Bubbles of air are formed at the bottom of the flotation tank, and as they rise the non-charged surfaces of the coal stick to the surface of the bubbles. The coal rises and the shale particles sink. The bubbles and coal particles form a froth at the top of the tank and the froth is skimmed off and dried. The unwanted material, the tailings, can be pumped out of the tank.

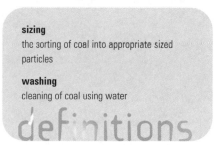

sizing
the sorting of coal into appropriate sized particles

washing
cleaning of coal using water

definitions

a A shaking table

b Dense medium bath (cross-section)

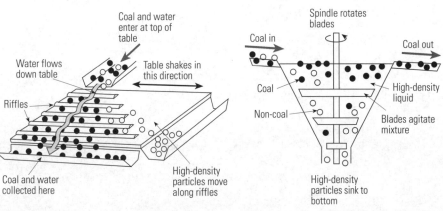

Figure 5.4.2 Two methods of physically washing coal.

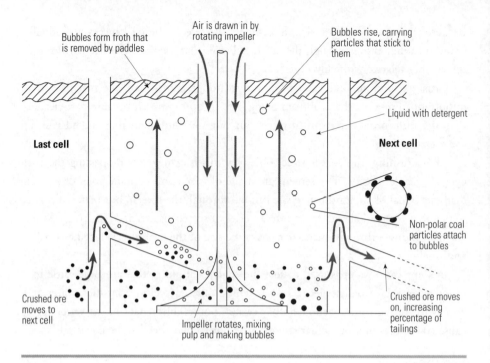

Figure 5.4.3 Froth flotation.

Dewatering

Dewatering is the process in which excess water is removed from the coal. Water, which has been introduced during sizing and cleaning, is removed in order to reduce the weight of the coal. In Australia, as in many other parts of the world, coal is usually used close to where it is mined. If the coal is stockpiled for any length of time it may absorb moisture from the air. In some cases the coal may be wet down with sprinklers to reduce the risk of fires caused by spontaneous combustion within the coal.

Even when coal has been dried by mechanical methods, some water remains locked with the coal. This water is referred to as inherent moisture and it is a property that is used in determining the quality, or grade, of the coal. The amount of inherent moisture affects the energy production of the coal when it burns. The more water that is present, the lower the heat that is produced during combustion.

Blending

Before sale, coal loads with different chemical properties may be combined to meet a customer's requirements. By blending coals, for example, the sulfur content may be adjusted to acceptable levels.

REVIEW ACTIVITIES

1
Describe the steps involved in preparing coal for use.

2
Contrast the processes used in coal washing with the processes used in sizing the coal.

3
Account for the need to blend coal before it is provided to buyers.

EXTENSION ACTIVITIES

4
Outline, using Figures 5.4.2 (page 261) and 5.4.3, how a shaking table and a froth flotation tank work.

5
Discuss the features of coal that are used in its preparation.

THE REFINING OF PETROLEUM

Petroleum is a complex mixture of organic compounds, and during refining these compounds are separated and modified. The liquid petroleum or crude oil is processed using physical and chemical techniques. The process has two outcomes. It separates one compound, or group of compounds, from others in the oil and modifies some molecules by either breaking them up or building them into new ones. The process of breaking up molecules is called cracking and the initial separation process is called distillation.

Distillation

In an oil refinery the primary separation of the crude oil is by the process of distillation. Distillation is a process that uses the different boiling points of chemical compounds in a mixture. Compounds with a low boiling point boil and escape from the mixture as the temperature increases. Compounds with higher boiling points remain in the liquid. The gaseous compounds can then be recovered by cooling and collection. In oil distillation the components recovered are referred to as **fractions**, and the process is referred to as fractionation.

Two types of distillation are used to refine crude oil. Atmospheric distillation occurs at normal atmospheric pressure and occurs over a temperature range of 350–400 °C. The other type of distillation, called vacuum distillation, occurs at low pressures. By lowering the pressure above the oil, the boiling point of compounds in the oil is lowered. This means that compounds, which normally boil at very high temperatures, are easier to distil. It also means that the compounds are separated without the chemical alteration that occurs at high temperatures.

Atmospheric distillation occurs in a structure called a distillation column. The crude oil is first heated and then placed in the bottom of the column. As the vapour rises, it passes through holes in a series of trays. The vapour also passes through a gradually falling range of temperatures. At a certain temperature, compounds in the vapour condense on a tray and can be collected. Large molecules condense at higher temperatures towards the bottom of the column and smaller, lighter molecules form fractions towards the top of the column.

The liquid, called fractions, is removed from the collecting trays. Gases, which do not condense, are removed from the top of the column and sent to a gas fractionation plant. The thick residue that remains in the bottom of the column may be burnt as a fuel or processed by vacuum distillation into oils, waxes, bitumen and chemical feed stocks. The various fractions prepared by distillation are shown in Figure 5.4.4.

The distillation process does not result in finished products. Further processing involves cracking, reforming, treating and blending of the compounds derived by distillation. Cracking breaks molecules into smaller pieces, while reforming creates larger molecules from smaller ones. Modification is used to alter the structure of molecules, producing chemicals with particular properties.

Cracking

Cracking uses heat and/or catalysts to break up the compounds in a petroleum fraction. (See Figure 5.4.5.) Thermal cracking is the oldest cracking technology and is used to treat the residue of vacuum distillation. Catalytic cracking converts large molecules into the smaller ones found in petrol and light fuel oils. Heat causes large molecules to break up. This is similar to the process that naturally occurs in petroleum maturation. Catalytic cracking uses a catalyst to break up compounds. A catalyst is a substance that takes part in a chemical reaction but is not altered by the reaction. In catalytic cracking the catalyst is silica-alumina or silica-magnesia.

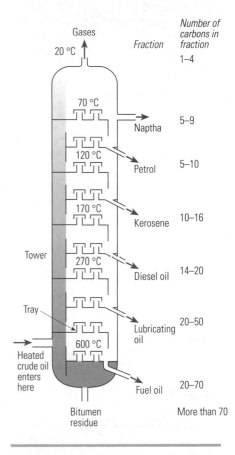

Figure 5.4.4 Atmospheric distillation and its products.

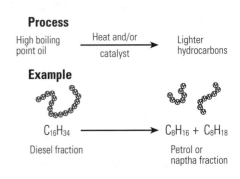

Figure 5.4.5 Cracking.

As pellets, these substances withstand the temperatures and pressures used in the process. Two types of catalytic cracking are fluid catalytic cracking and hydro-cracking. (See margin box.)

Fluid catalytic cracking shows the conditions under which cracking occurs and the way the catalyst is reused. The catalyst used in this process is often a fine powder consisting of alumina (aluminium oxide) or silica (silicon oxide). Recently, artificially produced minerals called zeolites have been used to carry out the process more quickly. In fluid cracking the catalyst is mixed with steam so that the catalyst acts like a fluid. When the steam–catalyst mixture comes into contact with the feed stock hydrocarbons, the hydrocarbons vaporise and the new mixture is carried to a reaction vessel. The feed stock is heavy gas oils. Here, under conditions of high heat and pressure, the hydrocarbons are broken up. The catalyst and the hydrocarbons are then separated and the catalyst is reconditioned. The products of the reaction are diesel oils and petrol. Carbon films on the catalyst are burnt off and then the catalyst is returned to the process. The process occurs at a temperature of 530–540 °C.

Hydrocracking is a process that is used to increase petrol yields and produce light distillate oils from crude oil. The process involves the use of hydrogen. It occurs at lower temperatures and higher pressures than those used in fluid catalytic cracking. Oils are converted to petrol and kerosene. At the same time, the hydrogen removes sulfur and nitrogen from the hydrocarbons. These elements are impurities and the process increases the quality of the products being produced.

The sulfur removed from the hydrocarbons may be converted to solid sulfur in a sulfur recovery plant. The removal of the sulfur also ensures that the equipment in the plant does not corrode quickly.

Reforming

Reforming, also called unification, produces compounds that are used in making petrol and as feed stocks for other chemical industries. Like catalytic cracking, reforming involves high temperatures, high pressures and catalysts. The difference is that rather than breaking up the hydrocarbon molecules, the process generates larger molecules. Metals such as platinum and rhenium act as catalysts and relatively small, straight chain hydrocarbons from naphtha fractions are turned into branched chain and aromatic hydrocarbons. (See Figure 5.4.6.) The addition of aromatic compounds to petrol modifies the rate at which the fuel burns and can increase the efficiency of the engines that use it. Some of the chemical processes involved in reforming are alkylation, polymerisation and isomerisation.

Treating and blending

The materials formed by reforming are separated by distillation, but before use they are treated and the fractions may also be blended. Treating involves removing impurities from the hydrocarbons. This is done in a series of steps, beginning with treatment in a column of sulfuric acid. The acid removes hydrocarbons that contain double bonds. The acid also removes small pieces of solids (such as asphalt) and contaminants (such as nitrogen and oxygen). Next, the hydrocarbons are passed through an absorbing column where water is removed. Finally, a sulfur treatment removes any sulfur present.

Blending involves mixing fractions of various compositions to produce products with specific properties. You may be aware that oils can be purchased in a range of grades. They are produced by mixing, or blending, oils with different properties. Petrol is also blended. The addition of aromatics to the liquids alters their burning characteristics, and other additives may also be introduced to produce particular characteristics.

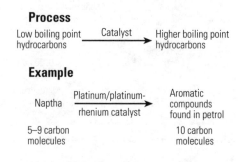

Process

Low boiling point hydrocarbons →(Catalyst)→ Higher boiling point hydrocarbons

Example

Naptha →(Platinum/platinum-rhenium catalyst)→ Aromatic compounds found in petrol

5–9 carbon molecules → 10 carbon molecules

Figure 5.4.6 Reforming.

The refinery as a system

At this stage it should be obvious that a petroleum refinery involves a number of processes. These processes are located within specialised parts of the refinery, and the number and type of treatment plants in a refinery will depend on the types of crude oil they use.

While a modern refinery is a very efficient place, it may affect the environment. The refinery monitors the air, water and land within and around the refinery. Government agencies, such as the Environment Protection Authority (EPA), also monitor the site.

Air monitoring aims to prevent the emission of hydrocarbon vapours, smoke, sulfur oxides and strong-smelling compounds. Water monitoring aims to restrict the occurrence of contaminated water. This means that water that comes into contact with the hydrocarbons must be treated, that rainwater must not be allowed to carry pollutants off the site and that no hydrocarbons must be allowed to leak into waterways. Land cannot be allowed to be contaminated by the refinery's products or wastes. A 'slops system' recycles wastes through pipes and tanks so that the material does not leach into the soil. Wastes are recycled, treated or disposed of at EPA approved sites.

REVIEW ACTIVITIES

1
Explain, using a flow chart, the process of distillation.

2
Outline the processes involved in distillation and catalytic cracking.

3
Define the role of catalysts in petroleum refining.

4
Describe the environmental monitoring that goes on in and around a petroleum refinery.

5
Distinguish between the functions of various parts of a petroleum refinery.

EXTENSION ACTIVITIES

6
Summarise how catalytic cracking may be applied to coal. Predict whether the products produced would be worth the energy expended in producing them:
a today
b 200 years in the future.

7
An oil contains hydrocarbons that contain molecules with between five and thirty carbon atoms. Identify the products that could be distilled from this oil and explain how those products could be converted into lighter and heavier fractions.

COAL AND OIL USES IN INDUSTRY

The uses of fossil fuels were described in Chapter 5.1 and summarised in Table 5.1.2 (page 228). Only 6% of Australia's energy needs are supplied by things other than fossil fuels.

Coal as a fuel presently provides more than 90% of the electricity in New South Wales. A little less than ten times as much is exported to twenty-five countries in areas such as Europe, North America, the Middle East and Asia. The quality of our coal provides good-quality coke for iron and steel, both in Australia and overseas.

Cement manufacture consumes a lot of coal, but coal products are also used in products as diverse as dyes, drugs, disinfectants, resins, acetylene and synthetic fibres. In addition to international income from exporting the coal itself, Australia also generates $1 billion exporting our coalmining and processing expertise.

Oil and natural gas also play a vital role in our lives. Oil supplies 52% of Australia's energy needs. Approximately 46% of the energy use in New South Wales is from oil. During the late 1990s about 25 billion litres of oil was produced in Australia and a little less than half that amount was exported.

Natural gas in New South Wales is sourced from Moomba in Queensland and a pipeline from Longford in Victoria will provide natural gas from Bass Strait.

In assessing the uses of fossil fuels we must acknowledge the costs as well as the benefits. Pollution from fossil fuels is an issue and while industry continues to make fuel use cleaner and more efficient, gases such as carbon dioxide are affecting the Earth's environment. Mining fossil fuels is a hazardous business. Australia has the best safety record of any coal-producing country in the world, but one person dies mining coal for every 10 million tonnes of coal we produce. Worldwide, approximately 11 000 workers in the coal industry die each year and millions are injured. In the next chapter we will examine our use of, and dependence on, fossil fuels in more detail.

REVIEW ACTIVITIES

1
Summarise and evaluate the uses of coal and oil in our society.

2
Compare the costs and benefits of fossil fuel use.

EXTENSION ACTIVITY

3
Evaluate how your life would change if the energy you utilise derived from fossil fuels was to be reduced by half (without another source being available).

SUMMARY

- Fossil fuels are prepared in particular ways to produce products optimised for specific uses.

- Coal preparation improves the energy value of the coal, standardises the range of particle sizes and removes fine dust.

- Coal treatment involves crushing and sizing, cleaning, dewatering and blending.

- Liquid petroleum is processed using physical and chemical techniques that have two functions: they separate one compound, or group of compounds, from others; and they modify some molecules by either breaking them up or building them into new ones.

- The process of breaking up molecules is called cracking and the initial separation process is called distillation.

- Coal and oil are important fuels.

PRACTICAL EXERCISE
The processes used in refining fossil fuels

In this exercise you will use information from Chapter 5.4 to draw diagrams showing the processes used in refining fossil fuels. You can create greater detail in your diagrams by referring to other resources, such as those listed in the resources list for this section. (See page 389.)

ACTIVITIES

Coal

1

In the material on coal preparation four steps were described. Starting with raw coal, list the steps and outline the features and product of each stage.

2

Identify the machines or processes used in each stage.

3

Use the information summarised from activities 1 and 2 to draw a flow chart that summarises the process. In your flow chart, use squares or boxes to represent materials or machines and use diamonds to represent processes. At the end of your flow chart you should have a series of products that are used in industry and the home.

Oil

4

Repeat the steps outlined at left for the refining of oil. Again, ensure that your flow chart ends in products that are used in industry and the home.

Comparisons

5

Compare the two flow charts you have produced and identify the similarities and differences between the two processes.

6

Assess the effect that removing the processing technologies we currently use would have on:
a the uses of fossil fuel
b the environment.

OUTCOMES

At the end of this chapter you should be able to:

- analyse and evaluate the types and effects of products of burning fossil fuels—gases, water, particulates

- analyse the dependence of modern society on fossil fuels and assess attempts to limit emissions

- describe and evaluate arguments concerning the greenhouse debate.

THE DILEMMA OF OUR DEPENDENCE ON FOSSIL FUELS

The wide range of materials we derive from fossil fuels indicates the degree to which we rely on them for our standard of living. A piece of coal, or a bucket of oil, does not have any value in itself. We place value on such things because we know how to use them to provide things we need. Fossil fuels provide energy as fuel and the raw materials for things we use every day, such as plastics and industrial materials.

Today, we face the dilemma of needing fossil fuels for the benefits they provide but knowing too that using fossil fuels causes harm. At this point it would be worthwhile reviewing Chapter 3.5 to remind yourself of the issues regarding air pollution. If something produces more costs than benefits, its value declines. Fossil fuels create problems in a number of areas, but perhaps the greatest problem we face in relation to fossil fuel use is that it is changing the Earth's climate.

The effect of burning fossil fuels, together with land clearing, has been to increase the temperature of the atmosphere. The temperature change is due to the increasing amount of global warming gases, termed greenhouse gases, in the atmosphere. Concern over the effects of such gases on global climate led members of the United Nations to agree to reduce greenhouse gas emissions to prevent further negative effects on the climate. The Kyoto Agreement, created in 1997, sought to have countries reduce their annual greenhouse gas emissions by an average of 5.2% relative to 1990 emissions.

The ability of a gas to increase the temperature of the atmosphere depends on its ability to capture energy. Figure 3.5.1 (page 163) shows the fate of energy as it reaches the Earth from the Sun. While the energy reaches the Earth's surface as light, it is re-radiated from the surface as infra-red radiation. It is in this form that the energy is captured and held by molecules such as water vapour and carbon dioxide. Such molecules are referred to as greenhouse gases.

The amount of heat held in the atmosphere depends on the type and amount of gases present. Of the gases that produce global warming, water vapour is the most important naturally derived greenhouse gas. It accounts for 75% of natural greenhouse effects. Other naturally occurring greenhouse gases are carbon dioxide, methane and nitrous oxide. Table 5.5.1 shows the main greenhouse gases generated by humans and their characteristics. Such gases are said to have an **anthropogenic** source.

definition

anthropogenic
of human origin

Table 5.5.1 The major greenhouse gases, excluding water vapour

Greenhouse gas	Chemical formula	Residence time (years)	Anthropogenic sources	Global warming potential	Pre-industrial concentration (ppbv)	Concentration in 1994 (ppbv)
Carbon dioxide	CO_2	Varies depending on sink processes	Fossil fuels Land clearing Cement production	1	278 000	358 000
Methane	CH_4	9–15	Fossil fuels Rice growing Waste dumps Livestock	21	700	1721
Nitrous oxide	N_2O	120	Fertilisers Industrial processes Combustion	310	275	311
CFC-12	CCl_2F_2	102	Liquid coolants Foams	6200–7100	0	0.503
HCFC-22	$CHClF_2$	12	Liquid coolants	1300–1400	0	0.105
Perfluoro-methane	CF_4	50 000	Aluminium production	6500	0	0.070
Sulfur hexafluoride	SF_6	3200	Fluid used in dielectric components	23 900	0	0.032

Note: ppbv = parts per billion by volume. For carbon dioxide 1 ppbv = 2.13 million tonnes of carbon.

Gases that contribute to global warming or cooling are said to show **climate forcing**. This means that they add energy to, or remove energy from, the surface of the Earth. Climate forcing is expressed in units of watts per square metre (Wm^{-2}). Carbon dioxide produces climate forcing values ranging from 1.3 to 1.5 Wm^{-2}. Particles produced by fossil fuel burning or forest clearing also cause climate forcing. Black carbon particles have a climate forcing value of 0.1 to 0.8 Wm^{-2}. Sulfate aerosols, on the other hand, cool the atmosphere and have climate forcing values of -0.1 to -0.3 Wm^{-2}. Models and measurements suggest that the global forcing due to human sources averages 2.43 Wm^{-2}. The most significant contributor to this heating effect is carbon dioxide (1.46 Wm^{-2}), followed by methane (0.48 Wm^{-2}) and **fluorocarbons**, such as CFCs (0.34 Wm^{-2}) and nitrous oxide (0.15 Wm^{-2}).

Another factor determining the effect of a greenhouse gas on the atmosphere is its **residence time** in the atmosphere. Table 5.5.1 shows the average time that greenhouse gases reside in the atmosphere. Some materials that cause warming remain in the atmosphere for short periods, such as tens of days. Carbon particles are an example. Methane, on the other hand, can remain in the atmosphere for as long as ten years. The human-manufactured fluorocarbons can remain in the atmosphere for more than 1000 years because they reach the atmosphere above clouds, which remove material from the atmosphere. It is difficult to measure the residence time of carbon dioxide in the atmosphere.

As part of the carbon cycle, carbon dioxide can be removed from the atmosphere in a variety of ways and at different rates. The movement, or flux, of carbon dioxide and major reservoirs of carbon are shown in Figure 5.5.1 (page 270). Note that, on balance, the terrestrial biosphere and the oceans remove more carbon from the atmosphere than they generate into it. Burning fossil fuels, however, puts twice as much carbon dioxide into the atmosphere as the amount removed to the deep ocean each year.

climate forcing
the effect of adding energy to, or removing energy from, the climate

fluorocarbons
a group of carbon compounds that contain the element fluorine

residence time
the time something remains in part of a cycle

definitions

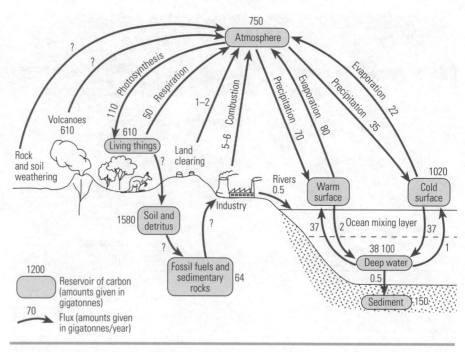

Figure 5.5.1 Global reservoirs and carbon dioxide flow.

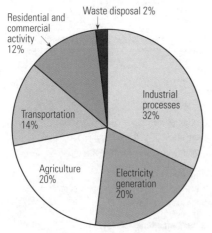

Figure 5.5.2 Sources of carbon dioxide from the use of fossil fuels.

While we face an enormous task to reduce the amount of carbon dioxide we add to the atmosphere, there are historical examples that show we can alter the emissions we cause. Over time we have seen air pollution legislation have a beneficial effect on the atmosphere. In the mid 1980s Australia introduced emission standards for cars, which has helped to reduce the emissions of carbon monoxide and unburnt hydrocarbons. Legislation has also driven the increasing efficiency of automobiles so that less petrol is burnt to provide the energy needed for transport. The development of clean coal technologies, particularly in the USA where coals have a relatively high sulfur content, has reduced nitrous oxide and sulfate aerosol emissions. At the same time, the furnace efficiencies of coal-fired power plants have been improved so that less coal is burnt. The two greatest sources of carbon dioxide are electrical generation and transport. The sources of carbon dioxide derived from fossil fuel use are shown in Figure 5.5.2. Improvements in efficiency in these areas will produce some of the most rapid reductions to emissions from fossil fuels.

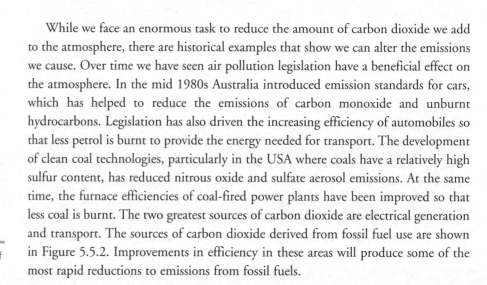

REVIEW
ACTIVITIES

EXTENSION
ACTIVITY

1

List the gases that promote global warming and indicate which of these are caused by fossil fuel use.

2

Identify the industries that produce carbon dioxide and methane.

3

Explain what climate forcing means.

4

Describe three steps that have been taken to limit greenhouse gas emissions.

5

Analyse the degree to which we are dependent on fossil fuels.

ENVIRONMENTAL IMPACT OF PRODUCING AND USING FOSSIL FUELS

The effect of fossil fuels in altering climate is based on evidence of changing carbon dioxide and methane levels derived from ice cores, direct measurements and climate modelling. Climates over geological history are known to some degree, but not with the accuracy needed to predict our future. It is known, however, that during the last interglacial period (up to 10 000 years ago) carbon dioxide levels seem to have averaged between 190 and 280 ppmv (parts per million by volume). The carbon dioxide concentration did not move beyond 280 ppmv until the industrial revolution, but by 1958 the level had increased to 315 ppmv. The concentration is currently about 370 ppmv and is rising at 1.5 ppmv per year. A graph showing recent changes in carbon dioxide emissions and temperature since 1860 is shown in Figure 5.5.3. Note the increasing temperature and emissions rate since 1940. Currently, global emissions from fossil fuel use are increasing at 0.6% per year and this is less than the rate used in some climate change models. (See Figure 5.5.4, page 272, and Table 5.5.2, page 273.)

It is hard to be sure to what extent climate change is affected by non-human processes. There have been regional fluctuations in temperature through recorded history that could have very little to do with fossil fuel burning. In Europe, records show that the period from 1100 to 1300 was quite warm, while from 1400 to 1850 conditions were very cool. We know that during each of these times the average temperature changed by only 1.5 °C. The temperature differences between the two periods caused changes in a wide range of human practices. Current models suggest that the globally averaged surface temperature will increase by 1.4 °C to 5.8 °C by 2100.

The accurate prediction of how climate will change depends on our understanding of how the climate system works. Figure 5.5.4 and Table 5.5.2 show some of the predictions made by recent models as to changes in carbon dioxide concentrations. Note the wide error envelopes for each model. This reflects the uncertainties that exist. The worst results reflect continuing population growth and a continuing reliance on fossil fuels to supply energy. (See S4.) The most rapid decline in carbon dioxide emissions depends on population growth peaking in 2050 and then declining, a simultaneous development of non-fossil fuel technologies and a move towards similar incomes across the globe (See S2.) The stabilisation of atmospheric carbon dioxide concentrations at about 450 ppmv will require global human-caused carbon dioxide emissions to drop below the 1990 levels within a few decades and continue to decrease steadily to a small fraction of current emissions. Under this model, emissions would peak in about one to two decades.

Uncertainties that affect the confidence in such models relate to assumptions that have to be made. These assumptions concern economic growth, technological development and change, population growth, and the efforts of governments to alter current emissions. An area of particular uncertainty is in modelling accurately the way the carbon cycle operates. In particular, the sensitivity of the climate to changes and the effects of climate **feedbacks** are not well understood.

Difficulties in modelling climate change, and the contribution of fossil fuels, is also affected by our knowledge of feedback in natural systems. Feedback occurs when something that is created by a change affects the thing that first caused the change. For example, the greenhouse gases absorbing heat directly cause 40% of the temperature change, but the remainder is due to feedback mechanisms. Increasing temperatures increase water evaporation, and the water vapour further increases

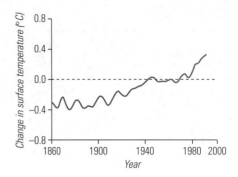

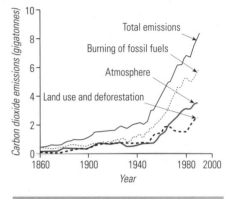

Figure 5.5.3 Changes in temperature and carbon dioxide emissions, 1860–1990

feedback
a change that alters the process that causes the change

definition

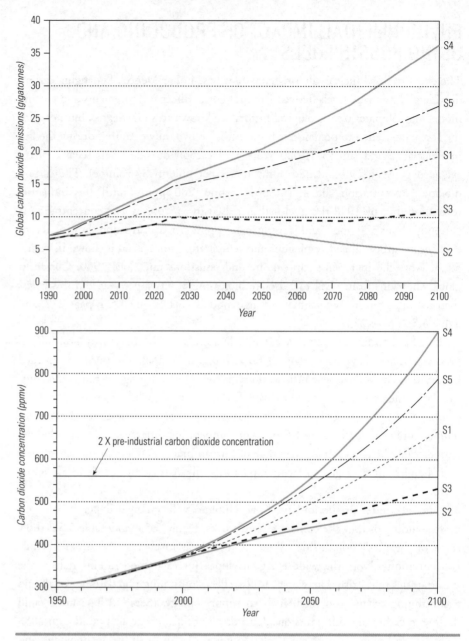

Figure 5.5.4 Recent models showing future changes in carbon dioxide emissions.

temperature by holding heat in the atmosphere. A decrease in the extent of icesheets due to warming reduces the albedo effect, that is, the amount of light being reflected from the surface rather than forming heat. This leads to more warming of the Earth's surface and re-radiation of energy that is held by the greenhouse gases. These two processes multiply the affect on temperature by as much as 250%. Changes in cloud distribution and humidity also affect the rate of temperature change.

The models do agree that projected climate change will create benefits and adverse effects for both the environment and human society. Some of these changes are shown in Table 5.5.3 (page 274). The adverse impacts of climate change are expected to affect different societies to varying degrees. In some areas the changes will be coped with because the societies have the wealth to adapt to them. It is in less affluent countries that the negative effects of climate change will be felt most. The three most important factors driving climate change are population, economic activity and technology. It is changes in these three factors that will allow a particular society to cope with, and help minimise, the changes we see.

Changes in carbon dioxide levels and temperature are not, and will not be, evenly distributed across the Earth. The rising carbon dioxide levels have led to an average

Table 5.5.2 Assumptions for the five scenarios shown in Figure 5.5.4

Scenario	Population	Economic growth	Energy supplies*	Other**	CFCs
S1	World Bank 1991: 11.3 billion by 2100	1990–2025: 2.9% 1990–2100: 2.3%	12 000 GJ conventional oil 13 000 GJ natural gas Solar costs fall to $0.075/kWh 191 GJ of biofuels available at $70/barrel	Legally enacted and internationally agreed controls on sulfur oxides, nitrogen oxides and non-methane volatile organic compound emissions. Commitments by many OECD countries to stabilise or reduce carbon dioxide emissions.	Global compliance with scheduled phase-out specified by Montreal Protocol
S2	UN medium–low case: 6.4 billion by 2100	1990–2025: 2.0% 1990–2100: 1.2%	8000 GJ conventional oil 7300 GJ natural gas Nuclear costs decline by 0.4% annually	Legally enacted and internationally agreed controls on sulfur oxides, nitrogen oxides and non-methane volatile organic compound emissions	Same as S1
S3	Same as S2	1990–2025: 2.7% 1990–2100: 2.0%	Oil and gas same as S2 Solar costs fall to $0.065/kWh 272 GJ of biofuels available at $50 barrel	Emission controls extended worldwide for carbon monoxide, nitrogen oxides and sulfur oxides. Halt deforestation. Capture and use of emissions from mining and gas production and use.	CFC production phase-out by 1997 for industrialised countries. Phase-out of HCFCs.
S4	Same as S1	1990–2025: 3.5% 1990–2100: 3.0%	18 400 GJ conventional oil Gas same as S1 Phase-out of nuclear by 2075	Emission controls (30% pollution surcharge on fossil energy)	Same as S3
S5	UN medium–high case: 17.6 billion by 2100	Same as S1	Oil and gas same as S4 Solar costs fall to $0.083/kWh Nuclear costs increase to $0.09/kWh	Same as S2	Partial compliance with Montreal Protocol. Technological transfer results in gradual phase-out of CFCs in non-signatory countries also by 2075.

*All scenarios assume coal resources up to 197 000 GJ. Up to 15% of this resource is assumed to be available at $1.30 per gigajoule at the mine.

**Tropical deforestation rates (for closed and open forests) begin from an average rate of 17 million hectares per year for 1891–90, then increase with population until constrained by availability of land not legally protected. S3 assumes an eventual halt of deforestation for reasons other than climate. Above-ground carbon density per hectare varies with forest type from 16 to 117 tonnes per hectare, with soil carbon ranging from 68 to 100 tonnes per hectare. However, only a portion of carbon is released over time with land conversion, depending on the type of land conversion.

Note: GJ = gigajoule; kWh = kilowatt-hour

rise in atmospheric temperature of 0.4–0.8 °C over the last 100 years, but this varies from place to place. Changes in ocean temperatures have also occurred and average out to an increase of 0.05 °C for the upper 3 km of the ocean. Arctic ice volumes appear to be decreasing at 3% per decade, but the effects in the Antarctic seem less pronounced. Changes in the rate of warming are not uniform. The periods from the early 1900s to 1940 and from 1980 to 2000 have been the periods of greatest temperature increase at the Earth's surface.

Inertia in climate, ecological and societal systems is a major reason why working to anticipate change is important. Inertia describes the fact that systems do not respond immediately to changes. For example, a rapid move to preventing carbon dioxide emissions will not stop temperature increase. Research suggests that when carbon dioxide levels become constant, atmospheric temperatures will continue to rise a few tenths of a degree for at least a century. The ocean responds even more

definition

inertia
the property of a system that describes its resistance to change

Table 5.5.3 Projected changes to climate due to global warming in the twenty-first century and resulting impacts

Projected changes	Probability	Examples of resulting impacts
Higher maximum temperatures More hot days and heat waves over nearly all land areas	Very likely	Increased: • death and serious illness in older age groups and poor living in cities • heat stress in livestock and wildlife • risk of damage to a number of crops • electric cooling demand and consequent reduced energy supply reliability Changes in tourist destinations
Higher minimum temperatures Fewer cold days, frost days and cold waves over nearly all land areas	Very likely	Decreased: • cold-related human illnesses and death • risk of damage to a number of crops but increased risk to others • demand for energy for heating Extended range and activity of some pest and disease vectors
More intense precipitation events in many areas	Very likely	Increased: • occurrence of natural hazard events, such as floods, landslides, avalanches and mudslides • soil erosion • flood run-off, which could increase recharge of some floodplain aquifers • pressure on government and private flood insurance systems and disaster relief
Increased rates of summer drying over most mid-latitude continental interiors Associated risk of drought	Likely in mid-latitude interior of continents	Increased: • damage to building foundations caused by ground shrinkage • risk of forest fire Decreased: • crop yields • water resource quantity and quality
Increase in tropical cyclone: • frequency • maximum wind intensities • average and maximum precipitation intensities	Likely in some coastal areas	Increased: • risk to human life • risk of infectious disease epidemics • coastal erosion and damage to coastal buildings and infrastructure • damage to coastal ecosystems, such as coral reefs and mangroves
More intense droughts and floods associated with El Niño events in many different regions	Likely	Decreased: • agricultural and rangeland productivity in regions that are prone to drought and floods • hydro-electric power generation potential in drought-prone regions
Increased variability in Asian summer monsoon rainfall	Likely	Increase in flood and drought magnitudes and the damage they cause, which will affect both temperate and tropical Asia
Increased intensity of mid-latitude storms	Unknown: wide range of outcomes among models	Increased: • risks to human life and health • property and infrastructure losses • damage to coastal ecosystems

slowly and the sea level may rise for centuries after the carbon dioxide levels become constant.

It is therefore important that our society not only works to develop better technologies that avoid adding to greenhouse emissions but that it also works to remove barriers that slow the adoption of low-emission technologies. Emission inventories are one way of working towards a better understanding of how a company or industry contributes to the production of greenhouse gases. Such inventories can be used for scientific models and policy development and by regulatory bodies to establish compliance of companies to laws regulating emissions. In the next two parts of this chapter we will look at the basics of greenhouse gas production and the sinks into which such gases are removed.

Greenhouse gases

Greenhouse gases were discussed in Chapter 3.5 and will not be covered in detail here. Burning of fossil fuels and other materials can give rise to a range of products. These include gases, such as carbon dioxide and unburnt hydrocarbons, water and particles.

The particles may vary in size from large ones that settle out of the atmosphere quickly through to small particles, which are referred to as fly ash. The types and amounts of the materials produced during burning, or combustion, depend on the heat of the process, the nature of the fuel and the amount of oxygen present.

When a fuel burns it may show either complete or incomplete combustion. Complete combustion occurs when all the carbon-based fuel is converted into carbon dioxide and water. The equation of the process is shown in Table 5.5.4 for a hydrocarbon found in petrol. Similar patterns of reaction can be found in coal burning.

In example A, the octane hydrocarbon is converted to carbon dioxide and water. Note the relatively large amounts of carbon dioxide produced. This does, however, generate the maximum amount of heat from the reaction. As the amount of available oxygen falls (reactions B and C), there is an increase in the amount of carbon and carbon monoxide formed. Carbon monoxide is highly poisonous and can react in the atmosphere to form carbon dioxide. When very little oxygen is available, hydrocarbons are left unburnt. These chemicals can form photochemical smog, which irritates eyes and the membranes lining the lungs.

Note that the oxygen preferentially combines with hydrogen rather than carbon. Coal and oil can be converted into coke, which is almost pure carbon and used as a fuel. This process produces one molecule of carbon dioxide for each atom of carbon in the coke, assuming complete combustion. It does not, however, produce a great deal of heat; only 32 kilojoules per gram of coke. An increase in the number of carbon atoms in a fuel molecule does not cause a corresponding increase in the energy produced by that molecule. (See Table 5.5.5, page 276.)

In addition to carbon dioxide, impurities in the fuel may produce gases that cause environmental problems. Sulfur is a contaminant of coal and petroleum. During combustion it forms sulfur dioxide, which is a major component of air pollution. In the air, the sulfur dioxide is converted to sulfuric acid by reacting first with oxygen in the air and then with water. Nitrogen in fuels is converted to nitrous oxide during combustion and then to nitrogen dioxide in the air. Nitrogen dioxide gives smog its brown colour.

Emissions of the various greenhouse gases differ between sectors. (See Figure 5.5.5.) Electrical energy production by fossil fuel burning produces more carbon dioxide than methane. However, agriculture (in particular, livestock) produces more methane than fossil fuel use does. Figure 5.5.5 shows that carbon dioxide is the largest contributor to greenhouse gas emissions in Australia.

The particles produced by the combustion of fossil fuels are formed by incomplete combustion or as impurities in the fuel. In coal, particles of sand and clay do not burn up in combustion and these particles form the larger particles that settle

Table 5.5.4 Examples of reactions in burning fuels

Complete combustion
A Adequate oxygen
Octane + oxygen → carbon dioxide + water
$2C_8H_{18}(l) + 25O_2(g) \rightarrow 16CO_2(g) + 18H_2O(g)$
B Less oxygen
Octane + oxygen → carbon + carbon monoxide + carbon dioxide + water
$2C_8H_{18}(l) + 18O_2(g) \rightarrow 4C(s) + 6CO(g) + 6CO_2(g) + 18H_2O(g)$
Incomplete combustion
C Even less oxygen
Octane + oxygen → carbon + carbon monoxide + water
$2C_8H_{18}(l) + 14O_2(g) \rightarrow 6C(s) + 10CO(g) + 18H_2O(g)$
D Very little oxygen
Octane + oxygen → unburnt hydrocarbons + carbon + water
$2C_8H_{18}(l) + 6O_2(g) \rightarrow 2C_2H_6(g) + 8C(s) + 12H_2O(g)$

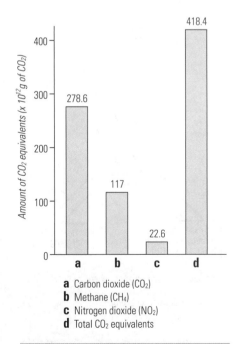

a Carbon dioxide (CO_2)
b Methane (CH_4)
c Nitrogen dioxide (NO_2)
d Total CO_2 equivalents

Figure 5.5.5 Greenhouse gas emissions for Australia, 1990.

Table 5.5.5 Heat generated for a standard amount of fuel

Fuel	Atoms/ molecule			Energy/ gram (kJ/g)	Molar mass (g/mol)	Heat of combustion (kJ/mol)
	Carbon	Hydrogen	Oxygen			
Straight chain alkanes						
• methane	1	4	0	55.6	16.0	889
• ethane	2	6	0	51.7	30.1	1 557
• propane	3	8	0	50.3	44.1	2 217
• butane	4	10	0	49.5	58.1	2 874
• octane	8	12	0	47.8	114.2	5 464
• hexadecane	16	34	0	37.8	282.5	10 687
Alcohols containing a single alcohol group (-OH)						
• methanol	1	4	1	22.7	32.0	725
• ethanol	2	6	1	29.6	46.1	1 364
• propanol	3	8	1	33.5	60.1	2 016
• butanol	4	10	1	36.1	74.1	2 677
Molecules containing a single double bond						
• ethene	2	4	0	50.1	28.1	1 409
• propene	3	6	0	48.8	42.1	2 056
• butene	4	8	0	48.4	56.1	2 715
• octene	8	16	0	47.3	112.2	5 305

out of the atmosphere relatively rapidly. Very small particles of carbon or fly ash may stay suspended in the air for some time until rain removes them or they clump together and settle out. Because of their size, small particles absorb heat and scatter light, which contributes to the haze visible in the air.

Carbon cycle sinks

When we burn a fossil fuel we are returning carbon that has been stored within the Earth back into the carbon cycle. The carbon locked up, or sequestered, in fossil fuels was once part of the marine or terrestrial life. These living systems, together with other parts of the carbon cycle, can be thought of as reservoirs, or sinks. A reservoir is a place where something can be found in large quantities. It can also be thought of as something that takes up something from another part of a material cycle. A sink can also act as a source for materials if processes remove materials from them.

Carbon sinks and sources take a variety of forms. Figure 5.5.6 is a simple model of the carbon cycle that shows four major parts of the carbon cycle: the atmosphere, the biosphere, the ocean and industry. We saw that, using the movement between the four parts, humans are adding some 6300 million tonnes of carbon to the atmosphere each year. However, to model the system more detail is needed. A better model, but not a complete one, is shown in Figure 5.5.1 (page 270). It shows that the deep ocean is the largest reservoir for carbon. It also shows that more carbon is removed from the deep and intermediate ocean to the surface than occurs in the other direction.

It is at first attractive to think of modifying parts of the carbon cycle to remove carbon from the atmosphere. Increasing the uptake of carbon dioxide by land or

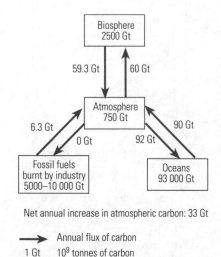

Net annual increase in atmospheric carbon: 33 Gt

→ Annual flux of carbon
1 Gt   10⁹ tonnes of carbon

Figure 5.5.6 A model of carbon movement in the carbon cycle.

marine vegetation would seem to be an advantage. Recently, scientists have carried out experiments to see the effect of seeding the ocean: adding nutrients to speed up the uptake of carbon by plankton. The problem is that we still have some way to go in understanding how feedback systems work in, and between, different reservoirs. To alter one part of the carbon cycle may lead, in time, to unexpected adjustments in other parts. Consider for a moment what would happen if cold, upwelling water no longer reached the surface of the oceans. Less carbon dioxide would enter the atmosphere, but the nutrients used by plankton near the surface would decrease. Less plankton, and changes to surface currents, would create new impacts affecting other parts of the system.

REVIEW ACTIVITIES

1
Describe the types of products produced by burning fossil fuels.

2
Analyse the effects caused by gases, particulates and water vapour from fossil fuels.

3
Summarise the areas in which we need fossil fuels in our society.

EXTENSION ACTIVITIES

4
Summarise the relationship between fuel composition and heat produced shown in Table 5.5.5. Explain why these fuels are described as clean fuels.

5
Summarise the information shown in Figure 5.5.5 (page 275).

6
Lung disease from cooking fires affects many people in the less developed world who use wood or dung for heating. What combustion products from the fire would most likely cause the disease and what processing of the fuel could reduce the problem?

SUMMARY

The burning of fossil fuels can produce a range of materials, including particulates; oxides of carbon, nitrogen and sulfur; water; and fly ash.

Incomplete combustion occurs when a fuel is not burnt to carbon dioxide and water. The products of such combustion include particulates, carbon compounds and carbon monoxide.

Greenhouse gases alter the temperature of the climate and are produced by the combustion of fossil fuels. The gases include carbon dioxide, water, methane and nitrous oxide.

The effect of a greenhouse gas on the climate depends on the ability of the gas to cause climate forcing, its residence time in the atmosphere and the amount of gas in the atmosphere.

Climate change due to greenhouse gases is predicted by models, which suggest that there will be costs and benefits if the climate warms.

Our society has a high dependence on fossil fuels. While some success has been made in limiting the increase in combustion emissions, their level of production does appear to be affecting our climate.

The products of combustion

In this chapter you have learnt that combustion produces gases, including carbon dioxide and water, as well as particles. In this experiment your objectives are to design an experimental method that allows you to measure the products of complete and incomplete combustion.

Designing the experiment

Consider the following aspects of the experiment and design a procedure to achieve the objectives listed above.

A POSSIBLE PROCEDURE

Here is a possible way to carry out the experiment. If a small piece of coal, or other fuel, is ignited in a deflagrating spoon and placed in a gas jar, it will burn until it runs out of oxygen or it is put out. If the spoon is removed carefully when it is still burning well, the products of full combustion should be left behind. Perhaps leaving the spoon in the jar for varying times will produce different products. Think about the variables you need to control as well as the materials you need. Perhaps you need to trial your method in order to check that it works well.

SAFETY AND CHOOSING A FUEL

For safety reasons some fuels, such as petrol, will not be acceptable for this experiment. Such fuels have a low flash point (the temperature at which a fuel will produce explosive mixtures with the air). Coal, wood or ethanol-soaked cotton wool are suitable materials to use. Ensure that only small amounts of fuel are used and store the bulk of the material well away from flames and where you are working. Consider other safety precautions and check this aspect of your method with your teacher before proceeding.

MEASURING THE PRODUCTS

Water vapour can be identified when it condenses on the side of a container. Think about how the amount of water may be measured.

Particles of carbon will vary in size and number. How could they be collected and examined? If collected on a sticky surface (such as petroleum jelly smeared on a glass slide), perhaps they can be seen using a microscope.

Carbon dioxide can be measured using limewater or by looking for a pH change when it dissolves in water. Decide on the best method to use by discussing this with your teacher and using reference books.

Fuels produce heat. How could the heat produced in this experiment be measured?

Write up a procedure for testing for these products.

Conducting the experiment

When you are confident of your method and have discussed it with your teacher you can proceed to gather data. Remember the importance of planning how you will record your data and make sure you record observations, including measurements, as you go. If time allows, consider whether repeating parts of the procedure will lead to better quality results. If you are working with others, consider making individuals responsible for particular measurements.

Analysing and presenting the results

Your teacher may ask you to write up the whole task formally. In any case, you need to compare the results you obtained and try to identify relevant trends and differences.

Write a discussion of what your results mean and how your learning may lead you to follow a different procedure if you repeated the task.

5.6 The search for alternative sources of energy

OUTCOMES

At the end of this chapter you should be able to:

- identify and discuss alternative sources of energy (solar, wind, hydro-electric, nuclear, synthetic oil, ethanol, wave) and evaluate the relative importance of each as an alternative energy source for the local community now and in the future

- describe and evaluate methods of conserving energy, including architectural design.

ALTERNATIVE ENERGY SOURCES

There are many alternatives to the use of fossil fuels. The use of nuclear energy and hydro-electric systems for generating electricity accounts for 12.6% of world power production. In addition to these two sources we will look at five others: wind, solar, synthetic oil, ethanol and wave energy. For each, we will examine how and where it works, its costs and benefits and its potential future.

Wind power

Wind power has been used as a means of propulsion for a very long time, but its use in generating electricity only dates back 100 years. Today, wind power accounts for 0.04% of global energy production. Since 1980 the production of wind generators has increased by 5% every year.

Wind is caused by the Sun. Unequal heating of the Earth's surface by the Sun causes convection and the movement of air. As hot air rises it is replaced by cooler, denser air, which flows across the Earth's surface to take its place. The winds produced in this way are not constant in their strength or direction. They do not blow all the time.

It is the variability of the wind that is the greatest challenge to the more widespread use of wind power. Not all areas are suitable for harnessing the wind, but in some areas there are places where strong winds occur regularly. Wind turbines are machines that convert kinetic energy caused by the wind into electrical energy. Before a wind turbine is built, a long period of measurements is usually undertaken to understand exactly how much energy is to be converted in an average year.

Many suitable places are found along the coast, but inland areas away from the coast may be suitable too. Figure 5.6.1 (page 280) shows some current sites of wind generation in Australia. Note, in particular, the large wind farms in New South Wales. Crookwell and Blayney, when complete, will together produce 14.8 MW of electrical energy.

Wind is converted into electricity by wind turbines. (See Figure 5.6.2, page 280.) As the wind spins the blades of the turbine, a generator transforms the kinetic energy

Energy units and terms

The amount of energy in something is described in terms of joules. One joule is a small amount of energy. For example, it takes more than 4 joules (4 J) of energy to raise the temperature of 1 mL of water by 1 °C. The energy content of something, such as food, is usually expressed in kilojoules (kJ) or megajoules (MJ), that is, thousands and millions of joules, respectively.

In most practical situations it is not the amount of energy that is important but the rate at which energy is transformed. The term used to describe this rate is power. Power is measured in terms of watts, a watt being a conversion of 1 J/s. A light bulb, for example, rated at 100 watts (100 W) converts 100 J of electrical energy into light and heat every second.

Just as joules are described in multiples of thousands or millions, so is work. A kilowatt (kW) is 1000 W, a megawatt (MW) is 1 million W and a gigawatt (GW) is 1000 million W. To describe the amount of energy used, the power is multiplied by the time over which the work occurs. Using seconds as the basic measurement of time is difficult in everyday applications, so hours are used. When you see a measurement in kilowatt hours (kW.h) it is describing the amount of energy used.

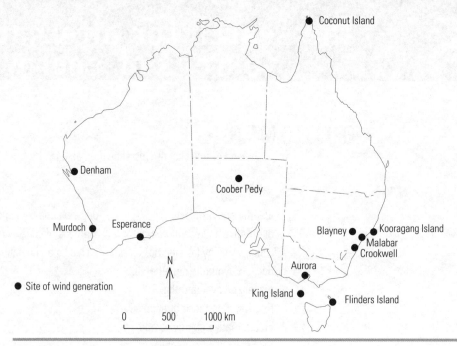

Figure 5.6.1 Current sites of large-scale wind generation in Australia.

of the spinning shaft into electrical energy. A number of mechanisms are present to protect the generator from damage in strong winds. Brakes and ways of changing the angle of the turbine blades ensure that the structure is not stressed and that the electrical system is not overloaded. The height of the tower is also important. Wind turbines are usually placed 12–30 m above the ground so that the turbine uses strong winds rather than the slower, gusty winds found at the surface.

Wind turbines come in a range of sizes. Some are suitable for supplying the energy needs of individual houses, but others are large enough to produce electricity for commercial purposes. At present, large turbines that are capable of generating hundreds of kilowatts are imported from overseas; although Australia does manufacture smaller units.

A number of criticisms are made of wind power. Wind turbines are noisy and they need to be placed in the open so they can capture as much wind as possible. Some people object to their presence in the scenery and see them as ugly. It may be possible to place wind farms on the ocean. They would probably be more efficient there, but the problem then arises as to how to transmit the power to users.

Some people argue that the number of birds that are killed by wind turbines is unacceptable. In Denmark, where wind power supplies 9% of the country's electrical generation, making it the world leader in wind power, 30 000 birds die each year as a result of collisions with the windmills. While this sounds a lot, it may be trivial when compared with the number killed by cats, planes and cars in the same period of time.

The cost of manufacturing a wind turbine is sometimes used to argue that they are uneconomic. The energy used in metal mining, processing and fabrication is an issue, but for large turbines the energy is recovered within several months of operation. For servicing remote locations, wind turbines can be located close to the places that use the power. The cost savings in transmission offset the cost of production.

Solar power

Each day the solar energy falling on the Earth is about 7000 times our global energy consumption and in Australia we receive 20 000 times the energy that we currently

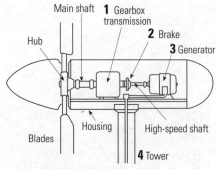

1 Gearbox speeds up rotation of shaft

2 Brake prevents damage in high winds

3 Generator converts kinetic energy to electrical energy

4 Tower elevates generator above slower ground winds

Figure 5.6.2 Structure of a wind turbine.

use. With our current **photovoltaic** technologies the world's energy needs could, theoretically, be serviced by **solar cells** covering only 0.15% of the Earth's surface. However, transmitting the energy from one place would be impractical.

The amount of solar energy received by different parts of the Earth varies. Less solar radiation is received at higher latitudes than near the equator. Clouds reduce the amount of solar energy reaching the surface, and in equatorial and mid-latitude areas there are frequently clouds.

Within Australia not all sites are equally good for solar power. Different amounts of light fall in different areas and the quality of the light also varies. For example, a north-facing slope will receive more light than a south-facing slope. The power produced is proportional to the intensity of the light falling on a solar cell and inland areas, where clouds are rare, are more productive than some coastal places. For example, in Tasmania an average of 111 J of energy falls per square centimetre, but in Port Hedland the average is over 181 J per square centimetre. The amount of light also varies with the seasons. In Sydney there are a little fewer than 200 hours of sunshine during July, but there are 250 hours during January.

TYPES OF SOLAR TECHNOLOGIES

There are three different technologies that use radiation from the Sun to generate heat or electricity. Each has a place as an alternative source of energy.

Thermal solar power involves the direct utilisation of heat from the Sun. Heating water uses 26% of the average energy use in our homes, and a further 42% of our energy use goes towards heating our homes. Thermal solar systems work by absorbing radiation and then preventing the heat from being re-radiated. The change in the temperature of the absorbing surface can be used to heat water, cook, warm an area or dry materials. In a domestic solar hot-water system, water is passed through tubing or between sheets in an enclosed box with a transparent surface. (See Figure 5.6.3.)

Thermal electric methods use structures to focus light from the Sun into a small area and generate high temperatures. The high temperatures can be used for cooking or for generating heat. An example of such a system is the parabolic dish concentrating system. This system uses dish-like mirrors with a parabolic shape to focus solar radiation onto a receiver that is positioned at the focal point of the dish. (See Figure 5.6.4.) The receiver contains water, which is heated to very high

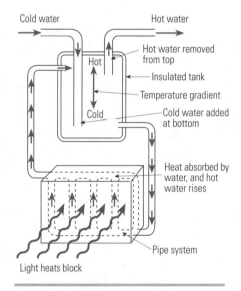

Figure 5.6.3 Structure of a solar hot-water system.

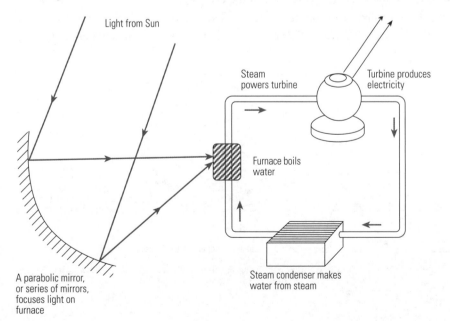

Figure 5.6.4 Basic design of a thermal electric system.

temperatures and then used to generate electricity in a small engine attached to the receiver. Parabolic dish systems are the most efficient of all solar technologies. They generate temperatures in the receiver of up to 750 °C and efficiencies approaching 25%. This compares quite favourably with other solar thermal technologies, which approach 20% **efficiency**.

Large-scale systems that generate commercial electricity involve the use of power towers. These are thermal electric systems that use banks of mirrors to focus light onto a receiver mounted in a tower. The Stanwell Solar Power Station in Queensland has a 5 MW solar thermal power plant and was fully operational by the end of 2001. The power plant uses a compact reflector system to produce steam, which is then used to operate conventional steam turbines.

Photovoltaic cells convert light energy directly into electricity. Each cell consists of thin layers of silica (quartz) that have been treated in a special way. It can be thought of as being like a sandwich in which the bread represents layers of silica and the filling represents a special junction. Light striking the junction causes electrons to move to one side of the junction and positive 'holes' to form on the other. An electric field generated in the junction prevents the positive and negative charges combining, but by connecting an electrical circuit to the two sides of the 'sandwich' a current flows through the circuit doing the work. (See Figure 5.6.5.)

The most efficient photovoltaic cells are cut from single crystals of silica. These are known as single crystal wafers. The technology is expensive because large crystals of silica are grown at temperatures near 1400 °C and are then cut into wafers. Less expensive technologies are polycrystalline cells and amorphous silica films. Polycrystalline cells are composed of silica that is cast in a mould and then cut into wafers. The wafers contain many intergrown crystals. Depositing thin films of silica on a glass or plastic surface produces amorphous silica cells. Deposition may be from a gas and it is easier and cheaper to process such cells than the other technologies.

The size of the cell does not affect the cell's voltage but it does affect the current, or flow of electrons, that is produced: the larger the cell, the larger the current. Single crystal wafers can reach efficiencies of 25% and are limited to this performance because some wavelengths of light cannot cause the charges in the cell to separate. Polycrystalline cells and amorphous silica cells are less efficient (20% and 10%, respectively) due to energy losses within the cell.

In the last thirty years, photovoltaic cells have become three times more efficient and fifty times less expensive. This is partly due to the development of new and better

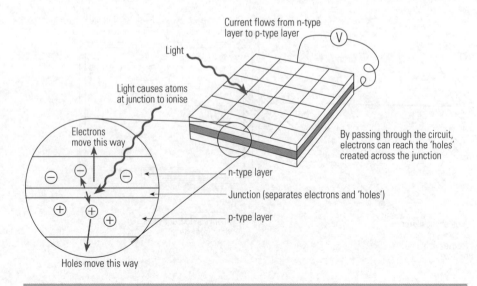

Figure 5.6.5 How a photovoltaic cell works.

technologies and it has been achieved by a growing demand fuelling the development. Photovoltaic cells have a number of advantages. For example, they can provide:

- convenient, silent and flexible forms of power in houses and products such as calculators and watches
- cost-effective, low-maintenance, long-lasting power suitable for remote locations
- non-polluting, renewable and sustainable power that does not generate greenhouse gases other than during its fabrication.

Worldwide, the solar industry is worth $500 million and its production currently doubles every three years. Australia has two small plants that produce photovoltaic cells capable of producing 10 MW of energy per year. Demand is growing at 30% per year and the technology is already cost-effective in remote areas where the cost of transmitting electricity outweighs the difference in generation costs. Worldwide, demand continues to reduce cost; current global predictions see the technology being competitive with fossil fuels within ten to thirty years. The cost of systems to store and transmit electricity from the panels is not declining as fast as the cost of the panels, however, and perhaps there is room for better efficiencies in this area.

Despite being a non-polluting and quiet form of electricity, solar power systems do have critics. One concern is the storage of energy. Photovoltaic cells only work during the day, and battery banks are needed to store the electricity. Such banks need to be stored properly and pose some risks. The panels may be damaged by hail and storms, but this disadvantage is similar to the damage to powerlines that affects people during such events.

Problems with shading of the ground and heating effects around the solar panels can produce minor ecological changes and some critics do not like the look of the panels or the use of land for panel arrays when it may be used for other purposes. The industry that makes the photovoltaic cells produces toxic wastes, but the semiconductor industry works with a set of standards to minimise pollution from manufacturing plants. Some people also see the variability in supply as a negative.

Synthetic oils

Synthetic oils are like other synthetic fuels in that they are produced from organic compounds. Previously in this section you read about the production of syngas from coal. The problem with many synthetic fuels is that they are produced from non-renewable resources, produce greenhouse gases and involve the use of energy in their manufacture.

The production of synthetic oil is an example of a biomass conversion technology. Biomass is plant or other organic material. It may be a waste product, or a by-product. It may be material that has been specially grown as feed stock for the process. Examples of biomass feed stocks are agricultural residues (such as rice husks or green vegetable matter), forestry by-products and animal wastes.

The biomass is placed in a chamber without oxygen and heated. The process is called **pyloris** and resembles the natural processes that produce petroleum. The heat and pressure cause the organic material to form three basic substances: an oil-like liquid, a solid carbon-rich residue and a hydrocarbon-rich gas. The carbon residue can be used as a fuel, as can the gas. The oil, or 'bio-oil', can be refined in the same way as natural petroleum to generate a series of products that can be used as fuels, lubricants or feed stocks.

The advantage of pyloris as a process is that it uses a sustainable raw material, if it is managed properly. It is important to remember that the production of the biomass involves water, soil and nutrients and sustainability involves managing these valuable resources well. Like fossil fuels, the combustion of the products produces greenhouse gases.

definition

pyloris
a process in which biomass is heated in the absence of oxygen

Ethanol

Ethanol is an alcohol composed of molecules containing two carbon atoms, six hydrogen atoms and an oxygen atom. It burns cleanly to produce carbon dioxide and water and can be produced relatively cheaply using the process of **fermentation**.

Fermentation occurs when a mixture of sugars, yeast and water is kept in a tank and oxygen is excluded. The sugars are broken down into ethanol, carbon dioxide gas and a liquid containing ethanol, water and other impurities. At this stage the alcohol content of the fermented product is between 10 and 15% alcohol. If wood pulp is used as the source of the sugars, a combustible pulp is a by-product of the process.

The feed stock for ethanol production can be a range of materials. Materials rich in sugar, such as sugar cane and sugar beet, produce relatively large amounts of alcohol. Starch-rich materials (such as maize, wheat and potatoes) are cheaper than sugar crops but require processing before fermentation. A third source of raw material is the lignin and cellulose making up wood. Sugars are released by treating the wood pulp with acids. Softwoods treated with strong acids produce the highest amounts of alcohol among the wood-derived feed stocks, but there is less alcohol produced than from sugar cane.

The product of fermentation is processed to produce pure ethanol in a number of steps. First, the ethanol is distilled to remove the solid wastes. Next, the alcohol is distilled again to remove the water. Chemicals may be added at this stage to help the efficiency of the distillation. At the end of the process the ethanol is nearly pure. A single-pass distillation can achieve a 95% by volume result, but the water present will affect the energy derived from burning the fuel.

The ethanol has the advantage of being a suitable fuel for cars. Mixed with some petroleum-derived petrol, it reduces the amount of fossil fuel needed. Volume for volume, ethanol produces about two-thirds the energy produced by petroleum-derived petrol. By blending alcohol and petrol we reduce the amount of petroleum needed for one of the largest areas in which it is consumed: transport.

It is sometimes argued that the greenhouse gases produced by ethanol can be thought of as being recycled. The carbon was removed from the air into the plant first, and burning the ethanol releases the carbon back into the atmosphere to be recycled again. This argument assumes that only the carbon locked in the plant material is released in the process, but energy is needed for transport, growing the crop and distilling the fuel. It has only been in recent years that scientists have found how to make ethanol so that it produces more energy than is used in its production.

Ethanol has the disadvantage that it is not as efficient as petrol in producing energy. A properly prepared, otherwise conventional car using only ethanol can achieve 80% of the distance that a similar amount of petrol produces. The raw material displaces other crops, and the need to conserve soil, water and fertilisers are also issues to be considered.

Wave and tide energy

WAVE POWER

The ultimate source of much of the energy in the ocean is the Sun. In addition, gravitational energy from the Moon and the Sun provide energy in the form of tides. By capturing some of the energy in moving water it is possible to generate electricity without generating greenhouse gases.

Wave power systems involve either fixed or floating devices. An example of a fixed device is shown in Figure 5.6.6. Each type of device has advantages and disadvantages. Fixed devices are often easier to service but the environments where they are most efficient are rare.

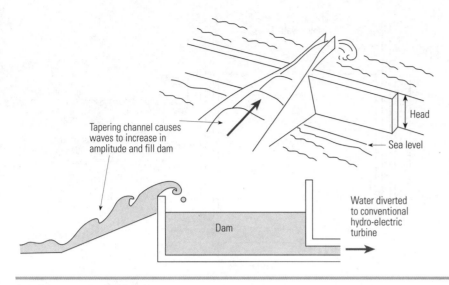

Figure 5.6.6 A tapered channel (TAPCHAN) wave energy device.

The disadvantage of fixed systems is that they will only work well in specific conditions. Deep water, regular waves moving in a consistent direction and relatively small tidal changes in water levels are needed for best results. Because these conditions are only found in relatively rare sites there are limitations on such systems. The building of these systems will also alter the water movement near the instillation and may alter the ecology of the area.

The major disadvantages of floating devices are difficulties in servicing them and in transmitting the energy they generate. Like the fixed systems, floating systems will also benefit from environments where waves of adequate energy are frequent. Both systems may also be damaged by infrequent storm events unless they are adequately engineered.

TIDAL POWER

As the Earth spins, its oceans are affected by the gravitational pull of the Moon and the Sun. On average, twice a day the sea level at a spot on the coast rises and falls. By capturing the energy of the water as it flows, or by using the gravitational potential energy of high-tide levels, renewable energy can be generated.

A simple way of generating electricity from waves is by using a **barrage** across a bay or estuary. A barrage is like a dam but has ways of allowing water to pass in both directions. (See Figure 5.6.7, page 286.) A series of barrages is used along the Murray River to regulate water heights. As the tide rises, the barrage is opened, allowing water to flow behind the wall. At high tide the barrage is closed, allowing the water to escape through an exit, which contains a turbine.

A recent development in using tidal energy to generate power is the tidal turbine, which resembles in many ways a wind turbine. The turbine is mounted on a pole attached to the sea floor. As tidal currents flow past the turbine, the blades of the turbine spin and electricity is generated. A vane ensures the blades are directed into the tidal flow. The system will work with both incoming and outgoing tides.

While tidal systems conserve fossil fuels and produce no greenhouse gases, they do affect the environments in which they are built. Fixed systems, such as barrages, in estuaries will alter sediment movement and the salinity distribution in the estuary. This will affect some parts of the marine ecosystem. Tidal turbines produce fewer impacts than barrages. A tidal power system is currently being considered for Derby in Western Australia. The area has a **tidal range** of 15 m and the area has the potential to generate as much as 300 MW of power.

barrage
an artificial barrier placed in a river to control water levels

tidal range
the distance between low tide height and high tide height

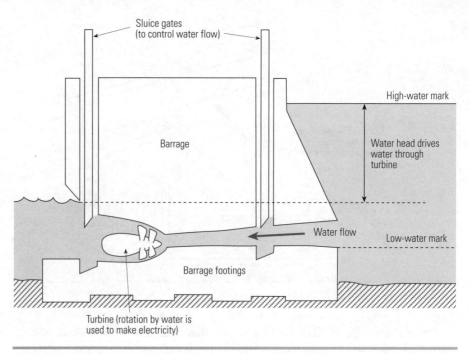

Figure 5.6.7 A barrage used to generate electricity.

Hydro-electric power

Hydro-electric power is electricity produced by the movement of water. The water is found in dams or rivers and the energy that is converted into electricity is the kinetic energy of the water as it flows from a high point to a lower point. Hydro-electric power stations in Australia range in size from the massive units making up the Snowy Mountains Hydro-Electric Scheme, which generates 3800 MW of power, to small units used in creeks on farms, which may generate tens of kilowatts.

The ideal conditions for a hydro-electric system are a deep dam or steep-sided valley and a reliable flow of water. Dams provide both the **head** and volume of water required for power generation.

The greater the head of water, the greater is the pressure entering the turbine and the greater is the power that can be generated. The amount of electricity produced also depends on the volume of water that flows through the water turbine. The greater the quantity of water that flows through the turbine, the greater is the amount of power that can be generated.

Converting kinetic energy into electrical energy generates the electricity. Water flows through a water turbine and spins a turbine shaft by pushing on turbine blades. As the shaft spins, it powers an electrical generator, which turns the kinetic energy into electrical energy. Figure 5.6.8 shows the way such a system works. Different types of turbines are used depending on the situation. In some places around the world, the water is stored below the turbine and later pumped back up to the dam above the turbine. The energy used to do this comes from coal or nuclear power stations and occurs when demand for the electricity is low. This procedure allows water reuse and provides power at peak times, that is, times of high demand.

Worldwide, hydro-electric power provides about 6.6% of energy production. In sixty-six countries such power generation supplies 50% or more of each country's needs. It is thought that at present only 32% of economically feasible hydro-electric power generation has been developed. Currently Asia is the area with the fastest growing hydro-electrical industry.

In Australia our hydro-electric capacity is currently 7.6 GW. There are thirty operating hydro-electric facilities in Australia. The majority of these are in South-eastern Australia, with the seven facilities making up the Snowy Mountains Hydro-

head

the distance from the top of a water storage to the point where the water is used to make electricity

definition

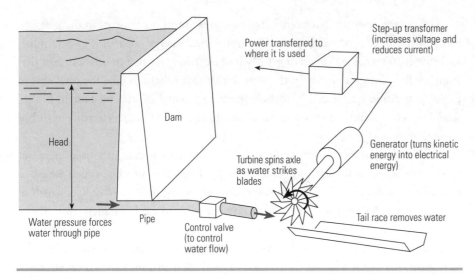

Figure 5.6.8 How a hydro-electric system works.

electric Scheme producing 50% of Australia's capacity. Next in terms of size is the contribution of the Hydro-electric Corporation of Tasmania, which generates 30% of the national capacity.

While hydro-electric generation does not generate greenhouse gases or other forms of air pollution, it does have an environmental impact. The flooding of large areas destroys or displaces a range or organisms. Changes to the flow of rivers affects areas both upstream and downstream of the dam. Increased erosion and thermal pollution by the cold bottom waters from the dam are two examples of the negative impacts of such systems. Large water bodies, such as dams, also alter local microclimates by modifying local temperatures and humidity.

In New South Wales the Snowy Mountains Hydro-electric Scheme has contributed a great deal of electrical power, but this has been at the expense of the health of some river systems. Reduced flow into rivers such as the Snowy River has changed the rates and character of sedimentation. Dams reduce the frequencies of floods and affect ecosystems that rely on such events. Reduced river flows mean there is less water to dilute salt influxes and so river salinity and nutrient levels may be higher than they otherwise would.

There are potential advantages to having some hydro-electric capacity to meet our needs. One advantage of hydro-electric systems over other alternative energy sources, such as solar power, is that it can provide electricity on demand. The energy capacity of the hydro-electric system is stored in the water and does not have to be stored as electricity. During peak times, a hydro-electric system can move rapidly to provide extra power. Another advantage of small-scale hydro-electric systems, so called micro-hydro power, is that it can do the jobs currently done by diesel generators in rural and remote areas. This is of potential benefit to developing countries in Asia where water is plentiful. The availability of suitable conditions in Australia limits the opportunities to use such technologies.

Nuclear power

Nuclear energy is a source of power that has generated a great deal of debate since the first nuclear power plant went into service in 1957. There are now some 400 nuclear power plants around the world and they supply approximately 6% of the world's current energy needs. In countries that use nuclear power, the proportion of energy supplied may be as much as 20% of national supplies. Recent decreases in fossil fuel costs have made nuclear power less attractive in terms of cost, but it is a technology that has some real advantages over fossil fuels.

The nuclear industry has been heavily subsidised by governments in the past and so the true cost of producing electricity from nuclear energy is difficult to determine. In 1999 the total cost of generating 1 kW.h of electricity seemed to be about half that using coal. A European study in the early 1990s that sought to take the total cost of power generation into account, including environmental effects and human health, found the cost of nuclear power to be similar to the cost of hydro-electric power but more than that of wind power.

Two nuclear processes generate energy. The one that is currently used in power generation is called **fission**. (See Figure 5.6.9.) The other process, called **fusion**, is regarded as better than fission. However, despite tens of billions of dollars worth of research, we still appear some way off from seeing fusion used commercially.

Nuclear fission is the process in which a radioactive atom of uranium is split apart. In the process, some of the mass in the atom is converted to energy. It is the heat generated by such events that is used to produce the steam that powers electrical turbines. Not all atoms of uranium are suitable for fission. The isotope of uranium with a mass of 235 (uranium 235) is the one needed for fission. One gram of this material will generate the heat equivalent of 3 tonnes of coal.

Nuclear fusion involves the combining of atoms to form new ones. An atom of deuterium (hydrogen with a molecular mass of 2) is fused together with an atom of tritium (hydrogen with a molecular mass of 3) to form a helium atom. In the process, energy is produced. This process does not produce radioactive wastes and the hydrogen fuel is obtained from sea water. No greenhouse gases are produced and the fuel will last for a very long time. Although the overall cost of developing the technology is very large, the actual running cost of a fusion plant may be quite low.

Uranium 235 only makes up about 0.7% of the uranium ore that is mined and so the uranium has to be enriched before it can be used. While Australia is a major source of uranium, world reserves of uranium 235 will last at current rates for about 100 years. A nuclear technology called a breeder reactor allows scientists to convert the more common uranium isotope, uranium 238, into an element called plutonium, which can be used as a fuel. A number of problems are associated with plutonium. It can be used to build nuclear weapons and it is one of the most toxic materials we know of.

As a source of energy, uranium is clean in the sense that the operating plant does not produce greenhouse gases or other pollutants, such as sulfur or nitrogen oxides. The technology of a nuclear power plant is, however, very sophisticated and the cost of building such power plants is substantial. The process of building a nuclear power plant generates a substantial amount of carbon dioxide, as does the building of coal-fired power stations.

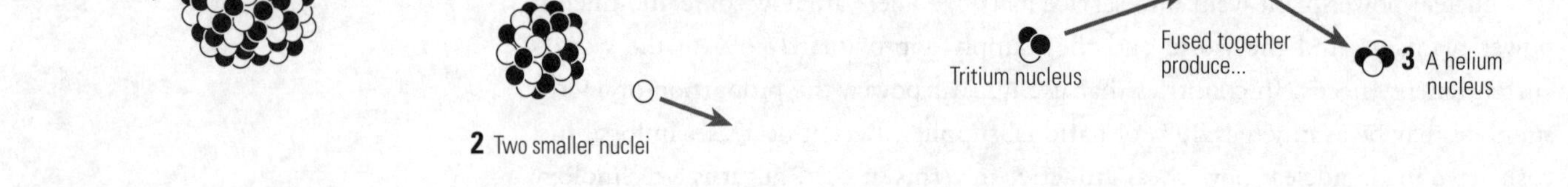

Fission

Fusion

Uranium nucleus

Fast-moving neutron   Leads to...

1 Energy (used to run turbines)

3 More neutrons

2 Two smaller nuclei

Deuterium nucleus

Tritium nucleus

Fused together produce...

1 A neutron

2 Energy

3 A helium nucleus

Figure 5.6.9 Methods of generating nuclear power.

A significant problem with nuclear power is that it generates radioactive wastes that must be stored and managed for millions of years. In addition to the spent fuel, materials in a nuclear reactor, both fusion and fission, become radioactive. The dismantling, or decommissioning, of a nuclear plant is therefore expensive in terms of disposing of radioactive components.

Public perceptions of nuclear power are generally poor. Major accidents, such as Chernobyl and Three Mile Island, have been caused by human error and such errors are hard to avoid. Potential health risks from radioactivity concern many people, but the health risks from coalmining still affect miners. Indeed, the mining of uranium presents the same environmental risks as other forms of mining. These include contaminated ground water, erosion and sediment contamination of waterways. Radioactive tailings, or mine wastes, do present greater problems than coal tailings if they are allowed to contaminate the environment.

REVIEW ACTIVITIES

1
Outline how wind power can be used to generate electricity.

2
Discuss the advantages and disadvantages of wind power.

3
Outline the ways by which solar energy can be used to provide for our energy needs.

4
Summarise how a photovoltaic cell works.

5
Describe the advantages and disadvantages of solar power.

6
Outline how biomass can be used to supply needs currently provided for by fossil fuels.

7
Compare ethanol and synthetic oil as petroleum substitutes.

8
Describe the advantages and disadvantages of wave and tidal power.

9
Hydro-electric power and nuclear power both provide some of the world's energy needs. For each energy source, outline the costs and benefits of the technology.

EXTENSION ACTIVITIES

10
Evaluate the alternative forms of energy that may be used where you live.

11
List the alternative energy sources covered in the text from most beneficial to least beneficial. For each, outline why it will not totally replace fossil fuels.

CONSERVING ENERGY

While alternative forms of energy generation will no doubt help reduce society's dependence on fossil fuels, there are things that can be done now to reduce our consumption of fossil fuel resources. More than three-quarters of Australia's energy consumption comes from three sectors: electricity generation, manufacturing and transport. Of these, transport and electricity account for 54% of energy consumption. If you recognise that electricity generation and transport exist as part of other sectors (such as mining, agriculture and residential), then you will see that electricity generation and transport are areas where we may achieve real changes if we reduce our consumption.

Homes: Energy-efficient architecture

Our homes are major consumers of energy. If we are to conserve fossil fuels, reducing our energy use in the home is a good place to start. Remember too that energy is not only used in processes such as lighting and heating but in the manufacture of building materials. The two activities that consume the greatest amount of energy in homes are heating space and heating hot water. We have already seen how a thermal hot-water system can supply many homes' hot water needs. Supplementary gas heating and solar heating are preferable to electrical heating on demand.

Energy-efficient buildings are the exception rather than the rule. But they do not have to be. While building a house allows the planning and use of good design principles, modifications can be made to existing buildings to improve their function. Insulating the roof, walls and floor spaces can significantly reduce heating and cooling costs. Shading by pergolas or trees can also reduce the heating of a house during summer.

Efficiency through a house's design can occur in a variety of ways. Figure 5.6.10 shows some of the features that can improve the energy efficiency of a house. **Insulation** and preventing air leaks are key features of an efficient design. The size, structure and placement of windows can play a role in both these areas. In cold conditions 10–25 % of a room's heat may be lost through them. In summer, light can enter a room through a window and remain trapped as heat. Today, window designs may involve multiple panes of glass, gas filling and special coatings that block infrared radiation.

Grouping living areas within a house prevents heat loss or gain. It also reduces heat loss from centrally located hot-water systems. By placing living areas on the north side of a house it is possible to use the Sun's heat for warmth in winter and keep sleeping areas cool and insulated during summer.

Passive heating and cooling can be achieved by using features with large **thermal masses**. Concrete slabs, earthen or brick walls, and water tanks can all absorb considerable amounts of heat and then re-radiate them when the air around them is cooler. During summer, a wall with a large thermal mass will absorb heat from the air and reduce the air temperature. In winter, the same wall can absorb heat from winter sunlight and re-radiate it at night, when the temperature of the air drops.

insulation
material that prevents heat exchange between two areas

thermal mass
the amount of heat that can be stored by a material

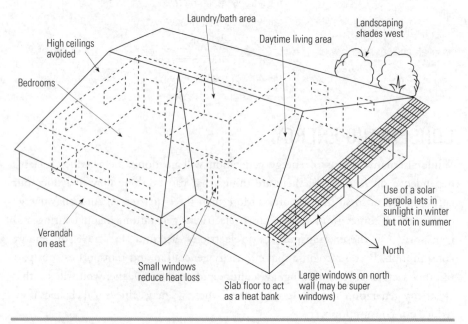

Figure 5.6.10 Features of an energy-efficient house.

REVIEW
ACTIVITIES

1

Draw a diagram to show the features of an energy-efficient room in a house.

2

Make a list of the features of an energy-efficient house.

3

Summarise the ways in which our society may conserve energy.

EXTENSION
ACTIVITIES

4

Make a list of the ways in which a person can increase the energy efficiency of their home.

5

Explain why a refrigerator adds to the heat in a kitchen. Try to think of a way that the heat may be removed from the room. Be energy efficient in your answer, if possible.

6

In 1999, 17% of people aged from fifteen to nineteen in New South Wales who were not in school were unemployed. Assess how high rates of employment may be used to conserve energy in our society.

SUMMARY

Greater efficiency, the development of alternatives and the increased efficiency of recovery are ways of conserving fossil fuels.

New technologies have improved on old ways of producing energy. These include hydro-electric and nuclear technologies and those that utilise energy from the Sun, wind, synthetic oil, ethanol and waves.

The success of an alternative energy technology depends on local conditions.

All technologies affect the environments in which they are built.

Changes in architectural design can lead to significant energy efficiencies.

A first-hand investigation of energy efficiency

The aim of this exercise is to perform a first-hand investigation to test the energy efficiency of various substances. In doing so, a number of problems will be presented and you will have to solve them in order to complete the task.

In this chapter you learnt about the importance of insulation in building energy-efficient homes. In this experiment you will test the insulation properties of four types of material. Before you begin, read the following outline carefully. There are many ways in which you may improve on the procedure. If you wish to do so, check with your teacher.

Caution: Be careful not to burn yourself, or others, with the hot water.

Materials

- Four similar aluminium soft drink cans
- Emery paper or sandpaper
- A sheet of white cardboard
- Enough natural fibre, such as wool or cotton, to wrap around a can
- A sheet of styrofoam
- A sheet of bubble wrap plastic
- A thermometer or data logger with a temperature probe
- A clock or stopwatch
- Hot water to fill the cans (hot water from a tap will do)
- A cardboard box with a lid, or capable of being closed

Procedure

1

Solve this problem: What sort of surface will reduce the can's ability to absorb heat from its environment? When you are sure of the answer to this problem, prepare the aluminium cans.

2

Each can is to be insulated with a different material. The depth of the materials is to be the same. The four materials are to be:

- air
- bubble wrap
- a natural fibre
- no insulation.

Prepare each can to meet the requirements listed above.

3

Prepare the box to hold the cans, using the styrofoam as a base to the box, and then discuss answers to the following questions with others in your class:

a Why is the box being used?
b Why is the styrofoam used?
c Does the size of the box matter?

Record your conclusions.

4

Use the same body of hot water to carefully fill each can. Transfer the cans to the box, and measure the temperature of each can. Measure the temperature of the air in the box, and then close it.

5

Leave the box for an hour and then measure the temperature of each can.

Results

Express your results in an appropriate form, or forms. A table of results is essential.

Decide whether it is worth pooling your data with other groups in the class. Before you do, consider what is different between the groups' methods and how the differences may affect your results.

Discussion and conclusions

Assess your results. Determine which is the best insulation material. If possible, make a hypothesis as to why some materials are better insulators than others.

Outline the nature of the problems you solved and the problems that arose during the experiment. Outline how the experiment may be improved if it were conducted again.

Mining and the Australian environment

6

Australia contains a great wealth of mineral deposits and the development of these mineral occurrences has helped shape our society and the environment we inhabit. The discovery of gold in 1851 and more recent discoveries (such as iron ore in Western Australia, porphyry copper deposits in central New South Wales and uranium in South Australia and the Northern Territory) have helped to shape our society through immigration, wealth generation and politics.

Australia's mineral resources continue to be an important part of our economy. The mining industry generates $6 billion a year and earns about $4 billion a year in exports. In recent years, exports of our mining technology and expertise have added another $1 billion to our national income.

Mining provides both benefits and problems. Mining operations are found throughout New South Wales and most of the 18 000 people directly employed by the mining industry live in rural and regional parts of the state. Other jobs generated by the industry support an additional 50 000 jobs. However, past ignorance and errors in terms of mining have led to environmental damage. While our ability to minimise harm to the environment has increased enormously in the last thirty years, public concerns still exist about the environmental effects of mining.

In this section we will discuss the issues associated with the science and technology of mining. We will also examine the way ores form, how they are located and how the economic viability of mineral deposits is determined. In addition, we will look at government policies that seek to make mining sustainable and some of the legal decisions that have shaped the way mining and exploration are conducted today.

CONTENTS

6.1 Why mineral deposits are not all ore deposits

At the end of this chapter you should be able to:

- identify renewable and non-renewable resources commonly used in society in terms of the processes and time required to generate them

- define ore deposits in terms of financial costs incurred in exploration, extraction and refining compared with market price, and in terms of grade

- distinguish between waste rock and ores in rock

- distinguish between ore minerals and gangue minerals in an ore deposit

- describe gangue minerals as those that must be removed to enrich the concentration and value of an ore deposit

- identify the relationship between tonnage and grade of deposit and the economic value of an ore deposit.

THE NATURE OF RENEWABLE AND NON-RENEWABLE RESOURCES

A **resource** is something that can be used to provide a human need. Freshwater is a resource and so are the plants and animals we use for food and clothing. Timber and other building materials are also resources. Metals are particularly important resources and ones that affect many parts of our lives. As technologies have developed new materials and processes, the amount and variety of metals we use have steadily increased.

Mineral resources are those naturally occurring materials that are used for industrial, agricultural and other purposes. In this section we concentrate on those materials composed of minerals from which we extract metals, but building materials (such as sand or limestone, which are used to make cement) are also mineral resources. Mineral resources such as diamonds and gold are sometimes found in a pure state. However, most of the metals we use are found in the form of **ore minerals** from which the metals are derived by the process of refining.

Some resources we use are replenished regularly and rapidly and others are part of natural cycles that take millions of years. (See Figure 6.1.1.) Renewable resources are those that can be replaced in a relatively short time. Solar energy, wind and rain are resources that are replenished on a time scale of seconds to years. The timber we use is harvested from trees, which can be replenished on a scale measured in tens of years. For some uses, however, timber needs to come from very large, old trees and the time taken for such trees to grow is measured in hundreds of years.

Non-renewable resources are those that take long periods to be generated. Mineral resources are produced by processes within the Earth that concentrate useful minerals. These events can take millions of years to occur. For example, porphyry copper deposits supply large amounts of copper ore and form within island arcs. An

resource
the amount of a commodity contained in all deposits—discovered and undiscovered

ore mineral
a mineral that contains metals that can be extracted

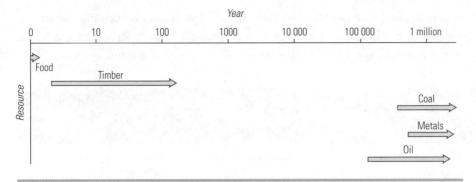

Figure 6.1.1 Estimates for the time needed to replace resources by natural processes. (*Note:* Uranium and banded-iron formations will not be generated in our oxygen-rich atmosphere.)

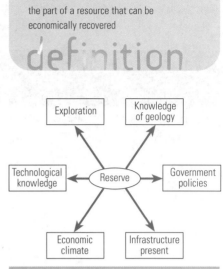

reserve
the part of a resource that can be economically recovered

definition

Figure 6.1.2 Factors affecting the size of a reserve.

island arc has a life measured in the tens of millions of years and it takes many more years before erosion brings such a deposit close enough to the surface to be mined. If we were to depend on such a process to provide the copper ores we use, our society would have to wait many times longer than the whole history of human life before the demand could be met.

Our mineral reserves must be thought of as finite and unrenewable. How long such reserves last is dependent on a number of factors. (See Figure 6.1.2.) The size of a **reserve** depends on the amount of the resource that is known to exist. Geologists have mapped and identified many deposits, some of which are currently uneconomic to develop. Demand may make these deposits economic by increasing the price of the material. New technologies may also increase the reserve by allowing new material to be mined. It is also possible that new discoveries may add to the reserve. Exploration is an expensive business and there are parts of the world where political events have restricted such activities. In some situations, new materials and technologies may replace the resource. The amount of a resource can also be affected by changes in the rate of its consumption. Asbestos, for example, is a material that has become less widely used due to health concerns, and this has affected the reserve size.

REVIEW ACTIVITIES

1
Compare renewable and non-renewable resources.

2
Name a renewable resource and describe the processes used to form it and the length of time over which it is renewed.

3
Name a non-renewable resource and describe how it forms. Explain why your example is non-renewable.

EXTENSION ACTIVITY

4
Consider an incandescent light bulb. List the materials that it is made of, their origin, and whether they are renewable. You may need to carry out research to complete parts of this activity.

THE NATURE OF ORES

Metals can exist in the Earth as pure metals (termed 'native metals') or as compounds (termed 'minerals'). Native metals include gold, silver, copper and platinum and exist as metals because of their low chemical reactivity. Most metals are found within minerals where one or more metals are chemically combined with non-metals. Such minerals are called ore minerals. An ore mineral is a mineral from which valuable materials can be extracted. (See Table 6.1.1.) Sphalerite, for example, is an ore mineral from which zinc metal is extracted. It has the formula ZnS, which indicates that the mineral consists of equal numbers of zinc atoms and sulfur atoms.

Table 6.1.1 Some common metal ore minerals

Metal	Typical minerals	Uses	Occurrence	Comments
Aluminium	Gibbsite Diaspore	Lightweight construction materials, packaging manufactured goods, etc.	As bauxite due to deep chemical weathering	Minerals are hydroxides and oxides
Chromium	Chromite	Steel making and chrome plating	Magmatic ores	Chromium and its compounds are highly poisonous
Copper	Chalcopyrite Chalcocite Bornite Malachite Cuprite Covellite	Manufactured materials, electrical wires and plumbing	Hydrothermal deposits, porphyry copper deposits and sedimentary deposits	Most often found as a sulfide, but malachite is a carbonate
Gold	Native metal	Economic trade, electrical applications, jewellery	Hydrothermal deposits and placers	Used in many more ways than as a precious metal
Iron	Hematite Magnetite Limonite Goethite	Steel making, manufactured materials, construction, transport, etc.	As banded-iron formations, contact metamorphic environments, and segregations within magmas	Found most often as oxide minerals
Lead	Galena	Batteries	Hydrothermal and sedimentary deposits	Most often found as a sulfide
Magnesium	Magnesite Dolomite	Lightweight alloys, raw chemical material and in insulators	Hydrothermal deposits, limestones, sea water	As a reactive metal it can be used to protect other metals
Mercury	Cinnabar	Electrical equipment	Hydrothermal deposits	Poisonous, but has important uses for which no substitutes exist
Nickel	Pentlandite	Stainless steel, alloys, catalysts, plating and coins	Sedimentary deposits formed by weathering	Sometimes found with copper and platinum metal ores
Platinum	Native metal Arsenide and sulfide salts	Chemical catalyst, alloys and electrical applications	Magmatic ores and placers	Ores associated with basic and ultrabasic igneous rocks
Rare Earth elements	Monazite	Electronics	Placers	Also found in rocks called carbonatites, together with phosphorus
Silver	Argentite As a solid in other metal sulfides	Photography, electrical equipment and chemistry	Hydrothermal deposits	Often found with copper, lead and zinc deposits
Titanium	Ilmenite Rutile	High-temperature alloys, and paint pigments	Magmatic ores and placers	Commonly sourced from heavy mineral sand deposits
Zinc	Sphalerite	Metal alloy	Hydrothermal and sedimentary deposits	Most often found as a sulfide
Zirconium	Zircon	Ceramics, castings and special glass	Placers	See comment for rare Earth elements above

While an ore mineral may contain a valuable metal, in order for the ore to be economically mined it must exist in a concentrated form. Tin, for example, has an abundance in the crust of only 0.0002%, but must be concentrated to about 0.5% in a deposit to be mined profitably. This means that tin has to be concentrated 2500 times before it can be profitably extracted and used.

Processes within the Earth can bring one or more ore minerals together with other minerals that are not of value. (See Figure 6.1.3.) The term 'gangue mineral' is used to describe the unwanted minerals with which the ore minerals are found. Two common gangue minerals in hydrothermal deposits are quartz and calcite. (Hydrothermal processes are those that involve hot, often acidic, water-based fluids.) They are not of great value in themselves, but add to the bulk of an ore. An ore is therefore a mixture of ore minerals and gangue minerals. During the processing of the ore, ore minerals are separated from the gangue minerals, producing an ore concentrate and tailings, which are the concentrated gangue minerals.

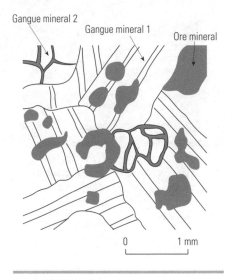

Figure 6.1.3 Parts of an ore.

REVIEW ACTIVITIES

EXTENSION ACTIVITY

1
Explain how an ore deposit differs from a mineral deposit.

2
Distinguish between ore and gangue minerals. Give examples of both in your answer.

3
Research a metal ore mineral and describe its occurrence and the gangue minerals found with it. Assess the metal's or mineral's importance to our economy.

HOW ECONOMICS DETERMINE MINE FEASIBILITY

An ore deposit is not the same as a mineral deposit because an ore deposit is something that can be mined economically. The resource is the concentration of material that is, or may be, feasible to extract. The net value of the ore body is the amount of ore that can be extracted economically. The term used to describe this quantity of ore is 'ore reserve' and the size of the ore reserve depends on the balance between the cost of production and the price people are willing to pay for the ore. The cost of producing the mineral, or metal, will depend on the cost of:

• exploring (finding and describing the deposit)
• extracting the ore (mining it and doing preliminary treatment on it)
• refining the ore (turning it from an ore into a metal or metal compound)
• delivering the ore to market.

The role of scientists in mining is not only to locate a mineral deposit but also to describe it so that its size and value can be determined, to determine how the metal can be extracted from the ore and how the environment can be maintained and rehabilitated. All these tasks involve a range of disciplines. In addition to the science related to ore deposits, a mine's development depends on a good understanding of economics. The science informs those who have to make economic decisions about the profit achievable from a mine.

Resources

In reporting the results of exploration and mining, companies use a set of terms to describe different aspects of the resources they are exploiting. Geologists and geophysicists generate information about the distribution, quantity and quality of the mineral deposit. Some of this information is known with confidence and some is estimated on the basis of other information that has been gathered. These resources are called **identified resources**. Note the difference between a resource and a reserve. A resource is a description of the mineralised material in a body and a reserve is the amount of the mineralisation that is recoverable.

Like any area of scientific inquiry, there is always some uncertainty in the measurements and calculations used to describe a resource. To account for this uncertainty, the resource may be described in terms of measured, indicated and inferred resources.

A **measured resource** is the quantity of the resource that has been calculated from detailed measurements. These measurements are obtained from surface mapping, subsurface drilling and mining. The quality of the ore is determined by detailed sampling and careful chemical analysis. On the basis of these data, the size, shape and distribution of the ore can be calculated with a reasonably high level of confidence. A term that is sometimes used in a similar way to 'measured' in terms of the size of a reserve is 'proven' and it gives a sense of the confidence that people may have in the description.

An **indicated resource** is determined in the same way as a measured resource, but the detail in the description is less than in that of a measured resource. If the spacing of measurements is far apart, or the number of drill holes is few, the confidence that the exploration geologists can have in the description of the resource will be lower. The term 'probable reserve' is used in a similar sense to 'indicated'. For an estimate to be an indicated resource the amount and quality of the data need to be enough so that continuity between points of measurement is assured. The sum of the measured and indicated resources is described as demonstrated resources.

Sometimes, assumptions about the continuous nature of an ore body cannot be supported by measurements or samples. Perhaps there are no good data to describe how deep an ore body extends, but previous experience in mining similar bodies may suggest how far such a body normally extends. An estimate based on such evidence is described as the inferred resource. New research of known ore deposits may alter inferred resource calculations.

Figure 6.1.4 shows how demonstrated and inferred resources may be described in a particular area. As the spacing between drill holes increases, the quantity of mineralisation becomes harder to judge. The shape of the body also becomes harder to estimate. Note that as the costs or price of the mineral being extracted changes, the reserves will change too.

a Widely spaced drill cores do not intersect much of the resource

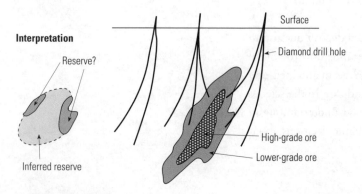

b More closely spaced drill cores define the shape and composition

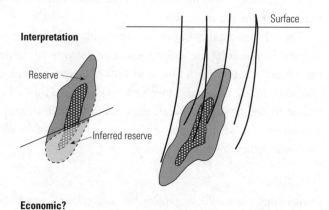

Figure 6.1.4 How the data affect resource and reserve estimates.

Classes of reserve

The part of a reserve that has been determined to be economic to recover is called the economic reserve. This is determined using the scientific knowledge of the deposit and economic considerations. The economic issues may involve the present and future price of the mineral, the investment that can be expected and possible models for the development of the mine. Parts of the reserve that do not meet the economic criteria for recovery are termed 'subeconomic reserves', but a third important category is the marginal reserve.

Marginal reserves are those that border on being economic. It is usually a case of economic uncertainty that leads to part of a reserve being defined as marginal. Exploration has shown that the material exists in a particular location, but it is not economic unless financial conditions change or new technologies occur. For example, ore in a prospect may occur in a form that is expensive to process. The cost makes the ore more expensive to produce than the income achievable under current commodity prices. Economic conditions may change. Perhaps the demand for the metal will increase, which pushes up prices, or political instability in other countries closes mines that also produce the metal. Technology may provide an answer. Perhaps a new way of treating the ore will reduce costs so that the processing of the ore becomes economic.

Grade and tonnage

In order to accurately determine the size of a reserve, the concepts of **grade** and **tonnage** are used. The grade of an ore is the amount of ore minerals within the ore. In the Broken Hill ore body, for example, lead, silver and zinc minerals occur. The zinc grades range from 5% to 20% and the silver grades range from 20 to 300 grams per tonne. In general, precious metal grades are expressed in grams per tonne and other metals are expressed as percentages. Grades are also expressed as parts per million (ppm). Different parts of an ore body will have different grades and the role of exploration geologists is to determine the distribution and variations of grades within the ore body. Samples of the ore are analysed to determine concentrations. This process is called assaying. Many samples will be assayed so that the grade distribution can be determined. Economics will then lead to a determination of the lowest grade that can be mined, that is, the **cut-off grade**. Another type of grade you may read about is the head grade. When the ore is mined, some of the surrounding rock will inevitably end up in the mined ore. This dilutes the overall grade. The head grade is the grade measured on arrival at the mill where the ore is processed.

Tonnage is the amount of ore that can be extracted from a mine. The total value of the ore depends on both tonnage and grade. If a zinc deposit has a zinc grade of 8% and the average price for zinc is $1500 per tonne then the value of the ore is $120 a tonne. A mine such as the Elura Mine, north of Cobar, contains a resource of just under 23 million tonnes. At a grade of 8% and a tonnage of 23 million tonnes the value of the resource is $2760 million. Note, however, that if the price of zinc halved, or tripled, the value of the resource would change.

Other factors affecting costs

Besides grade, tonnage and commodity prices there are a number of other factors that will affect the economic viability of a potential mine. One is the by-products of the mine. In some cases, more than one economic mineral occurs in a deposit. The sale of one may assist in financing the cost of mining another. Two examples of this in Australia are the Elura mine, mentioned above, and Olympic Dam in South Australia. At Elura the presence of silver has helped to make the zinc and lead mining more economic. At Olympic Dam the major metals are uranium and copper, but

grade
the quality of an ore; the amount of metal that can be extracted

tonnage
the measure of the size of a deposit

cut-off grade
the lowest grade at which mining is profitable

gold, silver and rare Earth elements are important by-products that help to make the deposit more valuable.

The characteristics of the ore affect the value of a deposit. The form of the ore minerals may affect the cut-off grade of the ore body. Sulfide ores, for example, are easier to process than silicate ores. In nickel deposits, silicate ores need to assay at grades three times that of sulfide ores to be economic.

The grain size of the ore also affects the economics of mining. Some of the ore is always lost during processing, particularly very fine particles. A loss of some percentage of the reserve needs to be weighed up against the cost of additional processing to reduce the losses.

Sometimes unwanted substances (such as arsenic or mercury) may occur, which may produce health and environmental problems. Unwanted substances may also complicate processing methods and reduce recovery of wanted metals.

A final issue related to the ore is the type of material in which the ore is found. Mineral sand deposits are often found in unconsolidated sands and their recovery is fairly easy. If an ore is found in hard rock, the cost of blasting and crushing will affect the cost of extracting it.

The size and shape of a deposit will affect the cost of mining. Large deposits close to the surface can be mined as open pits. This means that large amounts of ore can be processed, which may make lower grades economic. As the pit becomes deeper, however, the amount of **waste rock** that has to be removed increases. When an open pit reaches a waste to rock ratio of about 2:1, the pit may become uneconomic. Deep ore bodies or small, high-value deposits may be mined by underground methods. Such methods are more expensive than open-cut methods, but less waste rock needs to be dealt with.

waste rock
rock that lies above or around a ore body and has no immediate value

definition

REVIEW ACTIVITIES

1
Describe what the grade of an ore deposit is.

2
Draw up a table to show the factors that add value or costs to the mining of an ore deposit.

3
Compare measured, indicated and inferred reserves.

4
List the characteristics of a mineral deposit that affects its economic value.

5
An ore has a lead grade of 7.2%. If lead is worth $1000 a tonne, calculate the value of the lead in the ore.

6
The Peak Goldmine near Cobar had an estimated resource in mid 2000 of 7.9 million tonnes of ore, assaying at 5.33 grams per tonne of gold.
a Calculate the amount of gold in the resource.
b Calculate the value of the resource if the price of gold is $19 per gram.

EXTENSION ACTIVITIES

7
Write a word equation to show the relationship between grade, tonnage and value of an ore deposit. In point form, outline the correct way to use the equation.

8
Outline the mineral characteristics of an ore that may affect the economic viability of mining the ore.

9
Look at Table 6.1.3 (page 303) and calculate how a gold price of $280/oz would affect the value of the proven reserve. (This price occurred during the mid 1970s.)

SUMMARY

Renewable resources are those that can be replaced in a relatively short time, while non-renewable resources are those that take a relatively long time to form.

Ore deposits are described in terms of grade and the financial costs incurred in producing the minerals compared with market price.

An ore mineral contains metals that can be extracted, while waste rock contains minerals from which metals, or minerals, cannot be extracted economically.

Gangue minerals must be removed from an ore to enrich the concentration and value of an ore mineral.

Tonnage and grade of a deposit both affect the economic value of the deposit.

PRACTICAL EXERCISE
Classifying resources

Cement, aluminium, iron, copper, gold, nitrogen and zinc are major commodities used in industry. In addition, timber, water, food and gas for fuel are essential for our wellbeing. In this exercise you will use library resources, or resources supplied by your teacher, to identify information that will help you to classify these resources.

Procedure

For each material listed on the left:

a identify the source, or sources, from which we obtain the material

b list the uses of the material

c identify how the material is formed

d classify the material as renewable or non-renewable, using the definitions in the text.

ACTIVITIES

1
List the non-renewable resources.

2
List the renewable resources.

3
If a material is renewable, does this mean it is easy to produce?

4
If a material is non-renewable, does this mean it is scarce?

5
Summarise the characteristics you used to identify renewable resources.

PRACTICAL EXERCISE
Calculating the costs and profits of a mine

Tables 6.1.2 and 6.1.3 show details of a hypothetical, or imaginary, mine called Traflagara. It is a goldmine that starts as an open-pit mine and becomes an underground mine. The data in the tables illustrate some of the characteristic features of a mine's development. In this exercise you will process and analyse information from the tables to estimate the costs and conditions under which a mine operates.

As you complete the following activities, try to relate the costs you see to the information discussed in this chapter. Later, when you undertake your case studies, you will benefit by comparing some of the data recorded here with that of the mines you study.

ACTIVITIES

The measured reserve

1

Calculate the number of ounces in the measured reserve. Remember that the number of ounces will be the product of the number of tonnes of ore and the number of ounces per tonne. (*Note:* 1 kt = 1000 tonnes and 1 Mt equals 1 million tonnes.)

Mine production

2

Describe how the waste to ore ratio changes over time.

3

Explain why the ratio increases between years 6 and 8.

4

Suggest a possible reason why the ratio falls during year 9.

5

Describe how a high waste to ore ratio would affect the profitability of a mine.

Mill production

6

Compare the mill grades with the gold grade recorded for the measured reserve in Table 6.1.2. Suggest why the mill grades are higher.

Metal production

7

Mining costs increase during the period recorded here. List some of the reasons why this occurs.

8

Suggest why refining costs are high during year 7.

9

The production costs remain constant during years 7 to 9. Predict whether this will continue and give reasons for your answer.

Other costs

10

Production at the mine starts in year 4, but the exploration costs continue. Explain why this occurs.

11

Borrowing costs are the moneys paid on interest and fees on loans. For example, for all its mines, Newcrest Mining paid $27 million in the year ending 30 June 2000 with a net income of $293 million.

a Calculate what percentage of Newcrest's net income was spent as borrowing cost.

b Calculate the percentage of Traflagara's income in year 9 that goes towards borrowing costs.

c Predict what would happen if interest rates rose so that borrowing costs became higher than 15% of a mine's income.

Overall profitability

12

Make a graph to show income, expenditure and debt for the nine years of the mine. Describe what the graph shows.

13

Calculate the production cost of an ounce of gold using year 9 for the mine.

14

Predict when the debt carried by the mine will be repaid. List possible factors that could affect the debt carried by the mine.

Table 6.1.2 Reserves at the Traflagara Mine

Reserve	Measured	Indicated
Ore (Mt)	27.0	18.3
Gold (g/t)	4.74	3.4
Gold (troy oz/t)	0.15	0.11
31.2 g/troy oz		

Table 6.1.3 Productivity of the Traflagara Mine

	Year									Total
	1	2	3	4	5	6	7	8	9	
Mine production										
Open-pit ore mined (kt)	0	0	0	510	1627	1780	1900	1872	1420	9 109
Open-pit waste mined (kt)	0	0	0	2360	6002	4480	4983	5590	2640	26 055
Waste to ore ratio	–	–	–	4.6	3.7	2.5	2.6	3.0	1.9	18.3
Underground ore mined (kt)	0	0	0	0	0	0	0	0	420	420
Underground waste mined (kt)	0	0	0	0	0	0	0	0	32	32
Total ore mined (kt)	0	0	0	510	1627	1780	1900	1872	1840	9 529
Mill production										
Total ore treated (kt)	0	0	0	494	1446	1562	1879	1788	1796	8965
Mill grades (gold g/t)	0	0	0	4.10	3.30	4.54	4.91	5.20	6.82	28.87
Mill recoveries (gold %)	0	0	0	87.30%	83.40%	86.70%	87.00%	88.10%	88.50%	86.83% (average)
Metal production										
Gold produced (oz)	0	0	0	57 038	128 377	198 333	258 920	264 232	349 681	1 256 581
Gold price ($/oz)	0	0	0	558	692	712	752	665	558	656 (average)
Price received ($ millions)	0	0	0	32	89	141	195	176	195	828
Production costs ($ millions)										
Mining	0	0	0	6.5	16.3	17.1	17.4	18.4	18.6	94.3
Milling	0	0	0	6.3	17.4	19.2	20.1	21.9	21.7	106.6
Refining	0	0	0	6.8	21.1	20.4	23.1	21.0	21.1	113.5
Shipping	0	0	0	2.9	7.5	8.1	7.6	6.5	6.4	39.0
Royalties	0	0	0	0.6	1.9	2.0	2.5	2.3	2.3	11.6
Total production costs	0	0	0	23.1	64.2	66.9	70.7	70.1	70.1	365.1
Other costs ($ millions)										
Exploration	20.2	22.4	31.6	11.3	8.9	11.6	21.4	19.3	10.7	157.4
Borrowing	4.7	4.4	4.9	4.9	4.4	4.9	6.8	6.1	6.8	47.9
Capital expenditure	1.2	4.6	42.0	20.0	8.2	7.8	9.2	12.0	8.2	113.2
Balance ($ millions)										
Total income	0.0	0.0	0.0	31.8	88.8	141.2	194.7	175.7	195.1	827.3
Total expenditure	26.1	31.4	78.5	82.4	85.7	91.1	108.1	107.5	95.8	706.6
Profit before tax (negative values are losses)	−26.1	−31.4	−78.5	−50.6	3.1	50.1	86.6	68.2	99.3	
Total debt	312.0	332.26	344.47	344.07	320.83	275.95	206.92	140.17	62.976	

PRACTICAL EXERCISE
Examining ores

In this exercise you will examine some ore samples in order to identify the types of minerals and rock that are found together in an ore body. At the end of the exercise you should be able to distinguish between waste rock and ore, and between ore minerals and gangue minerals.

Materials

- A variety of ore mineral samples: galena, molybdenite, chalcopyrite, sphalerite and chromite
- Samples of ore containing some of the ore minerals: three if possible
- Hand lens or binocular dissecting microscope

Procedure

1

Begin by examining the ore minerals. You may need to review some of the terms used to describe minerals from *Earth and Environmental Science: The Preliminary Course* (pp. 127–32). Try to identify the following features:

- *Galena.* A lead sulfide with a metallic silver appearance when fresh. Its colour may range from silver to black, depending on how weathered the samples are. It forms cubes or triangular-shaped pits and produces a lead-grey streak when drawn across a white tile. It is soft (hardness 2.5) but very dense (7.5–7.6 g/cm^3). It is found with other sulfides (particularly spalerite) and gangue minerals, such as quartz, calcite, barytes and fluorite.
- *Molybdenite.* A molybdenum sulfide, it occurs as flakes or foliated masses. It is soft (hardness 1–1.5) and because of this and its single cleavage it feels greasy to the touch. It is found as small masses in granites and quartz veins. It has a green or bluish-grey streak. It has a density of 4.7 g/cm^3.

- *Chalcopyrite.* A copper iron sulfide that is a gold-yellow colour but produces a grey-green streak. It has a metallic lustre, but the crystals are often small and no cleavage is visible. The mineral is found with other copper and iron sulfides. It may weather to form copper carbonate and oxide minerals that are green and blue. It has a hardness of 3.5–4 and a density of 4.1–4.3 g/cm^3.
- *Sphalerite.* A zinc sulfide that may contain iron. Its colour varies (black, brown, yellow or white) and its lustre is similar to resin rather than a metal. It is harder than galena (hardness 3.5–4) but less dense (3.9–4.1 g/cm^3). Sphalerite forms tetrahedral crystals and has one good cleavage plane. It is found with other sulfides (particularly galena) and gangue minerals, such as quartz, calcite and metamorphic minerals.
- *Chromite.* Black with a submetallic lustre, chromite forms large, irregular, compact masses. It has a hardness of 5.5 and a density of 4.5–4.8 g/cm^3. It is found with serpentinites and peridotites.

2

Describe the appearance of each ore mineral and try to identify any other minerals present. If other minerals are present, try to determine whether they are ore minerals or gangue minerals.

3

Examine the ore rocks. Identify the ore minerals present and the gangue minerals. Try to draw a diagram to show the occurrence of the minerals within the rock. Can you identify waste rock and ore?

4

Compare the ore rocks. In what ways are they similar? In what ways are they different?

5

Reflect on the exercise. What aspects of it were interesting or difficult?

6.2 Mineral exploration

OUTCOMES

At the end of this chapter you should be able to:

- identify the main features of two Australian mineral provinces including a base/precious metal producing locality in an island arc terrane and one selected from an iron ore producing locality in an ancient continental area or an area of sedimentary ore formation

- discuss theories concerning mineral genesis related to sedimentary and tectonic processes responsible for the two minerals selected

- describe the exploration methods used to infer the size and grade, and indicate the presence, of one named ore deposit.

- outline exploration methods for the case study undertaken that may include geophysical and geochemical techniques, mapping, satellite imagery, and aerial photograph interpretation

AUSTRALIAN MINERAL PROVINCES

A mineral province is a specific region that contains significant concentrations of mineral ores. The ore deposits within a province show similar environments of formation and often contain deposits that formed within a particular period of time. An example of a mineral province is the Ordovician-aged volcanics of the Lachlan Fold Belt, which host a number of significant mineral deposits, such as Cadia-Ridgeway, the Cowal gold prospect and the Northparkes Mine. Another example is the Hamersley iron province in Western Australia, where Mount Tom Price and Mount Whaleback are located. The iron ore in the Hamersley province is of early Proterozoic age, approximately 2500 Ma old, and contains approximately 98% of Western Australia's iron ore reserves. The Hamersley province is one of four Australian iron provinces, each of which has its own particular characteristics.

A similar term to 'mineral province' is 'metallogenic province'. A metallogenic province is one in which, as the name suggests, metal ore deposits originate. While all metallogenic provinces are mineral provinces, the opposite is not true. Phosphate rock and gypsum rank ninth and tenth, respectively, in terms of world mineral production, but neither yields metals. Phosphate deposits can be either igneous or sedimentary in origin. The Kola Peninsula in Northern Europe contains significant phosphate deposits in alkaline igneous rocks of Late Palaeozoic age. Sedimentary phosphate deposits that are commercially exploited are of Phanerozoic age and form when ocean currents bring nutrient-rich water to the surface. Here the phosphates in the water are taken up by marine life, incorporated into sediments and then reworked to form phosphate-rich deposits. An Australian phosphate province is the Georgina Basin near Mount Isa. Of Cambrian age, the basin hosts some of the world's major

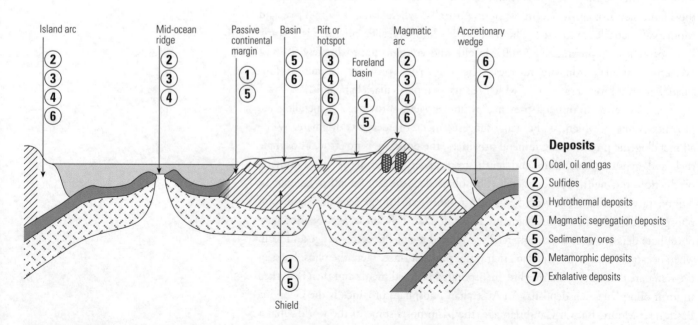

phosphate deposits. Bearing in mind such mineral provinces, we will concentrate in this chapter on mineral provinces that yield metals. Be aware that the Georgina Basin may well make a good case study of a sedimentary mineral deposit.

Tectonic settings

Mineral provinces arise in particular tectonic settings and give rise to **terranes**. You already know that plate tectonic processes shape the Earth's surface. It is the processes in particular tectonic environments that concentrate ore minerals so that we can mine them economically. There are six tectonic settings in which mineral deposits form:

- basins and rifts within continents
- oceanic basins and mid-ocean ridges
- passive continental margins
- subduction-related settings, such as island arcs and magmatic arcs
- continental collision zones
- strike-slip environments.

Examples of these environments and some of the deposits that form in them are shown in Figure 6.2.1.

In this chapter we will examine two particular tectonic settings that have given rise to mineral provinces in New South Wales. A discussion of the various forms of ore-forming processes will also be given so that you gain a broad knowledge of processes on which to base your case study later in this section.

ORDOVICIAN VOLCANICS OF THE CENTRAL WEST

The Ordovician volcanics of the Lachlan Fold Belt are an example of mineral provinces that form in an island arc tectonic setting. The volcanics, which host porphyry-style mineral deposits, formed during the Late Ordovician to Early Silurian as part of what is called the Molong Volcanic Arc. (See Figure 6.2.2.) During the Ordovician, the eastern seaboard did not exist. The edge of the continental part of Australia lay close to the current New South Wales–South Australian border. The Molong Volcanic Arc ran in a north–south direction through what is now Central New South Wales. To the west of it was a back-arc basin and to the east was a forearc basin and trench. Remains of the accretionary wedge can be seen at Narooma on the South Coast.

terrane
a geological unit with distinctive and diverse geology and structure

definition

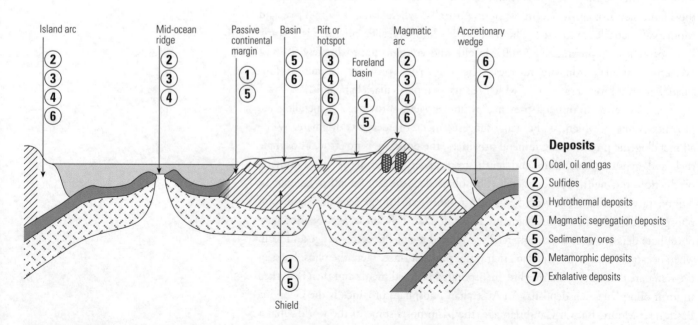

Figure 6.2.1 Tectonic settings and mineral deposits.

Deposits
1. Coal, oil and gas
2. Sulfides
3. Hydrothermal deposits
4. Magmatic segregation deposits
5. Sedimentary ores
6. Metamorphic deposits
7. Exhalative deposits

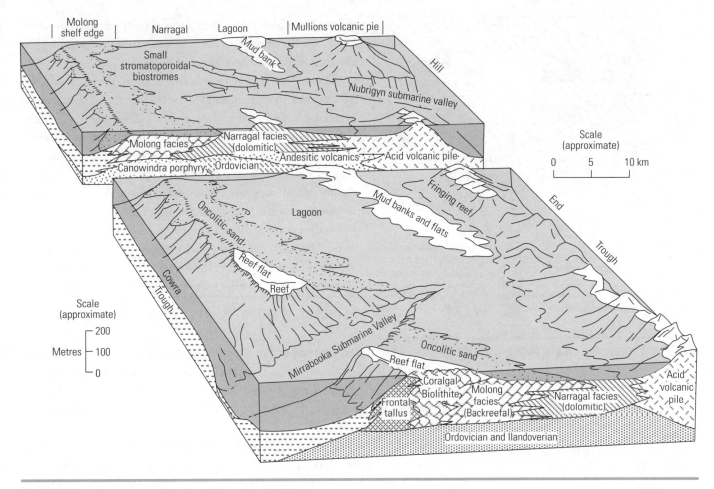

Figure 6.2.2 The palaeogeography of New South Wales during the Ordovician.

The Molong Volcanic Arc was relatively shallow due to the volcanic activity along its length. As a result, limestones were deposited in addition to volcanic rocks, **volcaniclastics** (such as tuffs) and igneous intrusions. The volcanic rocks range in composition from intermediate rocks (such as andesites and diorites) to more silica poor volcanics (such as basalts). The processes that formed the mineral deposits occurred within these rocks. Some of the volcanics and intrusions are classified as

volcaniclastics
rocks formed from fragments created by volcanic eruptions

definition

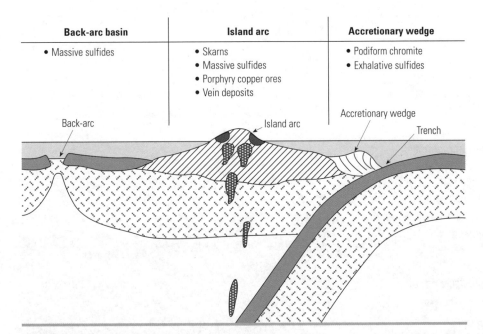

Back-arc basin	Island arc	Accretionary wedge
• Massive sulfides	• Skarns • Massive sulfides • Porphyry copper ores • Vein deposits	• Podiform chromite • Exhalative sulfides

Figure 6.2.3 Mineral deposits formed in island arc environments.

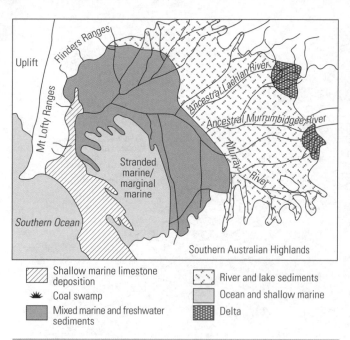

Shallow marine limestone deposition
Coal swamp
Mixed marine and freshwater sediments
River and lake sediments
Ocean and shallow marine
Delta

Figure 6.2.4 The Murray Basin during the Pliocene.

shoshonitic
describes igneous rocks with a high potassium to sodium ratio

epithermal
volcanic-related deposits formed near the surface

skarns
altered areas formed between carbonate rocks and igneous intrusions

definitions

shoshonitic, which means they formed from magmas generated late in the life of the arc. Later, tectonic compression disrupted the arc and gave rise to additional mineralisation.

The styles of mineralisation found in the rocks of the Molong Volcanic Arc include many of those found in other past island arc environments. (See Figure 6.2.3, page 307.) These styles include porphyry-related deposits, **epithermal** deposits and **skarn**-hosted deposits. These types of deposits are described in more detail later in this chapter. Copper and gold are major metals mined in these environments. In general, monzonite intrusions gave rise to fluids as the magma crystallised. This gave rise to disseminated mineralisation, and mineralisation in quartz veins. Mines located in the rocks of the Molong Volcanic Arc include Cadia-Ridgeway, Brown's Creek (a skarn deposit), Northparkes (Goonumbla), Peak Hill and Lake Cowal.

PLIOCENE SAND DEPOSITS OF THE MURRAY-DARLING BASIN

Another mineral province within New South Wales is the heavy mineral sands of the Murray Basin. The province is very much younger than the Ordovician deposits of the Central West. The heavy mineral sands of the Murray Basin were deposited during the Pliocene (5–1.8 Ma ago) and are of sedimentary origin. (See Figure 6.2.4.)

During the Cainozoic (65–1.8 Ma ago) the Murray Basin formed as a result of east–west compression. The Pliocene was a period of marine transgression and regression during which the basin filled with the sea and then emptied again. During a regression (sea retreat), medium to coarse beach sands containing the heavy mineral sands were deposited as barrier dunes along the coast.

It is worth noting that a number of other economic resources formed in New South Wales during the Cainozoic. The alluvial gold deposits that gave rise to the gold rushes during the 1850s, diamond and sapphire deposits from Tertiary volcanics and alluvial tin deposits all formed during this period by weathering, erosion and deposition.

REVIEW
ACTIVITIES

EXTENSION
ACTIVITY

1
Outline the characteristics of an Australian mineral province formed in an island arc environment (terrane).

2
Summarise the characteristic features of an island arc terrane.

3
Outline the characteristics of an Australian mineral province formed in an area of sedimentary ore formation.

4
Summarise the six tectonic settings in which mineral deposits form.

5
Describe the tectonic settings in which hydrothermal deposits form.

6
Contrast an ore deposit formed in an island arc environment with one formed on a continent in a sedimentary basin.

ORE GENESIS

There are many ways in which ore deposits form. The origin of ore deposits is referred to as **ore genesis** and in this part of the chapter we will look at important types of ore genesis found in New South Wales and other parts of Australia.

There are many ways in which Earth processes can concentrate ores. Igneous, sedimentary and metamorphic processes all contribute to ore genesis. In this text we will use a simple classification system for ore genesis and divide deposits into two large groups:

• those formed due to processes operating within the Earth
• those formed by surface processes. (See Figure 6.2.5.)

Internal processes

In relation to internal processes we will look at magmatic, metamorphic and hydro-thermal processes.

MAGMATIC PROCESSES

Magmatic processes are those that occur within magma bodies. Some ore deposits form simply by crystallisation: diamonds are an example. Processes that occur within the magma chamber or subsequent intrusion form other ores. Such processes include fractional crystallisation and **liquid immiscibility** and the formation of pegmatite. All occur as the temperature and composition of a magma changes during cooling. (See Figure 1.1.12, page 12.)

Fractional crystallisation separates ore and non-ore minerals according to their crystallisation temperature. As early crystallising minerals form they incorporate certain elements, some of which are metals. These crystals may settle onto the bottom of the intrusion, concentrating ore minerals there. Chromite and magnetite are ore minerals that form in this way.

Sulfide ores containing copper, nickel or platinum may form by a process called liquid immiscibility. As a magma changes, parts of it may separate from the main body of the magma. Two liquids that will not mix are called immiscible; oil and water are an example. In magmas, sulfides may separate and sink below the silicate-rich part of the intrusion or be injected into the rock surrounding it. These deposits are found

ore genesis
the way in which an ore deposit forms

liquid immiscibility
the situation in which two liquids will not mix with each other

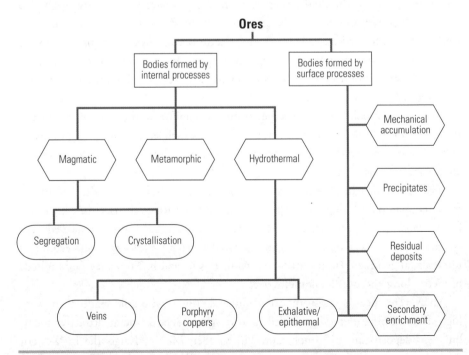

Figure 6.2.5 A classification of ore genesis processes.

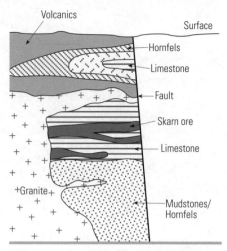

Figure 6.2.6 An example of a skarn deposit.

in mafic rocks (such as gabbros) and ultramafic rocks (such as komatiites and peridotites).

Pegmatites are igneous rocks formed from the last remaining magmas in an intrusion. The magma is enriched in water and other low melting point materials. As a result, the pegmatites have large crystals and may contain high concentrations of minerals containing beryllium, tin, tungsten, tantalum and niobium.

METAMORPHIC PROCESSES

Metamorphic processes produce many of the mineral deposits on which we rely. Talc, corundum, asbestos and graphite are all formed by contact or regional metamorphism. An important type of metamorphic deposit from which metals are produced is the skarn.

Skarns are altered areas that often form at the boundary between igneous intrusions and carbonate host rocks, such as limestones. (See Figure 6.2.6.) Copper, zinc, lead, tungsten and iron have all been mined from skarns, but most skarns do not contain economic deposits. Following initial metamorphism of the host rock by the intrusion, the skarn may be injected by hydrothermal fluids from the intrusion. Mineralisation results from deposition from the fluids. The Brown's Creek Mine near Blayney is an example of a skarn formed within an island arc environment. Hosted by Late Ordovician limestones and volcanics, the gold, silver and copper mineralisation formed from fluids derived from later Silurian granites.

HYDROTHERMAL PROCESSES

Hydrothermal processes involve hydrothermal fluids, that is, hot liquids that contain dissolved minerals. The water in the fluids may be derived from ground water or from an igneous intrusion, which is the source of the heat. In this text we will only touch on four types of deposits formed by hydrothermal processes; five if you include skarns.

Veins

Veins form when hot fluids, called hydrothermal fluids, deposit minerals in spaces within rocks. The fluids are composed of water and dissolved minerals. Heat and the presence of hydrochloric acid and hydrogen sulfide help to dissolve minerals from the rock through which the fluid passes. Most rock types have the capacity to give up metal ions to hydrothermal fluids, and the fluid itself can increase the surfaces to work on through a process called hydraulic fracturing. In this process, pressure in the fluid causes the rock to fracture, making new spaces for the fluid to move through. The fluids may be derived from igneous intrusions, metamorphic processes or from heated ground water.

As the fluids cool, the minerals they contain crystallise on the walls of the cavities that the liquid fills. Metal ores—from which metals such as gold, silver, tin, tungsten and uranium are extracted—are deposited with gangue minerals. Quartz and calcite are the major gangue minerals.

The spaces in which veins form may be fractures, faults and those within the fractured material making up breccias. Because of the variety of spaces in which they form, veins show great variation in form. Veins are often found as sheeted vein systems, that is, sets of closely spaced, parallel veins. Veins that host economic minerals range in size from centimetres to metres in width. They may be hundreds of metres long and over a kilometre deep.

Veins were once the most important mineral targets, with gold grades being very high. Bulk-mining methods have led to lower-grade deposits being attractive today, but vein deposits are still important. The Golden Mile at Kalgoorlie in Western Australia is an example of a place where vein-hosted gold is mined. The veins form in **shear zones** and the middle of folds. Fractures above granite intrusions may also host gold-bearing veins. The vein deposits occur in strongly deformed greenstones of the Archaean age Yilgarn Block and, while individual deposits are relatively small, hundred of deposits occur within a relatively small area. During the 1890s, when the deposits were first mined, gold grades of forty parts per million were acceptable. Today, grades as low as two parts per million are mined in open pits.

Porphyry deposits

Porphyry deposits are low-grade, high-tonnage deposits that are mined for copper, molybdenum and tin. The copper-rich deposits may also contain smaller amounts of gold and silver. The ore minerals are found in host rocks that have been intensely altered by hydrothermal fluids. The ore may be disseminated (scattered) throughout the host rock or be found in tiny quartz vein networks, called **stockwork**, running throughout the rock. The texture is due to fracturing and subsequent emplacement of ore minerals. (See Figure 6.2.7.)

Porphyry deposits derive their name from the rock type that forms the centre of the intrusions with which they are associated. You might already know that a porphyry is an intrusive igneous rock containing crystals of two different sizes: one coarse grained, reflecting slow crystallisation; and the other fine grained, reflecting rapid crystallisation. Surrounding the porphyry core is often a shell of coarse-grained rock that lacks a porphyritic texture. The porphyrys, which host mineralisation, are relatively rich in quartz and feldspar and range from intermediate to felsic compositions. They include granites, granodiorites, diorites and monzonites. The intrusions form close to the surface; they reach within 2.5 km of the surface.

The distribution of porphyry deposits and the chemistry of the intrusions indicate that the deposits form in subduction-related environments, both continental margins and island arcs. Within island arc situations, multiple intrusions occur and it is usually the youngest intrusions that host the mineralisation. The mineralisation may be found in the intrusion itself, partly within the country rock (the rock that surrounds the intrusion) or partly within both the intrusion and the country rock. The minerals that host the metals are sulfides; a pyrite-rich shell usually surrounds the ore body.

The hydrothermal solutions that cause the alteration and deposit the ore may have an origin in water from the intrusion magma or from circulating ground water. (See Figure 6.2.8.) If the water is magmatic in origin, the temperatures and salinity of the fluids are high and alteration occurs mainly within the intrusion. The mineralisation is derived from the intrusion itself. If the water is derived from convecting ground water, the fluid temperatures are lower and the mineralisation is most likely derived from the country rock. The mineralisation is consequently concentrated near the intrusion. In many cases the hydrothermal fluids are derived in varying amounts from both the magma and the ground water.

Within the intrusion, stockwork mineralisation suggests a period when sudden pressure increased within water in the cooling magma and shattered recently formed rock, both intrusive and country rock, to form what is called crackle breccia. Circulating hydrothermal fluids later seal the cracks with ore minerals. Perhaps the escape of the fluids during the formation of the crackle breccia removes the heat that leads to the porphyry forming. The nature of stockwork and veins within porphyry deposits suggests the occurrence of often repeated fracturing and sealing.

Compared to other hydrothermal ore bodies, porphyry deposits are extremely large. Tonnage may range from tens of millions of tons of ore up to giant deposits larger than 1000 million tons of ore. The low grade of porphyry copper deposits (0.4–1%), and their large sizes and relatively shallow emplacement mean open-pit

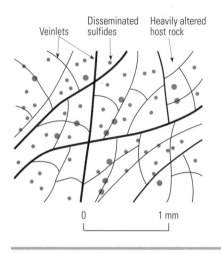

Figure 6.2.7 Stockwork and disseminated ores.

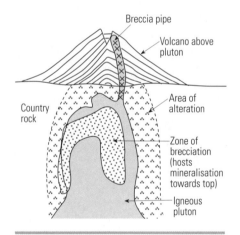

Figure 6.2.8 Structure of porphyry deposits. (*Note*: Not drawn to scale.)

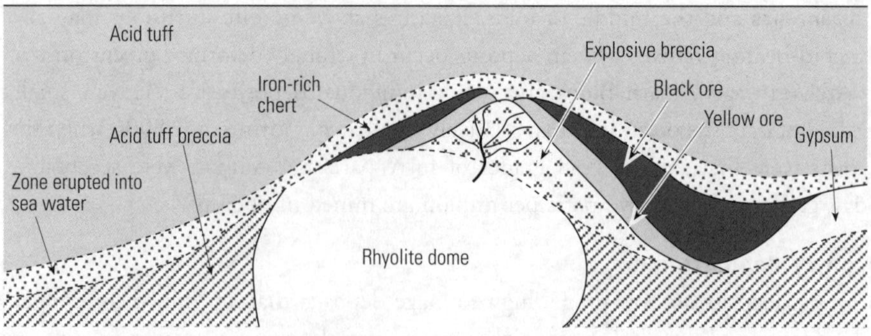

Acid tuff   Iron-rich chert   Explosive breccia   Black ore   Yellow ore   Gypsum   Acid tuff breccia   Zone erupted into sea water   Rhyolite dome

Order of deposition emplacement:
1 black ore
2 iron-rich chert
3 yellow ore

Figure 6.2.9 Structure of a massive sulfide deposit.

mines are economic ways to recover the ore. In some situations, however, underground methods are also used.

Volcanic exhalative and epithermal deposits

Volcanic-associated exhalative deposits (types of massive sulfide deposits) are formed at the boundary of igneous and sedimentary strata or between volcanic units. They are tabular (sheet-like) or lenticular (lens-like) in shape, and those that contain a number of metal ores show compositional zoning. (See Figure 6.2.9.) The deposits are marine in origin and the base of the deposit is a stockwork of veins, which were once the fractures through which the hydrothermal fluids reached the surface.

Like porphyry deposits, the hydrothermal fluids that build exhalative deposits may be derived from magmatic fluids or from water circulating in the country rocks. As the fluids dissolve elements from the rocks, they pass towards the surface. When the fluids reach the surface they mix with cold sea water and the minerals precipitate out of solution, forming mounds on the sea floor. This process has been observed at mid-ocean ridges where black smokers form, but such processes can also occur in back-arc basins near island arcs.

Exhalative deposits are sources of zinc and copper but lead, tin, gold and silver also occur. The most common mineral within the deposits is pyrite (iron sulfide). In zoned deposits a core of pyrite lies above the stockwork. A yellowish ore containing pyrite and chalcopyrite lies above the pyrite core and then a black ore of sphalerite, galena, baryte and pyrite overlays the yellow ore zone. The black ore is capped by an iron-rich chert.

Epithermal deposits are another type of mineral deposit formed by hydrothermal fluids in volcanic areas. These deposits form in both island arc and continental arc environments and they take the form of veins, stockwork or breccia-hosted ores. The deposits precipitate from hydrothermal fluids close to the surface and may deposit material in hot springs and fumaroles at the surface. The fluids cause alteration in the country rock and the deposits are often associated with subvolcanic intrusions and pyroclastics. They may also be associated with **calderas**.

Gold and silver are the main metals derived from hydrothermal deposits but antimony, arsenic and mercury are also derived from some epithermal deposits. Quartz and calcite are common gangue minerals. If the hydrothermal fluids are restricted to a few veins, very high grades of minerals may result. Such deposits may be relatively small in terms of tonnage but the grades may be measured in hundreds of grams per tonne. If the hydrothermal fluids work over a large area then large tonnage, low-grade deposits may occur.

calderas
volcanic collapse structures

Surface processes

There are five types of surface processes that you should be aware of. Four of them are due to sedimentary processes. The fifth process results in volcanic exhalative deposits, which have already been described in relation to hydrothermal processes. In general terms, mineral deposits formed at the surface can be divided into two groups:

- those that are transported to their site of deposition (often referred to as placer deposits)
- those that remain in their place of formation, including chemical precipitates and deposits formed by weathering (secondary enrichment and residual deposits).

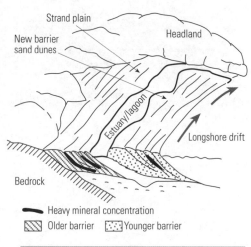

Figure 6.2.10 Placer deposits.

MECHANICAL ACCUMULATION: PLACERS

Placer deposits are formed by the mechanical accumulation of particular mineral grains. Agents of erosion, such as water and wind, move materials differently depending on the density of materials. Movement of heavy mineral grains requires more energy in the agent of transport than similar-sized, lower-density grains. Heavy, dense, hard and chemically resistant mineral grains are concentrated in places from which the agent of transport cannot carry them. In this way, barrier sand dunes, which separate estuaries from the sea, accumulate mineral deposits of heavy minerals. (See Figure 6.2.10.)

Mineral and gem deposits containing such resources as gold, platinum, cassiterite, chromite, ilmenite, uranium minerals, sapphires, rubies and diamonds are all found as placers. Sulfide minerals only rarely form placers because they chemically weather too easily. Uranium ores weather rapidly today, but during the Archaean, when the atmosphere had a low oxygen content, uranium placers also formed.

Most placers are recent and small. Large, ancient ones may be rich enough to warrant deep mining, but smaller deposits are less profitable to mine.

CHEMICAL PRECIPITATES

In some sedimentary environments, chemicals precipitate from water to form sediments that may become economic mineral deposits. The exhalative ores described earlier are an example of such precipitates and so are the banded-iron formations (BIFs) of Western Australia and South Australia. Phosphates, evaporates and manganese deposits can also form as precipitates.

The BIFs of the Hamersley Ranges formed between 1900 and 2500 Ma ago and were laid down over very large areas. Weathering of greenstones produced silica and iron oxides that were carried to the ocean. Cyanobacteria or iron-precipitating bacteria caused iron to precipitate when the iron reacted with the oxygen produced by the organisms. Consequently, metamorphism or leaching has removed some of the silica and enriched the iron oxides to a point where the grades make them highly economic ores.

SECONDARY ENRICHMENT

Secondary enrichment occurs when metal ores are carried from an area of low concentration to an area of high concentration. This occurs in sulfide ore bodies when water percolates through the ore, carrying dissolved ore minerals deeper into the ore deposit. The residual material left behind is called a gossan and gossans are features that exploration geologists keep an eye out for. The dissolved minerals stop travelling when they reach the watertable. There they form a zone of enrichment. (See Figure 6.2.11.)

The process operates in iron ores. As was mentioned earlier, leaching of silica from BIFs can significantly increase grades. At Mount Tom Price, original iron ore grades as high as 35% have been enriched to grades as high as 66% by secondary enrichment.

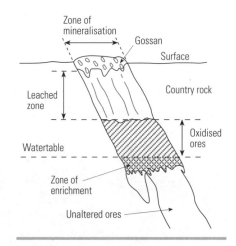

Figure 6.2.11 A section through a mineral deposit showing secondary enrichment.

RESIDUAL DEPOSITS

Residual ores are those formed by the leaching of soluble materials from rocks. The removal of such materials leaves the less soluble materials behind and enriched compared to their original concentrations. Intense chemical weathering, for example, creates laterites. In Australia, laterites consisting of almost pure aluminium oxides are called bauxite and are mined to produce aluminium. Nickel is another metal that is extracted from ores produced as laterites. In the case of nickel, the chemical weathering occurs in ultramafic rocks (such as peridotite) and metamorphic rocks (such as serpentinites).

REVIEW ACTIVITIES

1
Summarise the magmatic, metamorphic and hydrothermal processes that concentrate ore minerals.

2
Outline the characteristics of a porphyry copper deposit.

EXTENSION ACTIVITY

3
Summarise the geographical locations, climates and latitudes in which you would expect the sedimentary processes discussed above to operate.

HOW THE NATURE OF AN ORE BODY DETERMINES EXPLORATION METHODS

Chapter 5.3 outlines the characteristics of mineral exploration as it applies to petroleum and coal exploration. The same principles apply to the search for metal and other mineral deposits. The process of exploration moves from large area, broad detail studies to small area, highly detailed studies. Generally the exploration process will involve some, or all, of the following processes:

- a study of historical, technical, mineral-related data, which may include historical mine records and exploration reports
- the use of systematic regional geological, geochemical and geophysical surveys carried out by state geological surveys conducted by the Australian Geological Survey Organisation
- data derived from regional geological, geochemical and geophysical surveys in selected areas by mining companies or consortiums of companies
- the acquisition of exploration permits from state or territory governments or by acquisition of rights from other companies or individuals
- data derived from increasingly detailed, and expensive, levels of mapping and surveying, overburden and rock trenching, sampling of rock and/or mineral exposures, various types of drilling (percussion, reverse circulation and diamond drilling), stripping, bulk sampling, underground development and test milling.

A particular project will evolve through a number of stages before it becomes a functioning mine. Exploration and evaluation occur at each stage. (See Figure 6.2.12.) The following case studies of two mine projects in New South Wales will illustrate how the methods and stages occur.

Case studies

A DEPOSIT FORMED IN AN ISLAND ARC: THE CADIA-RIDGEWAY DEPOSIT

The Cadia-Ridgeway deposit is an upright mass of mineralised stockwork that caps a small (100 m diameter), intrusive stock of monzonite porphyry. The deposit is about 17 km south-south west of Orange and about 3 km north-west of the Cadia mine. A simplified geology map of the area is shown in Figure 6.2.13 (page 316). Both Cadia and Ridgeway are owned by Newcrest Mining Limited. The ore body is an example of porphyry copper mineralisation and it lies 500 m below the surface within Ordovician-age volcanics. The mineralisation contains rich copper and gold grades (2.6 g/tonne and 0.82 g/tonne, respectively), with the richest ore being located above the stock. Ore minerals include native gold, chalcopyrite and bornite. Importantly, magnetite is a major accessory mineral in the stockwork. Overall tonnage is estimated to be about 44 million tonnes.

The Cadia area has been known as a site of mineralisation for a long time. Copper, gold and iron were discovered in the area in 1851 and mining occurred there in an intermittent way until the end of the Second World War. Exploration of the area resumed in the mid 1980s, with the identification of possible targets using magnetic features in the area.

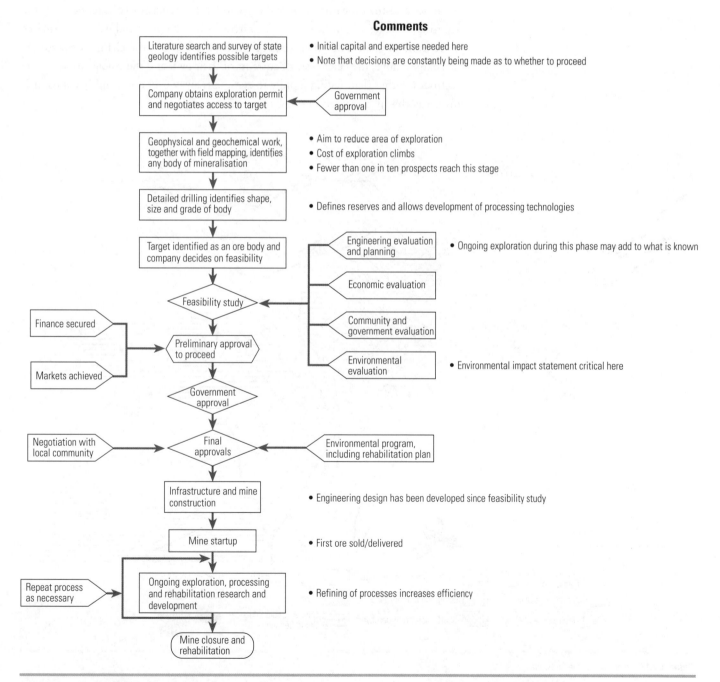

Figure 6.2.12 The development of a possible mine.

Magnetic exploration methods measure disturbances in the Earth's magnetic field due to the rocks in an area. Magnetite is magnetic, and rocks containing large amounts of it stand out in maps of the magnetic field. Magnetic exploration can be done by people on the ground, but it is more economic to use airborne techniques. The NSW Department of Mineral Resources has maps of the magnetism of most of the Lachlan Fold Belt and companies may also prepare more detailed magnetic maps of areas.

On the basis of the magnetic features in the Cadia area, a number of drill holes were made. Cadia Hill was discovered in 1992, but holes above the Ridgeway deposit did not penetrate the ore body. As more holes were drilled to define the Cadia Hill Mine, however, the mineralisation was found to trend in north-west and south-east directions. In early 1994 a regional reconnaissance of the area north-west of Cadia Hill was conducted using induced polarisation (IP) methods.

IP is a method that involves creating an electric current in the Earth between two electrodes and then measuring how the potential difference, or voltage, between the electrodes changes when the current is switched off. Charges build up on the surface of ore mineral grains and when the current is switched off, the charges disperse. As the charges move, they create small currents of their own and the voltage between the electrodes, rather than falling rapidly to zero, falls away over a few milliseconds. It is this decay that is measured in order to detect ore minerals.

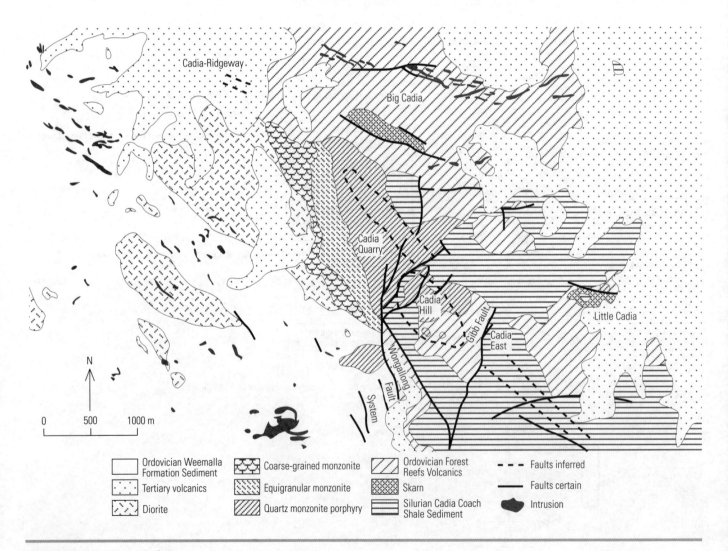

Figure 6.2.13 The geology of the Cadia area.

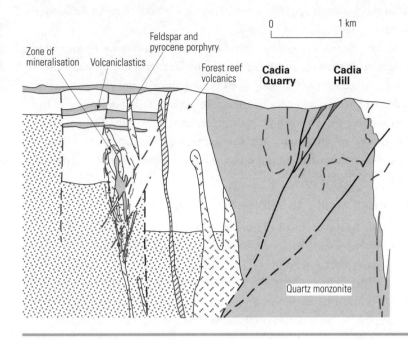

Feldspar and
pyrocene porphyry

Zone of
mineralisation   Volcaniclastics

Forest reef
volcanics   **Cadia
Quarry**   **Cadia
Hill**

0   1 km

Quartz monzonite

Figure 6.2.14 Geology Section: Cadia Ridgeway to Cadia Hill Deposits

Sulfide minerals, except for sphalerite, produce good IP responses. The IP survey over Ridgeway used spacings between the electrodes of 200 m and this allowed both a large area to be studied and, importantly, deep areas to be investigated. A regional reconnaissance is one that involves studies of a relatively large area and more closely spaced measurements. This would have been expensive and have given information only on a relatively shallow area. One anomaly, a high IP response, was detected and the geologists were encouraged enough by the result to drill in the area. Interestingly, it was later discovered that the IP anomaly lies above the Ridgeway deposit. The spacing of the measurements would have needed to be greater in order to detect the ore body.

At the start of 1995, drilling of the area commenced. Initially the IP anomaly area was drilled to a depth of about 200 m using percussion drilling. The first drill hole recovered gold and copper ore but, importantly, the set of results from all the holes suggested the halo of an ore body at greater depth. At the same time as percussion drilling was being conducted, a pattern of widely spaced core holes was being drilled and one hole, with a depth of 514 m, indicated increasing amounts of copper mineral-isation at depth. It was decided in late 1995 to deepen the hole.

In early 1996 the hole was deepened and at 610 m sheeted vein mineralisation was discovered. It was recognised as being similar to the mineralisation at Cadia Hill and so a pattern of deep core holes was drilled around the initial deep hole. The discovery hole, the one that intersected the ore body, was drilled 175 m to the west of the initial deep hole in November 1996. Its position was determined using a knowledge of the local geological structure determined from other holes. It encountered approximately 240 m of mineralisation, starting at a depth of almost 600 m and extending down to 821 m. Gold grades were as high as 7.4 g/tonne and copper grades varied between 1.20% and 1.27%.

The exploration that led to the discovery of the Ridgeway deposit showed a number of important principles in exploration. The area is in a known mineral province. The explorers knew the types of mineralisation to expect. Three different exploration methods were used. Magnetic mapping identified possible areas, the IP method focused the exploration, and drilling in a systematic way discovered the deposit. At each stage, knowledge from previous work was used to make decisions on what to do next.

In this case, drilling was very important. Magnetic methods were of limited use in the final stages because of surface basalt flows, which contain lots of magnetite. IP would not have worked on the ore body itself because it does not conduct. Gravity methods would not have worked because the ore is not very different in density from the rocks surrounding it. Down-hole measurements might have been useful in detecting the details of alteration, but the appropriate methods are yet to be developed for the area. Core drilling and assays of the recovered cores played the most significant role in the discovery.

A DEPOSIT FORMED ALONG AN ANCIENT SEA: GINKGO

The discovery and development of the Ginkgo Mineral Sands Prospect illustrates many of the techniques used in an exploration project. During the late 1990s a number of mineral exploration companies were working in the Murray Basin. The reinterpretation of the stratigraphy of the Northern Murray Basin, together with the discovery of high mineral sand concentrations in some exploration areas, led BeMaX Resources to apply for and obtain **exploration licences** that covered more than 12 000 km². The company recognised that in the Murray Basin they had discovered a Tertiary-age mineral sands deposit that is similar in origin and characteristics to modern-day deposits along the NSW east coast.

In choosing the exploration areas, BeMaX reviewed literature and data and used structural and stratigraphic models to determine likely areas of mineralisation. Criteria they used included:

- known occurrences of mineral sands identified by other companies in the past
- areas of the right topographic level, as indicated by models and research
- areas close to the ancient Darling River, which would have once brought heavy minerals from Central Western New South Wales and Southern Queensland
- areas that would not have been eroded by the ancient Darling River.

Another BeMaX prospect, the Massidon deposit, provided information that was used in the definition of the exploration targets. The sands of the Loxton Parilla Formation, which contains the deposit, are interpreted as a complex of high-energy marine, fluvial and back-barrier lacustrine deposits. Heavy minerals occur locally in the sand, with the best accumulations occurring where the sediments were deposited along beach shorelines and beach ridge environments.

The Ginkgo deposit contains over 7 million tonnes of heavy minerals in a measured and indicated resource of 205 million tonnes at 3.2% heavy minerals. (See Figure 6.2.15.) A cut-off grade of 1% heavy minerals was used in the measurement. It is one of the largest single concentrations of heavy minerals within the Loxton Parilla Formation. Mineralisation at Ginkgo occurs in clean, unconsolidated quartz sands. Impurities, called slimes, and oversize particles are relatively low (2.2% and 0.2%, respectively). These features enable low-cost mining techniques to be employed.

Preliminary success with air-core drilling led the company to carry out a ground magnetic survey. The purpose of the survey was to provide information on which to base a more extensive aeromagnetic survey. The subsequent airborne survey involved not only aeromagnetics but also radiometric measurements and a digital survey of the terrain. Using these data and combining them with a high-resolution seismic survey allowed the company to correlate existing drill-hole data with the geophysics data. This showed that high-sensitivity geophysics could detect the mineral deposits and also identified some previously unknown deposits. Another remote sensing technique

exploration licence
a government licence that outlines conditions a company must follow and which is required before exploration can occur

definition

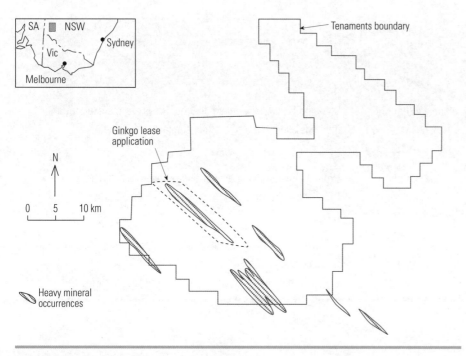

Figure 6.2.15 Location of the Ginkgo heavy mineral deposit.

that has been used in the area is the analysis of detailed satellite images to identify structures relevant to exploration.

Since 1998, three drilling programs have been undertaken. A total of 585 drill holes on forty-two lines and over 18 000 assays defined the deposit: a linear sheet some 14 km long and 2 km wide, averaging approximately 6% heavy minerals. The minerals occur as a series of **strandlines**. The company lodged a mining lease application in April 2001 following a pre-feasibility study and at the same time began compensation negotiations with the two pastoral leaseholders who would be affected by the development.

By September 2001, assays of the drilling results indicated a measured and inferred resource of approximately 7 million tonnes of heavy minerals. In October, a development application for the project was made, accompanied by an environmental impact statement. Using the 'right to negotiate' process of the federal *Native Title Act*, the company reached an agreement with the Barkandji people to protect the environment and cultural heritage as well as to provide employment opportunities. A binding Native Title Agreement was signed in December 2001.

A bankable feasibility study led to the decision to go ahead with mining. As part of the study, geological and geotechnical investigations demonstrated that the **lithology** of the ore body is ideal for dredge mining: a technique used in the mineral sand deposits along Australia's east coast. Although 25 m of overburden needs to be stripped before mining, the exceptional thickness of the ore body (15–30 m) results in a very acceptable overburden to ore ratio of only 1.3:1.

Commissioning of the Ginkgo Mine is expected to take place in 2003, with the first shipments of minerals occurring early in 2004. The estimated capital cost of the venture is $158 million and the estimated revenues, over fourteen years, is $1.2 billion dollars.

strandline
a line marking the location of a former shoreline from which the sea has receded

lithology
the general physical characteristics of a rock

definitions

1

Draw up a table to summarise the exploration methods used in a mineral province you have studied. Include the use of the technique and the way the techniques are conducted.

2

Evaluate the evidence that the Cadia-Ridgeway deposits formed in an island arc environment.

3

Evaluate the evidence that the Ginkgo heavy mineral deposit formed in a shoreline environment.

4

Compare Cadia and Ginkgo. In what ways are the exploration methods used in the two areas similar?

5

Explain why drilling is almost always used at some stage of an exploration project.

6

a Explain why the assessment of grade and size of a deposit is important.

b Using an example discussed in this chapter, describe the methods used to determine the presence of mineralisation and the size of the reserve.

7

Demonstrate how a mineral deposit's mode of origin can be determined using information within and around the deposit.

SUMMARY

Australian mineral provinces have formed in a variety of tectonic environments, including island arcs and basins located on continental shields.

The six tectonic settings in which mineral deposits form are basins and rifts within continents, oceanic basins and mid-ocean ridges, passive continental margins, subduction-related settings, continental collision zones, and strike-slip environments.

Ore genesis, the way in which an ore forms, occurs in a part-icular tectonic environment. Igneous, sedimentary and metamorphic processes are possible parts of ore genesis.

Ore genesis processes can be classified according to whether they occur within the Earth (internal) or at the surface (surface).

Internal processes that form ores include:
• magmatic processes
• hydrothermal processes
• metamorphic processes.

Surface processes forming ores include mechanical accumulation (forming placers), chemical precipitates and leaching (residual and secondary enrichment).

The characteristic of a mineral determines the techniques used to explore for it. Exploration methods include library searches; geochemical, geophysical and geological surveys; drilling; and other detailed methods.

As the detail of an exploration program increases so does the cost of the program.

Cadia displays the character-istics of a deposit formed in an island arc environment.

The heavy mineral sand deposits of the Murray–Darling Basin are character-istic of placer deposits formed by sedimentary processes.

Researching a mineral province

The aim of this exercise is to solve problems, identify data sources, and gather and analyse information from secondary sources to identify the locations and geological structures associated with a chosen mineral province. Before proceeding, be sure to read all of the task outlined here.

Procedure

1

Select one of the mineral provinces described in this chapter. The first problem to solve is the most likely source of information on the province. Consider its location and the state geological survey most likely to have information on it. Alternatively, consider the mining companies that may be involved. Thirdly, consider the towns in the area. Will an Internet search of the town name and 'mining' yield results? Another strategy is to concentrate on the minerals derived from the province and their origin, that is, their ore genesis.

2

Decide on, and describe, at least two research strategies and use the resources available to you to gather information on the province.

3

Carry out a search for information. Be sure to keep a record of the websites and other resources you use.

4

Consider the analysis. What is the most appropriate way to describe the location and structure of the mineral province? Have you gathered maps or cross-sections that detail the locations and geological structures associated with your chosen mineral province? Draw a simplified diagram to identify the location and geological structure of your chosen mineral province.

5

Ore deposits can be classified or described according to a number of different criteria, such as:

a the minerals contained within the deposit, for example, porphyry copper deposits

b the shape or size of the deposit, for example, stratiform and strata-bound deposits

c the host rocks, for example, shale-hosted deposits and breccia pipes

d the genesis of the deposit.

Analyse the information you have obtained and present information relevant to the categories above in a table.

6

Review the process you have used. Write no more than ten lines outlining how you carried out this task.

OUTCOMES

At the end of this chapter you should be able to:

- analyse the process of determining the feasibility of mining a named deposit, referring to the stages involved in its development from a resource to a reserve

- explain how local, state and federal government policies may affect the decision to mine

- assess the impact of installing infrastructure or using that which already exists on determining the feasibility of mining a named deposit

- outline the methods and technologies used in the extraction, concentration and refining of ore from a named deposit.

THE AIM OF THIS CHAPTER

The aim of this chapter is to canvas some of the issues that determine the feasibility and profitability of a mine. This chapter builds on issues from Chapters 6.1 and 6.2. At the end of this chapter you will have the basic knowledge to undertake a mine case study using a mine of your choosing. So far we have looked in some detail at Cadia-Ridgeway and the Ginkgo Mineral Sands Prospect. Both are suitable mines for the case study but others, particularly those you can visit, will make suitable subjects.

The importance of the case study for the HSC examination needs to be emphasised. As the syllabus stands, you can be asked to describe a range of issues for a named deposit. You may also be asked to outline how you conducted your study. Remember that your skills are as assessable as your knowledge of the course.

DETERMINING THE SIZE OF THE DEPOSIT

In Chapter 6.2, we described the methods that allow an exploration team to determine the size and quality of a mineral deposit. The most important method used to define the size of an ore body and the distribution of minerals within it is drilling. (See Figure 6.3.1.) Drilling allows exploration geologists to define the shape of the ore body and to sample the ore itself. The closer together the drill cores are, the better defined the ore body will be. This does, of course, have a cost. The larger the number of cores produced, the more it costs to drill and assay the ore body.

On the basis of the core results and geophysical data, two-dimensional and three-dimensional models can be built using computers. Such models allow miners to determine more accurately the actual size of the resource. The size of the deposit is important for a number of reasons. In the first instance, the grade and tonnage indicate the value of the ore that may be mined. The size of the deposit also influences the cost of production. Generally, the larger the deposit, the lower the cost of production per tonne of ore. Large, low-grade ore bodies can today be mined in

Figure 6.3.1 Drilling on land involves mobile drilling rigs.

such a way that the cut-off grade is reduced. The negative aspect of large ore bodies (particularly large, flat-lying sheets) is that the scale of environmental impact may be proportional to their size. Proximity to the surface is also an economic factor because it affects the cost of removing waste rock or overburden.

DEVELOPING THE DEPOSIT

Factors affecting mine feasibility have been described in Chapters 6.1 and 6.2. Once financial, community and government approval have been obtained, it remains for mine infrastructure to be built and the mine to commence production. As the mine progresses, further exploration may extend the size of the deposit or define the ore body more fully.

The case studies in Chapter 6.2 show that a mineral prospect develops with each stage being preceded by a review of data and decision making. During the life of the mine, further decisions are made. They may involve changes to processing or a decision to move from open-pit to underground mining. The quest for greater efficiency is a characteristic of mining.

REVIEW ACTIVITIES

1
Describe how exploration methods are used to determine the size and grade of a mineral deposit.

2
Outline the stages involved in processing ore.

3
Use a flow chart to show the stages involved in developing a resource into a reserve.

EXTENSION ACTIVITIES

4
Outline the methods used in the feasibility study for the mine you have studied.

5
Discuss the factors that affect the life of a mine.

6
Assess the circumstances in which wars pose difficulties for mining.

EXTRACTION, CONCENTRATION AND REFINING

Figure 6.2.12 (page 315) details some of the procedures used by miners to identify and extract mineral deposits. Essentially, there are four stages in turning the ore into a metal: crushing and grading, separation of the ore, mineral recovery, and smelting.

Liberation of the ore involves crushing and sizing the ore. Crushing reduces the size of the particles, and sizing sorts them into fractions that are suitable for processing. If particles are too large the ore minerals are hard to separate, and if the particles are too small the equipment used in ore separation can become clogged. Jaw crushers, sag mills and ball mills play a role in this stage of the process.

Separation of the ore minerals from gangue minerals can occur through physical or chemical processes. Gravity methods use differences in the density of ore and gangue minerals to separate them. Tables, screen cyclones and dense media can all be used to separate ore from gangue. Some minerals, such as some of the heavy mineral sands, have electrical or magnetic behaviour that allows them to be separated from other particles. Chemical methods include leaching and froth flotation. In leaching, a liquid is used to dissolve the minerals that are to be recovered. Gold is leached using cyanide solutions, and some copper minerals dissolve readily in sulfuric acid. Froth flotation is a very common separation method. Detergent is added to a tank and air is pumped through the liquid to form bubbles. Mineral grains adhere to the bubbles and form a shining froth at the surface of the tank, while gangue minerals sink to the bottom.

Recovery of the product often involves dewatering the ore concentrate. Removal of water reduces the weight of the concentrate and makes it easier to transport. Often it is the concentrate that is sold. At Cadia, copper concentrate is sold to Japan and refined there. Mineral sands are treated in a similar way. The water that is removed can be reused within the mine or treated before being disposed of.

Refining may involve smelting and/or electrolytic refining. Copper ores, for example, are smelted to produce blister copper. The sulfur produced is removed as sulfur dioxide and removed from the air by machines called scrubbers. Carbon is used to remove the oxygen from some metals. The carbon monoxide or carbon dioxide that results is a greenhouse gas. Electrolysis is often used to improve the purity of metals. This process, together with the refining of aluminium, requires large amounts of electricity. This, in turn, contributes to carbon emissions, which have an indirect cost on production.

REVIEW ACTIVITIES

1
Name a mine you have studied. Outline the methods used to extract and refine the ore at the mine.

2
Summarise the methods used in processing minerals into metals and identify the properties of the minerals used by each method.

EXTENSION ACTIVITY

3
Make labelled drawings to show how a sag mill and a froth flotation tank work.

GOVERNMENT POLICIES AND MINE VIABILITY

Government policy affects mines at three levels. At the federal level, a range of issues can affect a mine's viability. The Federal Government controls export approvals and administers international agreements that may affect a mine. It also determines taxation rates, legislates company law, and controls investment through the Foreign Investment Review Board and currency controls. Should the Federal Government wish to prevent a mine going ahead, it has the power to do so.

At a state or territory level, government controls the allocation of licences and approvals. It also monitors environmental impacts and takes royalties from the mine. The state or territory government has a great deal of power over how a mine is run.

At the council level, the role of local government depends on the size of the mine. Some infrastructure is controlled by council, and the council is the first point of contact for a company wishing to deal with the local community.

MINES AND MARKETS

Economic, political and environmental issues will also affect the value of a deposit. The **capital investment** is usually tens or hundreds of millions of dollars. Few companies have such funds on hand and so they must borrow the money. Repaying the capital and interest, in addition to government fees and the mine's running costs, has to be planned for before the mine proceeds. Accurate projections of the possible income from the mine rely on knowledge of currency changes and commodity prices. Locating a mine in a politically stable country with a well-managed economy allows the mine's costs and income to be more easily and reliably calculated. Global

capital investment
the money needed to develop a project, such as a mine

definition

economic conditions can affect commodity prices, and long-term forecasts contain some risk. Access to land and the nature of the legal system also play a part in mine feasibility.

Economic issues include not only markets but also the cost of infrastructure. The location of the ore body will have an impact on costs because it affects distances to transport, such as rail lines or ports. The closeness of the workforce also has an economic effect because it influences shift structures and potential productivity. An important factor in the feasibility of the Ginkgo Prospect was the proximity of Broken Hill. Road transport and treatment near the rail line was cheaper than treatment at the site because the infrastructure already existed in Broken Hill.

Mining treatment can be a major cost in a mine. As the mine develops, test processing often occurs to work out how cheaply the ore can be processed. On the basis of such trial results, a company can calculate the likely cost of processing plants and the nature of tailings needing treatment or disposal. Variations in ore quality can introduce technical risks into mine feasibility calculations.

REVIEW   EXTENSION
ACTIVITIES   ACTIVITIES

1
Discuss the economic and political factors that influence the viability of a mine.

2
Assess how the presence of necessary infrastructure, such as roads, will affect the feasibility of a mine.

3
Discuss three issues that were detailed in the environmental impact statement for the mine you studied.

4
Outline the costs and benefits of mines to the communities located near them.

SUMMARY

- Together with geophysical methods, drilling is used to determine both grade and tonnage of a deposit.

- Size, grade and proximity to the surface all affect the economic viability of a mine.

- Financial, community and government approvals need to be obtained before a mine is built.

- A variety of methods are used to extract, concentrate and refine metals. These methods may be physical or chemical processes.

- Methods used in concentrating ores include crushing, sizing, concentrating and dewatering.

- Governments affect mines at the federal, state and local government level.

- The Federal Government controls taxation, investment and export conditions.

- State and territory governments allocate licences and approvals to regulate how the mine runs and affects the environment.

- Councils may regulate aspects of a mine's infrastructure.

- The value and viability of mining a mineral deposit are affected by economic, political and environmental issues.

PRACTICAL EXERCISE
A mine case study

The purpose of this exercise is to plan and perform first-hand investigations and/or gather information from secondary sources to prepare a case study of a mineral deposit. In the study you will use available evidence to evaluate the methods employed in determining the feasibility of mining your chosen deposit. You will also gather information to describe the methods used to extract and refine the ore.

In this exercise you will also gather information from secondary sources or from a mine visit to build a case study of a mine. This is a particularly important exercise because it pulls together many aspects of the topic. You can also expect to be asked about your mine and its history in your final examination.

Procedure

Organise your information under the following headings:

- name of the mine
- company that owns the mine
- methods used to locate, gather and analyse information about the mine
- sketch map of the mine's location
- ore genesis of the deposit
- discovery
  - methods used in exploration and evaluation
  - history of the exploration program
  - structure of the ore body
  - grade and tonnage calculations
- evaluation of feasibility studies
  - methods used
  - results of the feasibility study
  - issues that arose
- methods used in mining
- methods used in extraction and refining
- current markets and commodity prices
- environmental impacts of exploration, extraction and processing
- rehabilitation plans
- sustainable development practices
- projected future of the mine.

Before you begin, decide on how you will obtain the information. If you are lucky enough to visit the mine you are studying, make sure you prepare questions and a checklist before you go. Prior preparation makes an enormous difference to what you will learn.

This is not an overnight task. Keep a logbook to track your progress and to keep your research together. Also review the syllabus Skills Outcomes, particularly H12.3. Use the points as a checklist to ensure you use a range of methods when gathering your information.

Present your study as a report. Consider how you will report your information succinctly. Use tables, flow charts and diagrams, where appropriate.

6.4 Mining, society and the law

OUTCOMES

At the end of this chapter you should be able to:

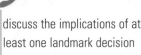

discuss the implications of at least one landmark decision on mining operations in Australia, such as Wave Hill, Mabo or Wik

outline the effect of at least one local, state and federal government policy on mining operations in the context of sustainability.

THE ISSUES OF ACCESS AND OWNERSHIP

Before a mining company can begin an exploration program it must obtain an exploration licence. An exploration licence applies to a specific area of land and is granted by a state or territory government for a certain period of time. This is usually two to three years, but may be as long as five years. The licence does not allow the company to mine, only to explore. An exploration licence contains a series of conditions that the company must follow. The company has to lodge a security deposit, which can be forfeited if the mining company does not follow its licence conditions. The security deposit in New South Wales is approximately $10 000 and officers from the Department of Mineral Resources check the program to see that the conditions are followed.

As part of obtaining an exploration licence, a mining company must reach agreement with the owner of the land over the terms of access. The company also must submit an environmental impact statement addressing the steps the company will take to minimise its impact on the area and how it will rehabilitate areas that are disturbed.

Since 1990, the *NSW Mining Act* has required a holder of a mineral exploration licence to advise the owner of the land they wish to explore of their intention to do so. They have to provide details of the activities they intend to carry out and the timing of their activities. The explorers must reach agreement with the landowner before they begin activities. The agreement is termed an **access arrangement**.

An access arrangement can cover a range of issues, including:
- the period during which the explorer has access to the land
- the areas of the property to be accessed and how they will be accessed
- the kinds of operations that will be conducted
- exploration licence conditions
- how disputes will be resolved
- compensation to be paid to the landowner
- how the arrangements may be varied during the exploration
- any other matters on which the two parties may agree.

If the explorer and the landowner cannot come to an agreement, an **arbitrator** is appointed by the minister responsible for mines and agriculture. The arbitrator looks at the cases put by the parties and endeavours to help the two parties reach an agreement. If the minister's arbitrator cannot resolve the issue, the mining warden for

access arrangement
an agreement between landowner and explorer as to the conditions of access onto private property

arbitrator
someone who is appointed to settle a dispute

definitions

the area determines the access conditions. Fewer than one in 100 cases require an arbitrator and the mining warden rarely has to make a decision.

Such agreements are important to landowners because they have to live with the result of exploration activities long after the exploration teams have left. (See Figure 6.4.1.) Remember that very few exploration projects result in a mine, but at any time in New South Wales there are many hundreds of exploration licences in effect. If a landowner cannot place some conditions on access the chance of pollution or the introduction of weeds or diseases may be significant.

The issue of land access is important to any landowner and is of particular importance to indigenous peoples. The land plays a central role in the lives of Indigenous Australians and there exist sites with particular spiritual significance in many parts of Australia. Since the start of European settlement, Indigenous Australians have been displaced from lands they traditionally occupied. Until the 1970s, a system of law and social understanding evolved in Australia that denied the claims of Indigenous peoples to ownership of land that they had occupied for thousands of years before European settlement.

REVIEW ACTIVITIES

1
Outline the characteristics of an exploration licence.

2
List the issues that are addressed by an access arrangement.

EXTENSION ACTIVITY

3
Outline how the mine you described in your case study is affected by state, federal and local council policies.

THREE LANDMARK DECISIONS

In 1971 the Yirrkala people of the Northern Territory sought through the courts to stop a mining company operating on their traditional lands. They failed because, while they could show they had a system of laws and practices that they applied to the land they traditionally owned, the court was unable to recognise the land tenure under Australian law as it stood. The effect of the decision was the establishment of the *Land Rights Act* in the Northern Territory in 1976. Under this act, traditional owners could claim land but the land they could claim was limited and often was only claimable on condition it was leased back to the government for public use.

In 1982 a court case began in the High Court of Australia that would change the way the law saw Indigenous Australians' relationship with the land. The result of the court case is known as the Mabo Decision. Eddie Mabo was one of five Torres Strait Islanders who sought in the case to establish the right of their families to land under Meriam Law. Mabo lived in the Murray Islands and belonged to the Meriam people. He and the other parties sought to show that their ownership of the land was based on a system of law that had existed for a long time within their culture.

It took ten years for a decision in the case, but the High Court found that the Meriam people had a right to possess and occupy their lands according to their laws and customs. Their right to the land was not based on colonial law, but on their distinct identity as the first owners of the land and the continuing system of law they practised. The High Court also found that the principle not only applied to the Meriam people but to all Indigenous peoples throughout Australia.

The effect of the Mabo Decision, together with the introduction of the *Racial Discrimination Act* (Cwlth) in 1975, was hotly debated by those with interests in land

Figure 6.4.1 Ranger uranium mine at Jabiru surrounded by Kakadu National Park.

ownership and access. The idea of extinguishment—that land title could not exist where a connection with the land could not be proved, or where an alternative title had been conferred—restricted the claims of Indigenous peoples. However, it still left available large areas of vacant crown land, forests, parks, foreshores and Indigenous lands.

State and territory governments and companies working in primary industries were concerned that their access and use of land would be restricted in ways they found unacceptable.

In 1993 the Federal Government introduced the *Native Title Act*, which provided some protection for **native title** at the same time as it provided validation for past land grants without the need for compensation. A consequence of this legislation was the recognition of Indigenous peoples' right to negotiate over dealings with their lands. As a result of negotiation, a number of agreements about access and use of land have been successfully concluded, but concerns by some parts of society have led to the Federal Government restricting native title recognition through what is known as the 'ten-point plan'.

definition

native title
the title to ownership of land by Indigenous peoples

REVIEW ACTIVITIES

EXTENSION ACTIVITIES

1
Summarise the features of the Mabo Decision.

2
Explain the implications of the three landmark decisions for access to land for mining.

3
Analyse the conflicting views on the desirability of native title. In particular, assess the effect of native title on the ability of mining companies to carry out their business.

4
Use secondary sources to find out what the 'ten-point plan' is.

REGULATION AND SUSTAINABILITY

The principles of ecologically sustainable development are:
- *Sustainable use.* This means that resources are not used to the extent that they are exhausted.
- *Efficiency of resource allocation.* Efficient resource allocation means that resources are not wasted and that the best use is made of the limited resources we have.
- *Equity.* This idea aims to ensure that minority groups and those with few resources have the ability to make their views heard even if other interests are more powerful.
- *Environmental quality.* This principle implies that nature has a right to exist in a state that does not depend on our sense of its usefulness. Given that we understand that the quality of the environment affects the quality of our life, the need to preserve and maintain the environment for itself is becoming more important.
- *Public interest.* This principle recognises that society as a whole has special needs with regard to the environment and that special treatment should be afforded the environment in all our interests.

Government regulations on mining and exploration aim to address these principles. A variety of government regulations were discussed in Chapter 3.6 of *Earth and Environmental Science: The Preliminary Course*. You are encouraged to review that part of the Preliminary Course.

The environmental impact statement is just one way in which a state or territory government can insist that mines are operated sustainably. Conditions of exploration and mining titles, together with Mining, Rehabilitation and Environmental Management Plans (MREMPs), help to ensure that environmental quality and public interest issues are managed well. **Audits** of mining operations and field inspections ensure that conditions are met and a range of regulatory mechanisms and sanctions are available to ensure that conditions are met. A mine's operation can be suspended by an inspector if the operations contravene the conditions under which the mine operates. The security deposit required under the NSW *Mining Act 1992* can be reviewed or forfeited. A company can also be prosecuted under the Mining Act or the NSW *Environmental Offences and Penalties Act 1989*.

In determining exploration or mining licences, the government also insists that a company shows it has the resources and expertise to carry out its role. This helps to ensure that unsustainable practices do not occur and that the resource will be mined efficiently.

The Department of Land and Water Conservation (DLWC) also plays a role in regulating the way a mine operates. Water is required in mines for many purposes and a water licence is required from the DLWC. The DLWC also regulates land clearing and is responsible for preventing soil erosion and degradation.

The importance of rehabilitation is addressed by the NSW Government's rehabilitation policy. This recognises that both exploration and mining are transient—they occur for a relatively short period of time. The policy lists a series of principles (fifteen for exploration and twenty-nine for mining) that aim to have land rehabilitated to a state where it can be used for other purposes, including use by the community.

audit
an examination of how procedures are carried out

definition

REVIEW ACTIVITIES

1
List three principles of ecologically sustainable development and match them up to legislation that regulates mining.

2
Explain the effect of three government policies on mining operations.

EXTENSION ACTIVITY

3
Research the NSW Government's mine rehabilitation policy. Which principles apply both to exploration and mining? Explain how the principles help achieve sustainability.

SUMMARY

An exploration licence is needed before exploration can occur, and the licence imposes conditions on the way exploration occurs.

Access to land and submission of an environmental impact statement are important steps in obtaining an exploration licence.

Ecologically sustainable development is encouraged by government instruments such as the EIS, the Mining Act and MREMPs.

The Mabo Decision and the Native Title Act have altered the rights of Indigenous peoples to control over traditional lands.

Government departments, such as the DLWC, play roles in monitoring the environmental effects of mines.

Summarising the Wik Decision

The aim of this exercise is to summarise the details and implications of a landmark decision in the form of a report.

Background

In 1996 the Federal Court reaffirmed the concept of native title in the case of the Wik peoples versus Queensland (1996). The Wik Decision, as it is more widely known, indicated that native title can co-exist with many forms of title. In particular, the Wik Decision dealt with native title and its relationship to pastoral leases.

Procedure

Use secondary sources to research the Wik Decision and write a short report on it. In particular, find out:

- the nature of a pastoral lease
- the reason for the Wik people bringing the court case
- the nature of the court's judgment
- how the judgment affects a mining company's access to land.

In your report, predict how the Wik Decision would impact on the mine you studied in your mine case study.

Alternatively, carry out this study for the Mabo Decision.

6.5 Mining and the environment

At the end of this chapter you should be able to:

assess the likely environmental effects of exploration, mining and processing methods for a named deposit

evaluate the purpose of the environmental impact statement for a named deposit in terms of protection of unique and endangered species, protection of sacred sites, community consultation and local habitat management

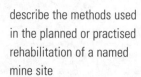

describe the methods used in the planned or practised rehabilitation of a named mine site

evaluate the relationship between mining methods and mine-site rehabilitation for a named deposit.

THE ENVIRONMENTAL EFFECTS OF EXPLORING, MINING AND PROCESSING

All mines have the potential to cause damage to the environment. Besides the mining processes at a mine site, it is often the case that processing plants are built at the site. Such plants may produce materials that affect the environment. The principles of sustainable development were described in the last chapter and they guide both government regulation and community involvement with mining. Some of the activities we expect of miners today include:

- minimising land contamination
- minimising impacts on archaeological, Indigenous and other cultural sites
- minimising impacts on natural ecosystems
- complying with government regulations covering other potential effects of mining and processing procedures.

Effects on land, water, air and living things

History has shown that a range of problems can be created by mining. The problems can be grouped according to how they affect land, water, air and living things.

Land can be affected in a number of ways. Changes to the shape of the land surface can increase rates of erosion and water run-off. Toxic substances can contaminate the soil. Minerals locked deep in the ground may be mixed with soils at the surface. Toxic chemicals, such as cyanide, from processing plants may also contaminate the soil. Vibration and subsidence can be issues for people who live near mines. Vibration most often results from the use of explosives within the mine. Subsidence occurs when ground levels change because of the collapse of parts of an underground mine.

Water is critical both to mines and the wider community. If toxic substances in water are allowed to enter streams they can create major problems for local ecosystems. The mine must ensure that such substances, or fine sediment called tailings, do not leave the mine site. Acid mine drainage is a problem associated with sulfide mines. Sulfide minerals react with the air, water and bacteria to produce sulfuric acids. The acidic water can release toxic elements from minerals, creating further problems. Acidic waters can be produced at every stage of a mine's life, from exploration to closure.

A particular problem associated with contaminated waters occurs when a mine closes. Without pumps to lower the water levels, flooding within the mine may result in surface discharges of contaminated water. Leaching of waste heaps by rainwater may produce similarly contaminated water. Ground water contamination can be as important an issue as leakage into waterways without adequate monitoring and planning.

Air pollution can arise from dust generated at the mine site or from smelter-derived gases and particulates. Dust is particularly dangerous when it contains toxic or harmful materials. When asbestos was mined in different parts of New South Wales early in the twentieth century, clouds containing cancer-causing asbestos fibres drifted across the countryside from mines, contaminating houses and farmland. When smelters were used to roast sulfides, sulfuric acid droplets caused acid rain, which killed vegetation around the mines. Queenstown in Tasmania shows the effects of such processes that began 200 years ago. (See Figure 6.5.1.)

Living organisms can be affected in a number of ways. The loss and alteration of habitats is a major issues in areas that are **strip mined**. Air and water pollution can also affect ecosystems in ways that reduce biodiversity. Animal or plant populations with limited distributions may be adversely affected if a mine is developed over most of their range. Also, the noise from mines affects both people and animals.

Figure 6.5.1 Environmental damage caused by smelting at Queenstown, Tasmania.

strip mining
a form of open-pit mining in which the pit is developed in one direction and filled in behind as the pit develops

definition

Limiting and monitoring environmental effects

Today, we recognise that any change to the environment can produce adverse effects. This is as true of urbanisation, farming, forestry and tourism as it is of mining. While the cumulative use of land for mining comprises 0.2% of Australia's land surface, miners have an obligation to develop their businesses in a sustainable way, that is, in a way that allows future generations to use the land after mining is finished.

The conditions associated with licences to mine and explore go some way to helping miners concentrate on their environmental responsibilities. The conditions under which a mine can operate often include limits on the production of harmful materials. In order to ensure the limits are not exceeded, monitoring of the environment is carried out by both government authorities and by the miners themselves.

In New South Wales the Department of Mineral Resources has a system of environmental audits for operating mine sites. The auditing process involves identifying key elements of a mine's operation that may affect the environment. The key considerations include surface water management, how monitoring is conducted, how ore tailings are processed and disposed of, and how waste rock is stored and disposed of.

Comparisons are also made between licence requirements and the mine's impact on the environment. If, for example, dust levels away from a mine pit exceed the levels mandated in the mining licence, action by the mine operator or regulating authority will occur. A third issue is modifications to the mine site. Such changes are checked to ensure that they conform to policy guidelines. How a mining company documents its conformation to the regulations is also examined.

Mining companies control levels of emissions in a variety of ways. Water leaving the mine site is monitored regularly. In some cases, no water may be allowed to leave the site. If water is allowed to leave the site, or sit open within the mine, a range of indicators will be measured. These include pH (acidity); the amount of suspended solids; the presence of metals, toxic chemicals, oil or other lubricants; and the amount of salts dissolved in the water. Geophysics and hydrology may be used to check dams for leaks and to ensure that water from the mine does not enter aquifers.

Noise is controlled in a number of ways. Scheduling of truck movements, controlled blasting, and walls or windbreaks can all reduce noise levels within and around the mine. In underground mines, primary crushing may occur underground. This not only reduces noise but can reduce dust emissions too. Occupational safety regulations will, however, require workers to be afforded protection if they work in noisy areas.

Other issues are controlled in a variety of ways. The nature of control programs depends very much on the particular situation of the mine. In your mine case study you will examine a particular example in some detail.

REVIEW
ACTIVITIES

EXTENSION
ACTIVITIES

1

Outline the features of an environmental audit.

2

Summarise the ways in which water quality in a mine site is monitored and controlled.

3

Compare the potential environmental impacts at a metal sulfide mine with those found at a mineral sands mine.

4

Mines use a great deal of energy, particularly electricity. Assess the environmental costs of supplying a base metal mine with the electricity it needs.

THE FUNCTION OF THE ENVIRONMENTAL IMPACT STATEMENT

We have already seen that an environmental impact statement (EIS) is a mandatory requirement of applying for an exploration or mining licence. An EIS allows both government and the public to evaluate a mining company's proposal. It also allows the government and the mining company to determine an acceptable set of targets for monitoring and rehabilitation.

The areas covered in an EIS describe what exists at the site, how the site will change and how it will be rehabilitated. First, the EIS outlines a description of the area. The description includes information on the geology, geomorphology, soils, flora and fauna that exist there. Particular attention is given to unique, threatened or endangered species that exist within the proposed mine area. The measures to be taken to protect unique and endangered species are described and attention is also given to how the local habitats surrounding the mine will be managed.

The human environment is also described. The location and nature of nearby towns, roads and other infrastructure are described and an outline of how the mine development will affect these areas is included. Such a description should include both positive and negative effects. Where possible, the EIS will also contain alternative proposals, including costings and implications. The protection of sacred sites and other sites of cultural significance are also considered.

Details of the control measures to be undertaken are also part of the EIS. For example, in areas where dust will be an issue dust limits will be set and the EIS will outline ways by which the mine will control the dust. For example, water may be used to damp down roadways, blasting may be avoided on windy days, and revegetation targets for areas may be undertaken.

Lastly, the EIS will contain a plan for mine decommissioning. This is the process of dismantling the mine and regenerating the area. The nature of such processes is described below.

REVIEW ACTIVITIES

EXTENSION ACTIVITY

1
Outline the issues covered within an EIS.

2
Assess the function of the EIS in ensuring care of the environment.

3
Create a model EIS for the mine you have studied.

REHABILITATION METHODS

On completion of mining, either of part of an area or at the end of a mine's life, the area has to be rehabilitated. Rehabilitation means returning an area to a condition where stable ecosystems can develop and where landowners can make use of the land. Figure 6.5.2 shows some of the features of a mine that may be left when a mine operation ends. Tailings may be stored in dumps. Mined-out pits may need attention. Changes to the shape of the land surface may need attention to reduce erosion and water build-up. As was indicated in Chapter 3.6, issues that need to be dealt with include:

- removal, relocation, or demolition of buildings and physical infrastructure
- closure of pits and shafts
- stabilisation of underground workings, soils and slopes
- treatment of tailings and associated waste
- treatment of waste dumps
- revegetation of the land.

Landform stability is a major issue for mine rehabilitation. Contouring of the surface to allow revegetation and control water run-off is only part of the process.

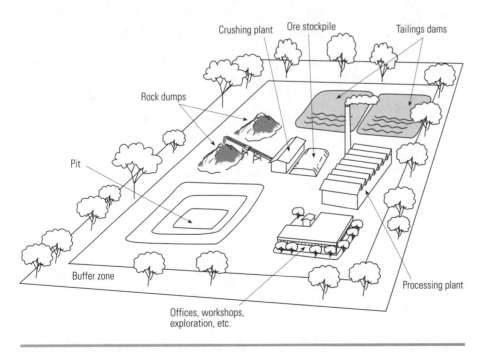

Figure 6.5.2 Features of a mine site. (*Note:* The arrangement of the features varies, but efficiency of movement is an important consideration. Safety is also a major design consideration.)

Slopes need to be of the correct angle so that landslips do not occur. Besides being dangerous to people and affecting vegetation, such an event may expose tailings to the air and allow acidic waters to develop. Burial of tailings that pose a hazard may entail using a clay seal before overburden and topsoil are distributed. Where water contamination is an issue it may be possible to reroute small streams around the mine site. In some cases embankments, trenches and drains are used to control water flow.

Revegetation returns the land to a useable state. The plants that are used may be pasture plants or native vegetation. If native vegetation is used for regeneration the mining company needs to plan ahead. Remnant vegetation is often fenced off in parts of the mine site and may be used to provide seed if the area is large enough. A thorough study of pre-existing vegetation and animal populations is important in setting goals for the revegetated areas. Strip mining may occur so rapidly that topsoil is stockpiled and replaced on the contoured land quickly enough that the seed within the topsoil germinates and starts a return to a vegetation similar to that which existed before. Sometimes this process produces unexpected results. If the area was stressed before the land was cleared, the vegetation that regenerates may differ from that which went before. Fencing of revegetated areas is important. By keeping stock off newly planted areas the chances of erosion are reduced and the plants are given time to grow. Control of introduced weeds and animals, such as rabbits, can also be an issue. Wildlife corridors into the mine site can also hasten the revegetated process.

Open pits can present problems for rehabilitation. Around cities such pits are in demand for landfill sites. If the minerals removed are not sulfides the pit may make a good water reservoir. Fencing and landscaping are carried out to make these sites safe and reduce their visual impact.

REVIEW
ACTIVITIES

EXTENSION
ACTIVITIES

1
Describe some of the methods that are used in rehabilitating a mine site.

2
Explain how rehabilitation methods prevent erosion and water contamination.

3
Explain how acidic mine drainage occurs.

4
Mines that have closed because they were uneconomic may be reopened as technologies allow lower grades to be utilised. Outline how a rehabilitation program could prevent a mine from reopening.

5
Draw up a table containing headings for sustainable development principles, monitoring, planning and rehabilitation for the mine you have studied. Then complete the following activities:

a Use the table to match practices and plans seen at the mine against the sustainable development principles.

b Assess the table. Analyse the most effective sustainable development practices used in the mine.

All mines can have an adverse effect on the environment unless they are well planned, monitored and managed.

Mines can affect land, water, air and living things.

Environmental audits involve identifying key elements of a mine's operation that may affect the environment: surface water management, how monitoring is conducted, how ore tailings are processed and disposed of and how waste rock is stored or disposed of.

An EIS details a number of issues, including how a mine will affect the environment and how regeneration will occur.

Rehabilitation and regeneration are important processes that occur when a mine ends its life.

PRACTICAL EXERCISE
Rehabilitation planning:
An investigation

The aim of this exercise is to describe the methods used in the planned, or practised, rehabilitation of a mine site. Ideally, you will research the mine you are using for your mine case study.

A central document in understanding the rehabilitation planned for an area is the EIS. Feasibility studies also contain information relevant to rehabilitation planning.

Procedure

1

Generate a set of search terms you may use to find information on the company that runs the mine.

2

Decide on a set of search terms that deal with rehabilitation practices or issues.

3

Consider organisations that may be a source of relevant information, such as the Environmental Protection Authority, the Department of Mineral Resources and Geoscience Australia.

4

Use the terms you have developed to search for information in a library or on the Internet.

5

On the basis of the information you find, write a point-form description of the methods planned for, or practised in, the rehabilitation of the mine site.

PRACTICAL EXERCISE
The environmental impact of mining

The aim of this exercise is to plan, gather and present information on the environmental impacts of a mine. In particular, you should aim to distinguish between the different environmental impacts produced by exploration, extraction and processing methods.

Ideally this exercise will be done on a mine visit. Otherwise you will have to treat the exercise as another secondary source exercise.

Procedure

1
Outline the features of the exploration, extraction and processing methods used at the mine you are studying.

2
For each phase of ore extraction or treatment, identify the major environmental impacts generated by the process operating at the phase.

3
Try to ascertain the limitations set under the mining licence for water quality, noise and dust within the mine site and in the area immediately surrounding the site.

4
Outline the area of the mine where the greatest potential for environmental impact occurs. Describe the methods used to contain the potential impact.

5
Summarise the environmental impacts occurring at the mine.

6
Summarise the nature, cause and possible remediation strategies of environmental impacts due to mining. Create the summary in the form of a table.

Oceanography

7

When viewed from space, the Earth looks blue. This is because almost 71% of the planet's surface is covered in oceans. The Earth is unique among the planets of our solar system because water is stable in all three forms (liquid, solid and gas) on the planet's surface. This occurs because the Earth is placed within the range of distances for this to be possible. If the Earth had formed too close to the Sun, the potential oceans would have vaporised and become a component of the atmosphere. If the Earth had formed too far away from the Sun, all the water would have frozen.

Much of the Earth's water is contained in large ocean basins or smaller seas whose floors are composed of basalt that has been generated at mid-ocean ridges. Away from the ridges, the sea-floor basalts are buried under a relatively thin covering of mostly fine-grained sediments that have settled through the sea water and come to rest on the sea floor.

Serious scientific investigation of the oceans began early last century. In this section we will investigate how the oceans are studied and find out many fascinating facts that have been discovered about them.

CONTENTS

7.1 Collecting data about the oceans

OUTCOMES

At the end of this chapter you should be able to:

- outline the range of data that can be collected by echo sounders and describe the principles involved in the collection of data

- describe the processes involved in the collection of sea-floor samples by dredges, grabs and core samplers

- assess the use of biological nets for gathering plankton and other organisms and relate this assessment to the need for continuous data on food chains in the oceans

- identify the information obtained by nansen bottles with electronic sensors

- identify the role of bathythermographs in terms of continuous surface temperature measurements

- identify the use of magnetometers for measuring magnetic intensity and the polarity of ocean-floor sediments and rocks

- describe the use of research submarines and deep-ocean drill ships in collecting information about the oceans and identify one example of the importance of using such research.

THE ROLE OF DATA COLLECTION

Our understanding of oceanic processes comes from the analysis and interpretation of a wide variety of information. Scientists call this information 'data' and they collect it using many different instruments and devices. Some measure physical properties, such as the temperature and **salinity** of sea water. Some are designed to take actual samples of the sea floor or of water from a particular depth. Other instruments use sound reflected off the sea floor and the oceanic crust below it to provide profiles and images of the sea floor and the underlying crust.

Over the last century, and particularly since the late 1970s, the sophistication of the equipment used to survey the sea has improved enormously. The quality of the information collected today is so good that many of the difficult deductions that were made in establishing the theory of plate tectonics, such as the magnetic polarity striping of the sea floor, seem almost obvious to modern-day ocean scientists. However, an important point needs to be made about this sophisticated equipment; the data that lead to new theories are always collected before the scientific explanations are made. Better data usually make it easier to work out which idea is the most plausible explanation.

Collecting oceanographic data is an expensive business. It requires a safe ocean-going vessel and specialised equipment. Examples of many of the tools, vehicles and instruments that are used to study the oceans, its creatures and the sea floor are described in this chapter.

MAPPING THE SEA FLOOR

Echo sounders are devices that are used to determine the depth of the sea floor. They emit brief pulses of high-frequency sound from a special underwater speaker, called a transducer. The echo sounder measures the time that it takes the sound pulse to travel to the sea floor, bounce off it, and then return to the surface. (See Figure 7.1.1.) The time taken can be converted directly to water depth if the speed of sound in the sea water is known. (See margin box.)

The first echo sounders were used on ships in the 1920s and provided mariners with an accurate estimate of the depth of water under the vessel. Echo sounders give sea-captains instant and continuously updated readings of water depth while their ship is under way. They are a vast improvement on the earlier technique of water-depth measurement. This technique, called lead-lining, consisted of stopping the ship, throwing a measured line (weighted with a lead sinker) over the side and waiting until it hit the bottom. Ships under way can't accurately use lead-lining because the drag on the line prevents it from descending to the bottom in a straight line. Until the late 1980s, most maps of the sea floor were made by contouring the data recovered from lead-line measurements or echo-sounder profiles.

Multiple-transducer versions of the echo sounder, called **multibeam echo sounders**, were invented in the 1980s and 1990s. Instead of emitting a single pulse, like the original depth sounders, these instruments generate a fan of many tightly directed beams that are orientated perpendicular to the ship's length. This technique produces maps of the sea floor directly underneath and on both sides of the ship. The area that these devices map is called a swath. Modern swath-mapping echo sounders produce highly detailed **bathymetric maps** of the sea floor, which allow geologists and geo-physicists to examine the intricate shape of submarine features.

Another device that uses the reflection of sound to produce images of the sea floor is the side-scan sonar. Sound can also be used to determine the shape and thickness of sediment and rock layers beneath the sea floor. This method is called seismic reflection profiling. Seismic-reflection equipment (see Figure 7.1.2, page 342) works like an echo sounder, but uses much lower frequency sound pulses. The pulses penetrate the sea-floor sediments and bounce off each successive layer of material. The profile that is generated is like a cross-section, and it shows the shape of the subsurface layers. One of these profiles is shown in Figure 7.1.3 (page 342).

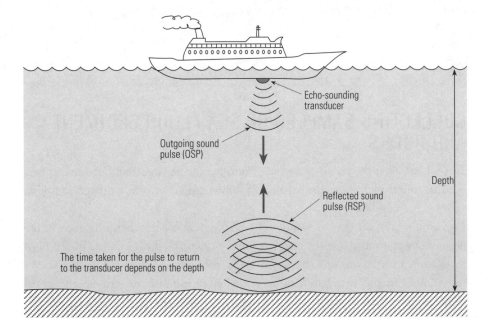

Figure 7.1.1 Echo sounding.

Calculating velocity using data

The speed of sound in sea water varies with temperature, depth and salinity, but is commonly approximated to 1450 metres per second (ms^{-1}).

The formula for velocity is

$$Velocity = \frac{Distance}{Time}$$

We know that the velocity of sound in sea water is 1450 ms^{-1}, that is,

$$Velocity = 1450 \ ms^{-1}$$

We can see from Figure 7.1.1 that the sound pulse travels from the ship to the sea floor and back again. So the distance travelled by the sound pulse is twice the water depth, that is,

$$Distance = 2 \times depth$$

So we insert these two pieces of information into the velocity equation, as follows

$$1450 \ ms^{-1} = \frac{2 \times depth \ (m)}{Time \ (s)}$$

We then rearrange it to make depth the object, as follows

$$Depth = \frac{1450 \times time}{2}$$

$$Depth = 725 \times time$$

---

echo sounder
a device that uses sound reflections to determine the depth of the sea floor beneath a specific point on the sea surface; sometimes called depth sounder

multibeam echo sounder
a highly sophisticated, multi-channel echo sounder that maps areas of the sea floor using a fan-shaped array of sound-beams

bathymetric map
contour map showing the shape of the sea floor

definitions

A marine magnetometer is towed behind a ship to measure the Earth's magnetic field. The magnetometer's signal can be processed to remove the dominant component, that is, the Earth's present-day magnetic field. The remaining component can then be used to determine the basaltic sea floor's remnant polarisation and magnetic field strength. This type of data is the same as that used to demonstrate magnetic stripping of the sea floor resulting from spreading at mid-ocean ridges.

Figure 7.1.2 Seismic-reflection equipment.

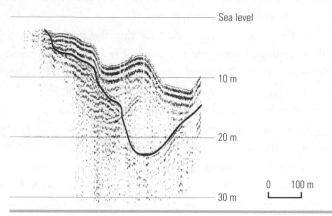

Figure 7.1.3 An interpreted seismic-reflection profile recorded by the equipment shown in Figure 7.1.2. It shows the location of a hard, rocky basement (indicated with a thick line) beneath layered, muddy sediments.

REVIEW ACTIVITIES

1
When were ships first equipped with echo sounders and what information do they provide to mariners?

2
Outline how an echo sounder works and indicate why these instruments are an improvement on lead-lining.

3
Explain why geologists and marine scientists prefer to undertake explorations of the sea floor using multibeam echo sounders rather than single-channel echo sounders.

4
Why was the invention of the marine magnetometer important for the theory of plate tectonics?

EXTENSION ACTIVITIES

5
Search the Internet for a high-resolution image of sea-floor topography from the ocean areas offshore from the Australian landmass. Name the method that was used to generate the image and investigate how this method works.

6
Find out how a side-scan sonar works and what it can be used for. Locate side-scan images on the Internet. Present your findings in a poster.

COLLECTING SAMPLES OF SEA-FLOOR SEDIMENT AND ROCKS

To find out what the sea floor is made from, marine geologists and oceanographers need to collect samples of the sea floor. Samples are taken with a dredge, sampling grab or core sampler.

Sampling the sea floor is wonderfully exciting, especially when it is done in the deep sea away from the main shipping lanes. When the dredge comes back full of rocks from a sea mount or a submarine slope, the crew knows that nobody has ever seen those particular materials before. Even when no samples are retrieved, the arrival of the sampler back on deck is often a cause of excitement for the crew. Corers and dredges frequently come back bent out of shape because the bottom was made from materials different from what the crew expected. Sometimes the gear is lost entirely because it is irretrievably snagged on the bottom.

Sampling devices are lowered to the bottom of the sea on a braided steel wire that is let out and then reeled back in by a winch. Various samplers have been designed and built to recover different sea-floor materials. The choice of sampler to use will depend on whether the crew expects mud, sand or rock to be on the bottom.

Dredges

Dredges are open-mouthed steel buckets that are dragged over the sea floor. (See Figures 7.1.4 and 7.1.5.) They are generally used to recover samples from rocky pavements and submarine cliffs or coarse sediments, such as cobbles and boulders, from the sea floor.

Some dredges are made from large pieces of circular steel pipe (0.5–3 m in diameter) with the bottom end sealed by a piece of steel. Another type consists of an open steel box, from which a bag made from steel chain hangs. The openings in the chain bag allow the finer material to be washed out so that only the large rocks are brought back to the surface.

Grabs

Grab samplers are generally used to recover surface samples of soft muds and sands. The grab consists of a spring-loaded scoop or a pair of jaws that are triggered when the sampler touches the bottom. The scoop digs through sediment or the jaws take a bite out of the bottom and then seal up tight so that the sediment cannot be washed out of the sampler as it is winched back up to the surface.

Corer samplers

A **core sampler**, or corer, is a long, thin cylinder of steel pipe (about 10–20 cm wide and up to about 20 m long) that has a sharpened cutting end at the bottom and many lead weights attached to the top. Corers are used for sampling soft, fine-grained sediments, such as sand and mud. They provide a short, continuous, cylindrical sample of the sediment layers that occur just below the sea floor.

The corer is lowered to near the sea floor by a cable and winch. (See Figure 7.1.6, page 344.) It is then dropped a short distance into the sediment and the sample is collected. The weights and the fall of the corer drive it down into the sea floor like a giant nail. In fact, you can think of a corer as being like a giant hollow nail. The hollow inside of the corer is lined with a length of plastic sample tube into which the sediment sample is forced. After the corer stops moving, the ship's winch is slowly reversed and the corer is gently pulled out of the sea floor and brought back to the surface. When it is recovered from the water, the corer is laid out on the deck of the ship and the cylindrical plastic core liner, which has the sediment inside it, is removed from the steel core barrel. The plastic liner and sediment are then cut in half. The samples can then be analysed and the relatively undisturbed sedimentary layering and structures can be examined and photographed.

Deep-ocean drilling

The Deep Sea Drilling Project (DSDP) began in 1964 and ran many scientific cruises aboard the research vessel *Glomar Challenger*. In 1969, one of the first cruises aboard this ship recovered progressively older pieces of basaltic crust in holes drilled further

Figure 7.1.4 Recovering a closed-pipe dredge and a small box dredge.

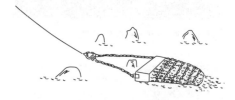

Figure 7.1.5 A box dredge in action.

dredge
a type of sea-floor surface sampling device that collects sediment or rock from the sea floor as it is dragged across it

grab
a device that recovers a sample of the sea floor at a single point

core sampler
a pipe-like device that is used to recover a thin cylinder, or core, of sub-bottom sea-floor sediment

Figure 7.1.6 Deploying a piston corer.

Figure 7.1.7 Recovering the drill bit aboard the ODP's ship *Resolution*.

and further away from the Mid-Atlantic ridge. This information proved that Harry Hess's more detailed model of sea-floor spreading was correct and was instrumental in providing conclusive proof to support the continental drift and plate tectonics theories.

The DSDP was followed by the Ocean Drilling Program (ODP), which began in 1985. These two programs will be replaced in about 2005 by the Integrated Ocean Drilling Program, which will coordinate a research program that will use a number of different drill ships.

Each year the ODP's geological research vessel, *JOIDES Resolution*, departs on six scientific expeditions. The *Resolution* is 143 m long and 21 m wide, and its derrick rises 61.5 m above the water line. The ship's drilling system can handle 9150 m of drill pipe, which is long enough for drilling in 99.9% of the world's oceans. A computer-controlled system regulates twelve powerful thrusters in addition to the main propulsion system, which keeps the ship almost completely stationary over the drill hole despite wind and waves. The ship can drill in water depths up to 8235 m.

Each expedition has a specific scientific goal that aims to work out the cause of a particular geological event, such as the K/T event. Hydraulic piston corers or rock drills are lowered from the ship and cut through layers of sea-floor sediment and rock. Cores of material from below the sea floor are collected in tubes. The cores are in segments up to 9.5 m in length. The cores are then returned to the ship's deck by a special wire cable that pulls the core up through the middle of the drill pipe. (See Figure 7.1.7.) When the 9.5 m long core segments are extracted from the core barrel onto the drill deck, they are immediately cut into shorter 1.5 m sections and taken to the ship's core lab, where they are permanently labelled. Information about each core section is entered into a database to enable data about it to be properly recorded. Then the core's materials are identified and many measurements made.

The *Resolution* is equipped with shipboard sediment and rock laboratories. The physical, mineralogical, chemical and biological characteristics of the materials in the core samples can be determined from the samples as they are recovered. (See Figure 7.1.8.) Upon completion of each expedition, the precious cores are transported to storage facilities for curation and future research.

Figure 7.1.8 Measuring the physical properties of a core.

REVIEW
ACTIVITIES

EXTENSION
ACTIVITIES

1
What is a dredge and how is it used?

2
Name the sorts of materials that dredges collect from the sea floor.

3
What is a grab and what is it used for?

4
What materials are corers used to collect?

5
Describe the operation of a simple corer.

6
Find a photograph or diagram of a dredge, other than a pipe or box dredge, and explain the advantages of its design.

7
In what circumstances would a marine geologist use a dredge in preference to a corer for sampling the sea floor?

8
Explain why the samples recovered by deep-sea drilling are an improvement on those recovered by dredges, grabs and corers.

STUDYING THE OCEANIC BIOTA

Living things that float or swim in the ocean can be caught in nets. Bottom-living creatures are sometimes caught in traps that are left on the bottom for a period of time. Also, metal or wooden stakes can be driven into the sea floor for encrusting and burrowing organisms to grow on or in.

Some surveys of the health and abundance of sea life are undertaken by direct visual inspection. Scuba divers can easily view shallow water sites, while deep water sites are investigated using special submarines, such as the Alvin, or submersible robots called ROVs (remotely operated vehicles) fitted with cameras and video equipment. (See Figures 7.1.9 and 7.1.10.)

Research trawls

Trawling, or biological netting, is the most commonly used technique for studying oceanic animals. Trawl is the name given to a single trawling operation. A trawl involves towing a trawl net behind a special boat that is, unsurprisingly, called a trawler. Nets of different designs and mesh sizes are rigged and used to sample the various forms of sea life. The choice of net design and mesh size depends on the size of the aquatic organisms that are the focus of the study. Small-mesh nets have holes so fine that only the tiniest creatures are trapped in them. These nets are used for studying **plankton**. Coarser-sized, open-mesh nets are used for catching bigger fish, squid and other large sea life that would probably damage the fine, stocking-like material used to collect plankton.

trawling
the act of catching marine life in special nets dragged behind a boat

plankton
free-floating and drifting sea life

definitions

Figure 7.1.9 The Alvin deploys a set of tube corers.

Figure 7.1.10 A remotely operated vehicle, the ROV Phantom, inspects a wrecked ship.

Trawls can be divided into three categories based on where in the water column the sample is to be taken from. The three categories are called surface-water, mid-water and bottom-water trawls. Surface-water trawls are made in the upper layer of the ocean, between the surface and 200 m depth, where the water is generally warm. Mid-water trawls are made in depths of 200–1000 m, which is a layer of the sea beneath the average depth of the bottom of the thermocline. In this layer of the sea, the water temperature rapidly decreases from a warm surface temperature to being quite cold (generally somewhere between 4 °C and 0 °C). Bottom-water trawls are made at depths greater than 1000 m in the lower, cold-water layer of the ocean.

Mid-water trawls are undertaken because this zone of the sea is inhabited by fast-swimming organisms, ranging from small shrimp and fish to whales, which we call the **nekton**. Nektic animals are important because they provide the majority of the world's seafood. A good understanding of the nekton and their environment is vital so that we do not overexploit them.

REVIEW ACTIVITIES

EXTENSION ACTIVITIES

1
What do the letters ROV stand for?

2
What is the Alvin, and why is this vessel so special?

3
Explain why the mesh size of nets that are used to catch plankton is much smaller than the mesh size of nets used to catch individuals in the nekton.

4
Outline why it is necessary for humans to use special submarines, such as the Alvin, if we wish to directly observe the deeper parts of the sea floor.

5
Research the use of trawls in improving our knowledge of the deep ocean's biota.

SAMPLING SEA WATER AND MEASURING ITS TEMPERATURE

The properties and composition of sea water are usually determined in one of two ways. First, they can be measured directly by taking an actual sample or a direct reading, such as a temperature reading. Secondly, the property of sea water can be measured indirectly. For example, by measuring the electrical conductivity of sea water it is possible to work out how **saline** it is. This is possible because the conductivity varies in a predictable way that is dependent on the amount of salt present in the sea water. Over the past 100 years or so, increasingly elaborate ways of making these measurements have been developed.

Nansen bottles

A **nansen bottle** is a special brass or stainless steel bottle that allows sea water to flow through it as it is lowered on a wire to a chosen depth. (See Figure 7.1.11.) Once the bottle is lowered to the correct depth, a trip weight (called a messenger) is dropped down the wire. When it reaches the bottle, the messenger triggers a release mechanism that causes the bottle to turn upside down. When this happens, special valves close and seal the bottle, trapping the sea water from that depth inside the bottle. This sample of sea water can then be brought back to the surface for analysis.

Nansen bottles are also fitted with a special thermometer that traps the mercury in them as they flip over. In this way, they record the seawater temperature at the time the bottle flips and takes its sample.

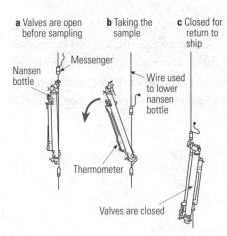

a Valves are open before sampling **b** Taking the sample **c** Closed for return to ship

Nansen bottle

Messenger

Wire used to lower nansen bottle

Thermometer

Valves are closed

To ship's deck and monitoring unit

Thin wire

Weight

Wire spool

Temperature sensor

Figure 7.1.11 Operation of a single nansen bottle. It is usual to use a large number of bottles in each hydrocast.

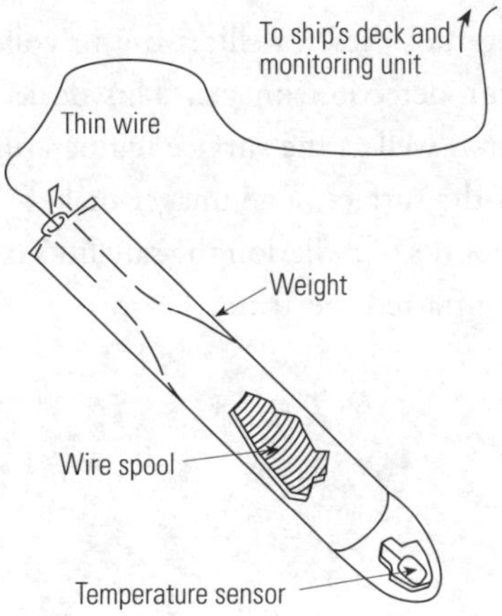

Figure 7.1.12 An XBT. These devices are used to take temperature measurements in the upper 500 m of the ocean.

In deep sea water it is common practice to attach a nansen bottle and a messenger to the hydrographic wire every 100 m as the wire is lowered. When the whole wire and fifty bottles are deployed, the first messenger is dropped down the line to trigger and flip the first bottle. When it flips it releases a second messenger. This travels down the wire to flip the next bottle over, which takes a sample and, in turn, releases yet another messenger. This process repeats down the hydrographic wire until all the bottles flip over and take samples from the depths at which they are deployed. This whole procedure is called a **hydrocast** and in water 5000 m deep it usually takes between five and seven hours to perform; if the sea state is favourable.

Bathythermographs

It is easy to understand that making a hydrocast is a time-consuming process that requires the hydrographic vessel to be stationary, which can be both difficult and dangerous in stormy seas. This limits the number of sites that can be surveyed in a given period of time. The bathythermograph was developed to make measurements of sea temperature at different depths while the ship is moving.

A bathythermograph consists of a weight, which contains a temperature sensor, and a spool of wire. One end of the wire is connected to the sensor and the other end is connected to a recorder on the ship.

Early models of these devices were deployed and then recovered back on board, but today expendable ones that look like small bombs are used. (See Figure 7.1.12.) These are called **XBTs** and they are thrown off the back of the ship while it is under way. They have a standard weight and size and their rate of sinking is well known, allowing them to give a temperature and depth record for the sea at the point they are released.

Remote sensing

Today, sea-surface temperature is measured remotely from satellites. This information is gathered and distributed by several organisations, including the US National Oceanic and Atmospheric Administration (NOAA). Its geostationary operational environmental satellites (GOES) and polar operational environmental satellites (POES) provide constant, worldwide monitoring of the sea-surface temperature, which allows daily and seasonal changes to be detected with ease. A GOES satellite is shown in Figure 7.1.13.

hydrocast
procedure of taking a continuous set of seawater samples at different depths at a particular location

XBT
expendable bathythermograph; a single-use, disposable device used to measure seawater temperature at different depths beneath a moving vessel

definitions

Figure 7.1.13 The GOES-3 satellite.

The tool these satellites use to collect sea-surface temperature data is a special radiation-detection imager. This device can determine the amount of cloud cover present as well as the surface temperature of the land and sea for 1 km^2 segments of the Earth's surface. The imager makes the temperature measurements by comparing the amounts of radiation the satellite's detector receives in different wavelength bands of the infra-red spectrum.

REVIEW ACTIVITIES

1
Outline how a nansen bottle works and state how many would commonly be used for a hydrocast taken in ocean depths of 5000 m.

2
What is an XBT and what data do this instrument provide oceanographers?

EXTENSION ACTIVITIES

3
a Name two different types of oceanographic data that modern satellite technology is now used to routinely collect, and outline why expensive satellite technology is used to do this.
b Describe at least one benefit that society receives from the collection of these data that makes the expense of collecting them worthwhile.

4
Search the Internet for some satellite images that show how sea-surface temperature varies during the course of a year. Outline some of the effects that the changes in sea-surface temperature may have on the weather in Eastern Australia over the course of a year. (NOAA provides an enormous amount of information on ocean science on its website <http://www.noaa.gov>.)

MEASURING CURRENTS IN THE SEA

Currents in the sea are bodies of moving water that flow from place to place. Two groups of instrument are used to directly measure current speed and direction: fixed current meters and drifters.

Fixed current meters

Current meters are instruments that are anchored to the bottom so that they remain stationary. (See Figure 7.1.14.) Their moving parts consist of a small propeller and a water vane, which measure the current speed and direction, respectively. The faster the water moves past the immobile current meter, the faster the meter's propeller rotates. The rate of the propellor's rotation is directly dependent on the speed of the water moving past it. This relationship is carefully determined by the meter's manufacturer. The current meter's vane swings around in the current in the same way that a wind vane does in a breeze. The current's direction relative to north is determined using a compass, which is also in the current meter. The speed and direction information is recorded on a device called a data logger, along with the time and date that the measurement was made.

Drifters

Drifters float on the surface and move with the water that makes up the current. Drifters work best when they move with water in the current rather than being moved around by the wind as well, so they need to be slightly submerged below the sea surface.

current
a stream of water flowing in a particular direction with a measurable speed

current meter
stationary or anchored device used to measure the speed and direction of a water current

drifter
free-floating device used to measure current direction and speed

definitions

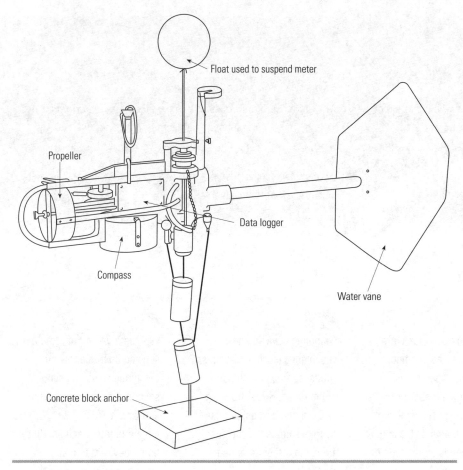

Figure 7.1.14 A current meter.

The first drifters used to measure ocean currents were simple floats whose motions were monitored and recorded by observers who followed them about. Large numbers of bottles were also released by the early oceanographers who were interested in ocean currents. Inside each bottle was placed a message that contained information about where the bottle had been released and by whom. The message also requested the bottle's finder to mail a letter to the oceanographer about where the bottle was found and the date of its recovery. Sometimes a small reward was offered as an inducement to help make sure the person who found the bottle delivered the required information.

More modern equivalents of the early floating drifters include dumping food dye into the water and releasing sophisticated floats with radio beacons that can be tracked. These drifters are released from a known spot, and some time later they are retrieved from the place they drifted to. Their position and the time they take to get to their final destination give an indication of the average current that they moved with.

Modern drifters contain data loggers, highly accurate global positioning system (GPS) devices, and transmitters that communicate their position, via satellites, to one of the big oceanographic institutes. (See Figure 7.1.15.) These drifters give accurate and instantaneous measurements of current speed and direction, allowing the bends in their drift path and changes in their speed of motion to be determined reliably. They are providing much more detailed information than was provided by their predecessors and this information is being used to determine the intricate details of current motion.

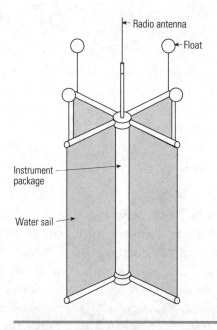

Figure 7.1.15 A modern Davis Drifter. This device sits just below the sea surface.

1

What two pieces of information do current meters always measure?

2

Outline how a drifter is used to investigate ocean currents.

3

Describe how the operation of a drifter is different from a current meter that is attached to the sea floor.

4

Why are modern drifters more effective than the first drifters used by oceanographers in the nineteenth century?

5

Explain why a conventional drifter is not generally used to measure subsurface water movements.

SUMMARY

Information about sea water, marine biota, the sea floor, and the sea in general is collected in a number of different ways. The choice of which instrument or technique to use depends on the type of information required.

Sampling devices and techniques (such as dredges, grabs, corers, drill ships, submarines, ROVs, trawling and nansen bottles) are used to take samples of the sea floor, marine biota or sea water.

We use a variety of remote-sensing instruments and techniques (such as echo sounders, seismic-reflection equipment, XBTs, magnetometers and satellite-borne devices) to make indirect observations of the sea, its biota and the sea floor.

PRACTICAL EXERCISE
Collecting information about our oceans

This exercise will improve your understanding of the way information about the oceans is gathered.

Procedure

1

Use the material in this chapter, along with resources on the Internet or in a library, to further your knowledge about the history of gathering information about the world's oceans. Focus on one aspect of

the study of the oceans that you find particularly interesting, for example, bathymetry, fisheries monitoring, sea-floor sampling, sea surface temperature monitoring or ocean-current monitoring.

2

Locate photographs and diagrams of the equipment used over time for gathering data about the oceans and/or examples of the samples or data gathered.

ACTIVITIES

1

Produce a poster that presents the information you have found. You should give brief descriptions of how the equipment works and how it is used to study the aspect of ocean behaviour in which you are interested.

2

Describe how the sampling or monitoring equipment has changed over time.

3

Indicate how improvements in the equipment used to investigate the oceans have led to a better understanding of the natural phenomena they measure.

7.2 The world's oceans

OUTCOMES

At the end of this chapter you should be able to:

describe the modern oceans in terms of
– average temperature
– mean depth
– average salinity
– average density

identify the area of the Earth covered by oceans and explain how this influences conditions on the Earth's surface

identify the probable origins of the oceanic waters

compare the evolution of the oceanic waters with the evolution of the atmosphere and explain how and why the two are linked

outline the reasons why the oldest sea floor present on the Earth today is generally less than 200 Ma old

identify the role of plate tectonics in maintaining the equilibrium between the area of sea floor and area of continental land present on the Earth

discuss the reasons for, and impacts of, possible shifts in the equilibrium between the area of sea floor and the area of continental land

describe evidence for the closing of former ocean basins in terms of the presence of deep marine sedimentary rocks in present-day continental mountain belts

identify the regions of the crust where new ocean basins are forming and where ocean floors are subducting

outline the types of evidence used to date ocean floors

assess the reliability of information used to date the age of ocean beds

outline the origin of salinity in the world's seas and oceans

explain examples of common processes that change the salinity and temperature of oceans and small, enclosed seas

relate the range of temperatures and salinities measured in selected areas of the Pacific Ocean to the distribution of specific species

discuss evidence that indicates that there are differences in the current and past distribution of oceans.

PLATE TECTONICS AND THE OCEANS

You probably know from your previous study of plate tectonics that the majority of the area occupied by the world's oceans and seas is floored by oceanic lithosphere that is at most 200 Ma old. These areas of dense crust form deep depressions in the surface of the rocky Earth. These depressions are called ocean basins. If we could drain the ocean basins, we would see vast higher-standing ranges of young mid-ocean ridges where oceanic lithosphere forms, as well as the deep ocean trenches where older oceanic lithosphere is subducted. The volume of crust created at the mid-ocean ridges is equal to the volume of crust subducted at the trenches. This keeps the system in balance.

Until the second half of the twentieth century the ocean basins were thought to be ancient and to have formed about 4 billion years ago when the Earth first solidified. We actually knew very little about the nature or age of the oceans and their sediments. Navies collected an enormous amount of information about the oceans between the 1950s and 1970s during the Cold War. They used many of the instruments and techniques described in Chapter 7.1. This research was done so that warships were better equipped to find and destroy submarines and mines if this became necessary.

Since the 1980s, research into the world's oceans has continued, with more of an emphasis on gathering scientific knowledge for its own sake, and on commercial purposes, such as communications cable laying and fisheries management. As time passes, the tools used to gather this information improve and so does our understanding of the oceans. We will now examine a few of the facts that these research efforts have discovered over the last century.

Distribution and morphology of ocean basins

Almost 71% of the Earth's surface is occupied by ocean, but this 71% is not evenly distributed around the planet. You can see in Figure 7.2.1 that the Southern Hemisphere is dominated by ocean, while the Northern Hemisphere is almost half land and half ocean. The mean depth of the ocean basins is 4.5 km and the average depth of the mid-ocean ridges is 2.5 km. The deepest parts of the ocean trenches lie 11 km below the sea's surface and their average depth is about 9 km.

There is 1.35 billion km^3 of sea water in the oceans. More than half is contained within the Pacific Ocean, making it by far the largest of the world's oceans. The world's oceans can have two different types of edge. The edges of the Pacific Ocean mostly consist of plate boundaries, that is, ocean trenches, where we know that oceanic lithosphere is subducted into the mantle. Sometimes this type of ocean boundary is called an active margin, in contrast to the other type of ocean edge, which is called a passive margin. The passive margins dominate the boundaries of the Atlantic, Indian, Southern and Arctic Oceans as well as many of the smaller seas

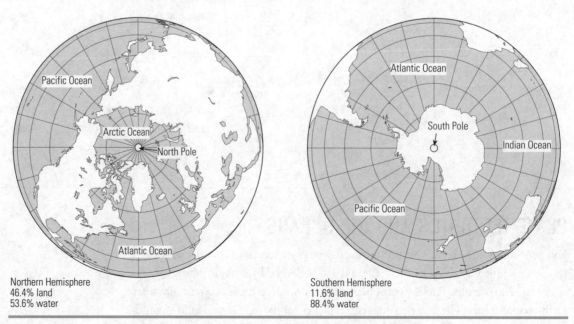

Northern Hemisphere
46.4% land
53.6% water

Southern Hemisphere
11.6% land
88.4% water

Figure 7.2.1 Comparison of the distribution of land and ocean in the Southern and Northern Hemispheres.

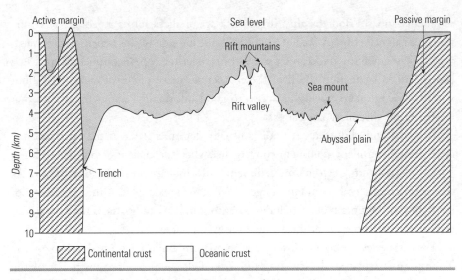

Figure 7.2.2 Cross-section of an ocean basin showing ocean margin types.

composed of oceanic lithosphere. An example is the Tasman Sea, which lies to the east of Southern Australia. These passive margins are sites well within the plates where continental crust joins oceanic crust. Figure 7.2.2 shows the general cross-sectional shape of an ocean basin and the two types of oceanic edge.

When new sea floor is created at the mid-ocean ridges it is hot and buoyant, which results in it having a relatively low density. This is one of the reasons why mid-ocean ridges are high-standing areas of the sea floor. As this newly formed lithosphere begins to cool, it gets slightly denser and sinks a little into the underlying lithosphere in a process called subsidence. With increasing age, the oceanic crust cools further and subsides some more. In this way, as the oceanic lithosphere gets progressively older and denser with age, its depth below sea level increases. The cooling of oceanic lithosphere to an equilibrium state takes about 100 Ma, and during this time it sinks from an average sea-floor depth of 2.5 km to an average sea-floor depth of 5.5 km.

The motions of the plates, the drift of the continents, the creation of new oceanic lithosphere and the subduction of old oceanic lithosphere all combine to constantly change the shape of the ocean basins and cause rises and falls in sea level. Because the Earth's overall mass and volume are not changing, the amount of oceanic lithosphere subducted into the mantle has to roughly balance the amount of new oceanic lithosphere created at the mid-ocean ridges. So, a major consequence of the plate tectonic cycle is that sea floor is subducted before it gets very old.

Evidence for ancient sea floors

It is possible to determine an approximate age of most oceanic crust that is less than 100 Ma old just by determining its depth below sea level. A graph showing the relationship between the age and depth of oceanic crust is given in Figure 7.2.3.

The oldest dated sea-floor basalt on the planet is located in the North-west Pacific Ocean near Japan. It was dated by isotopic methods and the age determined for the basalt was confirmed by dating microfossils that occur in sediment deposited above the basalt. This sea floor is about 180 Ma old, and will probably be subducted into the mantle in the next few million years. Geologists and geophysicists who study the formation and tectonics of ocean basins are of the opinion that the maximum age that sea floor could ever achieve is about 200 Ma.

During subduction, some subducted material is reprocessed to form magma, which erupts as lavas in the volcanoes of the overlying volcanic arc or is intruded beneath the volcanic arc. 'Slices' of oceanic crust are 'scraped' off the downgoing plate and accreted onto the overriding plate. These slices usually consist of sea-floor

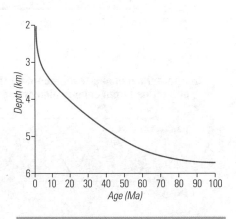

Figure 7.2.3 The depth–age relationship for oceanic crust.

sediments and sea-floor basalts. Together, a set of slices forms a geological feature known as an accretionary wedge on the volcanic arc side of the trench. Sometimes a very large segment of an oceanic plate is thrust over the top of continental crust in a process called obduction. Both of these processes place pieces of ancient sea floor within the continents, giving us access to good information about the oceans that existed in the distant geological past.

Accreted and obducted sea floor is usually deformed later in the plate tectonic cycle during continent–continent convergence when the ancient ocean finally closes. This often converts deep marine sediments into folded and metamorphosed rocks that form continental mountain ranges. Well-known examples of the conversion of sea floor into mountain belts include the European Alps and parts of the Himalayas, both of which formed after the ancient Tethys Ocean closed.

Similar cycles of ocean closing and mountain building formed much of the Thompson, Lachlan and New England Fold Belts, which make up much of the eastern third of the Australian continent. The accretion of these younger rocks onto the rest of the Australian Cratonic Shield took place between the Cambrian and the Triassic. This is why fossils of many extinct Palaeozoic and Mesozoic deep-sea creatures can be found throughout much of inland Eastern Australia.

Effect of sea-level change on the distribution of marine conditions

If you leave aside short-term sea-level changes (such as tides and storm surges) and consider an average daily or annual measure of sea level, it is evident that the level changes almost constantly. This constant change is especially clear if you consider the

a Approximated sea-level record suggested from seismic stratigraphy

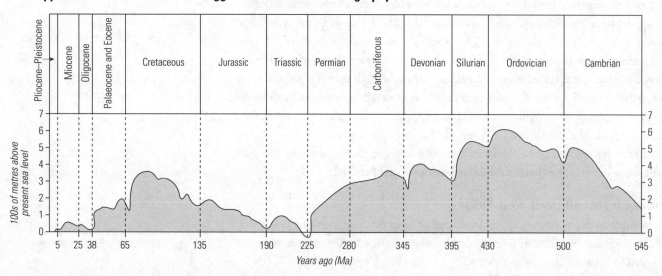

b Approximated sea-level record for the last million years suggested from sea-surface temperature record determined from composition of deep-sea sediment

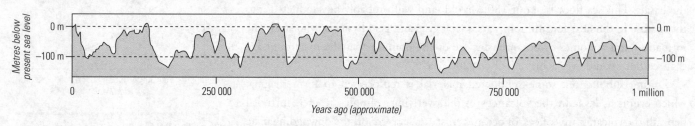

Figure 7.2.4 Sea-level change during the Phanerozoic (a) and during the last 1 Ma (b).

sea level from the perspective of geological time. The study of sea-level change is of great interest to many geologists. Figure 7.2.4a shows how sea level is thought to have changed over the past 545 Ma of Earth's history. Figure 7.2.4b provides a detailed portrayal of the rise and fall of sea level due to the cyclic advance and retreat of the polar icecaps over the last 1 Ma or so. It shows that, during this period, the level of the sea has fallen around 120 m then risen back to about its present position many times.

Figure 7.2.4a shows that 65 Ma ago the sea level was about 200 m higher than it is today. This is partly because the planet was warmer at that time, and so there was no ice at the poles; this accounts for about half of the 200 m. It is also partly due to the change in the shape and volume of ocean basins, which is mostly the result of plate tectonic processes (such as sea-floor spreading and continental collision).

The main effect of the rise and fall of sea level on the distribution of marine conditions occurs at the outside edges of the continents and over continental areas that are near sea level. For example, when the sea level fell by about 100 m during the last glacial maximum, much of the area occupied by continental shelves around the world was exposed and became dry land; this was particularly so on the narrow continental margin of New South Wales.

These sorts of fluctuations of sea level would have a relatively minor effect on the vertical distribution of marine habitats present in the *deep water* areas of the world's open oceans, and the geographical distribution of these environments would hardly change. This is because these open-ocean environments already occupy such deep water that fluctuations of sea level would merely shift the vertical position of the boundaries between the warm, well-lit, upper-ocean layer, the mid-water layer and the deep-water layer. Remember, though, that the distribution of currents and changes in climate resulting from such a change in sea level would have other effects that may have an impact on open-ocean communities.

REVIEW ACTIVITIES

1
Draw a sketch that shows the shape of the ocean basins and label it with their prominent features.

2
A sample of sea-floor basalt has been dredged from a site near an active mid-ocean ridge. The sample was recovered from a water depth of around 4.5 km. How old is this basalt likely to be?

3
Explain why it is unlikely that sea floor older than 200 Ma will ever be found.

4
What evidence have geologists collected for ancient oceans in Eastern Australia? Outline how this evidence may be interpreted in terms of the theory of plate tectonics.

EXTENSION ACTIVITIES

5
Draw a generalised cross-section through oceanic lithosphere that shows the different types of material that it is composed of.

6
Find a map that shows the age of oceanic crust. Where is the oldest known oceanic lithosphere located and how old is it?

7
Locate maps and diagrams that show how plate tectonic processes changed the distribution of the world's oceans during the last 200 Ma. Use this information to write a list of the places where you would expect to find ancient oceanic crust that has been incorporated into the continents.

WHERE DID THE OCEANIC WATER AND SALT COME FROM?

The Earth's oceans and most of the gases present in the atmosphere have been derived from material incorporated into the planet when it first formed. Volcanic eruptions during the Hadean expelled the material that formed the oceans from the young

Earth's interior, in a process called outgassing. Enormous amounts of water, carbon dioxide, nitrogen, sulfur and sulfur oxides, methane, hydrogen sulfide, hydrochloric acid and other volatiles were brought to the planet's surface by this initial outgassing. The large, permanent bodies of water formed when the gaseous water condensed and fell as rain to form bodies of water in the lowest lying areas of the planet. Permanent oceans were probably present on the surface of the Earth by about 4 billion years ago, when the heavy bombardment by asteroids, comets and large meteorites ended.

The composition of volcanic gases erupted from basaltic hot spot volcanoes, such as those of Hawaii, probably give us the best indication of the composition of the volatiles outgassed during Earth's early history. Gases expelled from Hawaiian volcanoes have sources located at depths of over 200 km. The gases that are expelled, and their proportions, are water (about 55%); carbon dioxide (about 20%); sulfur dioxide (about 13%); nitrogen (about 10%); and argon, chlorine, hydrogen sulfide, methane and other gases (about 2%). The water, sulfur dioxide and chlorine rain out and mix into sea water relatively quickly. This means that the first oceans were probably very acidic. The carbon dioxide, nitrogen and other gases became the early atmosphere.

The source of salt in sea water

The bulk of the salt in the oceans comes from elements leached from rocks and those released during the early outgassing (hydrochloric acid and hydrogen sulfide) and subsequent volcanism. It is thought that the concentration of salt in the ocean has been stable for at least the last 1 billion years. This stability is due to the equilibrium developed between input (mostly from rivers) and removal (mostly in sediments).

If the salt was extracted from the ocean today and dried, it would form a 45 m thick layer over the whole planet. Sea salt comprises 3.5% of the weight of sea water. Salinity is the total concentration of dissolved solids and is usually recorded in parts per thousand. The parts per thousand symbol is $^0/_{00}$ and you can see that it looks like

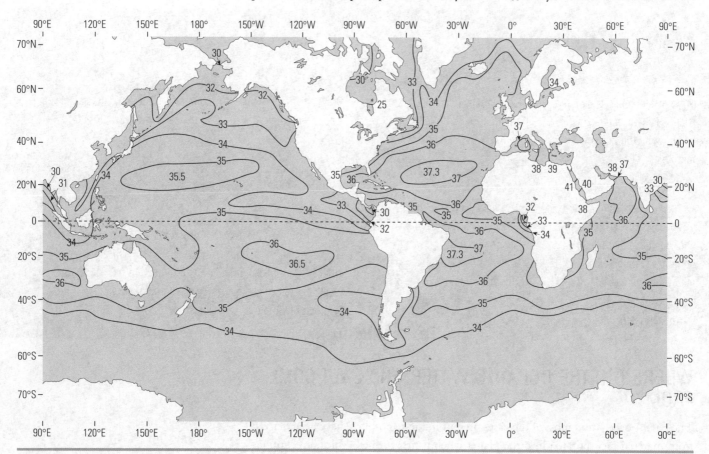

Figure 7.2.5 Average sea-surface salinity ($^0/_{00}$).

the per cent symbol. This notation allows us to easily indicate salinity as the number of grams of dissolved sea salt per kilogram of sea water. For example, a salinity of 35 ⁰/₀₀ means 35 grams of salt per 1 kilogram of sea water. While coastal waters can exhibit a wide range of salinity as a result of differences in freshwater run-off, most of the world's ocean lies in the narrow salinity range of 33.8–36.8 ⁰/₀₀.

Sea salt is composed of a number of components, the most abundant of which is chlorine, followed by sodium, sulfate, magnesium, calcium and potassium. (Almost all the chlorine is thought to have come from the volatiles mentioned earlier.) The ratio of these individual components is extremely constant, although the total amount of the salt in sea water can vary between locations. This means that all the water in the ocean must be constantly and quickly mixed; otherwise the ratio of particular components (say, the ratio of potassium to calcium) would vary significantly from ocean to ocean.

The variation in **sea-surface salinity** worldwide is shown in Figure 7.2.5. This variation can be due to a number of factors. Both evaporation of water and the formation of sea ice can increase the salinity of sea water. As water evaporates, the salt it contained is left behind, thereby raising the salt concentration of the water. The structure of ice cannot accommodate very much salt. Therefore, as the water freezes, salt is expelled and the surrounding water becomes more saline. Salinity decreases when freshwater from rivers, rain, snow and melting ice mixes with the sea water.

In the open ocean, the sea-surface salinity is fairly stable and most of the changes in its value are due to precipitation (dilution) or evaporation (concentration). Therefore, being able to monitor sea-surface salinity gives us information about the global hydrological cycle, global warming, changes in ocean circulation and sea surface–air interaction. In contrast to the open ocean, salinity in the coastal zone is highly variable, affected by seasonal changes in precipitation and run-off. There are currently not enough long-term salinity measurements of the coastal zone to discern changes reliably. However, changes in land-use patterns around the world, many of which result in greater run-off, suggest that coastal salinity would be affected. Many regions of the world's oceans have recently experienced a decline in salinity.

sea-surface salinity
the salinity measured at the sea's surface

definition

REVIEW ACTIVITIES

1
Name the mostly likely source for the water in the oceans and the process by which this water accumulated.

2
Where did the salt in sea water originally come from?

3
Why is sea water near the continental coast generally less saline than sea water found far away from land?

4
List the common materials that make up sea salt.

5
How much salt is there in a 1 kilogram sample of sea water if it has a salinity of 34.7 ⁰/₀₀? Give your answer in grams.

6
Outline two processes that change the salinity of sea water.

EXTENSION ACTIVITIES

7
What are the proportions of the different ions in sea salt?

8
Find a map of oceanic salinity on the Internet and compare it with the one given in Figure 7.2.5.

9
Invent a simple experimental procedure that would allow you to determine the amount of salt in a sample of sea water and could be done using kitchen implements.

10
Outline the process or processes that originally led to the world's early oceans becoming saline.

THE INFLUENCE OF OCEANS ON CLIMATE AND SURFACE CONDITIONS

The circulation of currents and winds in the ocean and atmosphere, respectively, are linked and are jointly responsible for distributing heat received from the Sun around the Earth. The atmosphere and the ocean act as insulating blankets that trap energy from the Sun and stop it from being radiated back into space. If it were not for the oceans and the atmosphere, the present-day average temperature of the planet's surface would be about −20 °C instead of the comfortable +14 °C we are used to.

Water has a high thermal inertia, or heat capacity, in comparison with other substances and, consequently, moderates the temperature of the atmosphere and the land near it. This means that small changes in circulation and sea-surface temperature can have a marked influence on the world's climate. Changing distributions of sea-surface temperature affect the global pattern of winds as well as rainfall patterns on the land.

Permanent or long-term shifts in ocean–atmosphere interaction (measured in decades or centuries) can cause climate change, which has major consequences for agriculture, habitats and natural and human communities. We will now examine an example of periodical cycling of the ocean–atmosphere interaction: the El Niño effect.

The El Niño effect

The **El Niño effect** is a consequence of changes in the ocean–atmosphere interaction causing a change in sea-surface temperature in the Pacific Ocean. Normally the trade winds blow towards the west across the Pacific. These winds push the surface water west, so that the sea surface is about 50 cm higher at Darwin than at Ecuador. This pile up of warm surface water results in a sea-surface temperature about 8 °C higher in the west than it is in the east. The westward movement of the warm water produces upwelling of cold water off South America's coast. Rainfall occurs due to air rising over the warmer water, resulting in the summer monsoon in Asia and the wet season in Northern Australia and the Western Pacific. At the same time, cooler and drier conditions dominate in the Eastern Pacific along South America's west coast.

During an El Niño event, the trade winds relax in the Central and Western Pacific, leading to a cooling of the sea-surface water in the Western Pacific and a warming of the sea-surface water in the Eastern Pacific. Consequently, rainfall is higher in the Eastern Pacific. This usually causes flooding and landslides in places such as Peru and California. The related cooling of the sea-surface water around Indonesia and Australia has the opposite effect: rainfall decreases and droughts result. During an El Niño event, there is an eastward displacement of atmospheric low pressure and rising air over the warmest water. This causes changes in the global atmospheric circulation, which, in turn, forces changes in weather patterns in quite distant locations.

The cyclic warming and cooling of the Eastern and Central Pacific can be recognised in changes in air pressure at sea level, which is measured by a figure called the Southern Oscillation Index. This is calculated by a numerical comparison between the air pressure measured at Darwin and the air pressure measured in Tahiti. The difference between the two air pressures is compared with their long-term averages and used to generate a number we call an 'index'. Changes in this index since 1850 are shown in Figure 7.2.6. When this index is a positive number, there is a La Niña (where there is a cooling of the ocean off Peru). This is the more common condition. When the number is negative, there is an El-Niño (where the ocean off Peru is warming), which is the less common condition.

> **El Niño effect**
> cyclic variation of ocean currents and water mass movements in the Pacific Ocean related to pressure changes in the atmosphere
>
> *definition*

Atmospheric and ocean scientists have, as yet, not fully explained what drives the El Niño–La Niña cycle. There is such a close relationship between the periodicity of their occurrence and the periodicity of a variety of solar phenomena that many scientists think there must be a causal link between the periodicity of fluctuations in the Sun's behaviour and the El Niño effect.

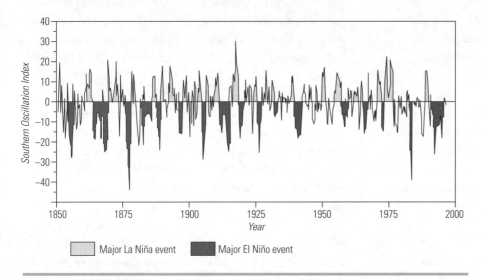

Figure 7.2.6 The Southern Oscillation Index since 1850.

REVIEW ACTIVITIES

1
What would be the average temperature of the Earth's surface today if it did not have oceans or an atmosphere?

2
Outline how the El Niño effect works.

3
Describe the two main climate regimes associated with El Niño that affect Eastern Australia.

4
What is the Southern Oscillation Index and how is it calculated?

EXTENSION ACTIVITIES

5
Explain how the sea-surface temperature changes that result from the El Niño–La Niño cycle cause changes in the amount of rainfall on the continental landmasses that border the Pacific Ocean. List some of the consequences that result from these changing rainfall patterns.

6
Investigate how the Southern Oscillation Index is calculated and explain how it is used to predict the severity of El Niño and La Niña events.

7
Find out the average density of sea water.

8
Investigate how changes in seawater temperature and salinity affect where sea creatures live. Why are there differences in the types of sea life found near Antarctica, the NSW coast and the Great Barrier Reef? Compare brackish water animals with oceanic animals.

SUMMARY

The ocean dominates 88% of the area of the Southern Hemisphere. In comparison, it dominates 54% of the Northern Hemisphere.

The average depth of ocean basins is 4.5 km. The average depth of mid-ocean ridges is 2.5 km and the average depth of ocean trenches is about 9 km.

The oceans are made up of different types of sea floor, such as shallow continental margins, ocean basins, mid-ocean ridges and ocean trenches.

Oceanic crust that forms at mid-ocean ridges is eventually subducted at ocean trenches or incorporated into mountain belts at convergent plate margins.

The sea level changes constantly over geological time due to climatic change and the changing shape of the global ocean.

The water in the Earth's seas and oceans contains, on average, 35 grams of salt per 1 kilogram of sea water.

Oceanic salinity varies slightly from place to place, but is generally restricted to a small range that lies between 33.8–36.8 parts per thousand.

The bulk of sea salt consists of sodium chloride, which has mostly been leached out of rocks by weathering.

Salinity decreases occur due to the addition of freshwater, while salinity increases result from evaporation and the formation of sea ice.

PRACTICAL EXERCISE
Variation in salinity and temperature in the Pacific Ocean

This exercise will improve your understanding of the variation in temperature and salinity in the Pacific Ocean.

Procedure

1

Find maps on the Internet or in a library showing the spatial variation in salinity and temperature of the surface waters of the Pacific Ocean.

2

Find profiles showing the variation in salinity and temperature with depth in the Pacific Ocean.

ACTIVITIES

1

In the Pacific Ocean:
a Where is the most saline water found?
b Where is the least saline water found?
c Where are the warmest waters found?
d Where are the coldest waters found?

2

Outline how sea-surface salinity and sea-surface temperature vary across the Pacific Ocean from:
a north to south
b east to west.

3

List any broad relationships between sea-surface salinity and sea-surface temperature that you can identify.

4

Refer to the sea-surface salinity and sea-surface temperature profiles you found. Select a particular place in the Pacific Ocean, then complete the following activities:
a With reference to the profiles, outline the variation in sea-surface salinity and sea-surface temperature with depth at the selected location.
b Try to explain why there is a difference in the salinity and temperature of the surface and bottom waters at the selected location.

7.3 Mass motion of the oceans

OUTCOMES

At the end of this chapter you should be able to:

describe the four types of mass motions of water
– surface currents
– deep circulation
– tides
– tsunami
and identify the energy source for each

explain how the oxygen supply on the ocean floor is renewed, making life possible

explain how long-lived materials, such as synthetic chemicals and heavy metals, that enter the sea in one place can be found thousands of kilometres away

discuss the implications of the movement of materials by ocean currents for the use of the oceans for waste disposal, including
– pollution
– ocean sewage outlets.

OCEAN CURRENTS

Ocean currents are movements of water in the oceans. They transport oxygen and dissolved materials throughout the entirety of the ocean, with the water travelling tens of thousands of kilometres. Currents also transport floating objects; fine, suspended sediments; and dissolved materials, such as oxygen and synthetic chemicals. Contaminants, such as heavy metals, are often absorbed onto the fine, suspended sediment and can also be transported around the ocean basins by currents.

Currents may affect small areas for a short period of time or they may be permanent features on a global scale. The large, permanent ocean currents move warm water from the equatorial regions towards the poles, as well as cool water from the polar regions towards the equator. These movements mix the upper and lower oceanic waters and the ocean's warm and cold water masses. This moderates their temperatures, so that equatorial waters are cooled and polar waters are warmed. Oxygen is replenished in the lower water by mixing with water at the surface. The redistribution of heat and substances by ocean currents in this way plays a major role in controlling the global climate and affects the marine and terrestrial habitats in which all the Earth's organisms live.

We will only examine a few of the important ocean water motions:
• wind-driven currents
• tidal currents
• gravity driven currents
• tsunami.
There are other ocean water motions, such as the Eckman flow, geostrophic currents, storm surges, rips, upwellings, downwellings and Langmuir circulation. However, these are more complex flows and are beyond the scope of what you need to study in this course.

WIND-DRIVEN SURFACE OCEAN CURRENTS

Mariners have used currents on the surface of the ocean since they started going to sea, so our need to understand these flows of water is obvious. Near-surface currents are driven by the wind and cause movements of oceanic water down to about 100 m below the surface. As wind blows across the surface of the ocean, it drags the water with it. Consequently, oceanic surface currents are closely connected to atmospheric circulation patterns and the dominant seasonal wind patterns. This means it is necessary to have some understanding of surface wind patterns in order to be able to understand the near-surface ocean current patterns.

Generating surface winds

Atmospheric circulation patterns are controlled by the difference in temperature at the equator and the poles, and the rotation of the Earth. These winds arise in response to heating and expansion of air at the Earth's surface by the Sun. As air becomes hot, it expands and rises, forming a low-pressure system at the surface. Air that rises to the upper atmosphere in this way moves away towards the poles, as a mass of cooler, denser and higher-pressure low-level air moves in to replace the warm air that rises. This causes high-level air to circulate towards the poles, and the low-level air to circulate towards the equator.

If the Earth did not rotate, the global surface wind pattern would be dominated by winds moving towards the equator. These would consist of two rotating cells of air, one in each hemisphere. (See Figure 7.3.1.) Air heated at the equator would move towards the poles, where it would descend and cool to become a dense, high-pressure, surface air mass that would flow back towards the equator. The Earth's size and rotation modifies this pattern, however, and produces three circulating cells of moving air in each hemisphere:

- an equatorial cell (called the Hadley cell) that brings surface air towards the equator
- a temperate-zone cell (called the Ferrel cell) that produces surface winds that move toward the poles
- a polar cell (called the Polar cell) that produces surface winds that move towards the equator.

The surface air that is forced to move north or south by the circulation of the Hadley, Ferrel and Polar cells is also deflected by the Coriolis effect. This results in the equatorward-driven Hadley cell surface winds being deflected towards the west, the poleward-driven Ferrel cell surface winds being deflected towards the east, and the equatorward-driven Polar cell surface winds being deflected towards the west. (See Figure 7.3.2.) A simplified map of the Earth's dominant wind patterns that results from the combination of the large-scale circulation of the atmosphere and the Coriolis effect is shown in Figure 7.3.3, page 364.

Ocean gyres

When the surface winds that drive the ocean currents encounter a continent they are not greatly affected by its presence; they simply flow over the top of the landmass. The water in the surface ocean currents cannot do this because the continents create barriers that block the currents and deflect them. The blocking action of the ocean currents by the continents also contributes to the way in which ocean currents move and helps to form large, closed circuits of moving surface water called **gyres**.

Each ocean basin has a gyre with its centre located at approximately 30°N or 30°S latitude. Figure 7.3.4 (page 364) shows the five major continuous gyres present in the oceans of the world today. There are two Northern Hemisphere gyres: the North Pacific Gyre and the North Atlantic Gyre. These gyres produce currents that move in

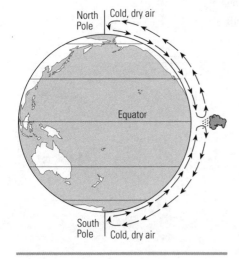

Figure 7.3.1 The pattern of global air circulation that would occur if the Earth did not rotate.

gyre
a type of ocean current that consists of a large, circulating circuit of moving surface water

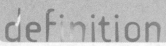

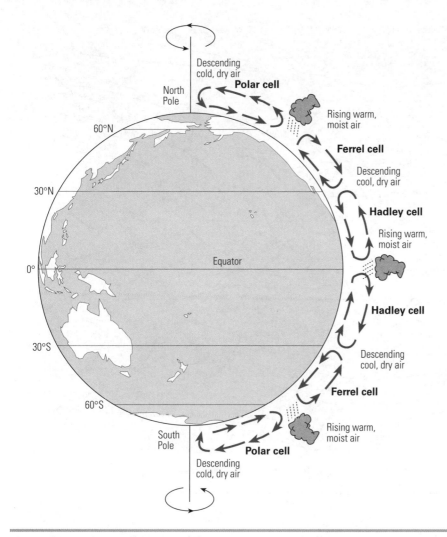

Figure 7.3.2 The Hadley, Ferrel and Polar cells.

a clockwise direction. There are three Southern Hemisphere gyres: the South Pacific, South Atlantic and Indian Ocean Gyres. These gyres produce currents that move in an anticlockwise direction. In addition to these currents, the east-moving winds of the roaring forties, howling fifties and screaming sixties drive the Circum-Antarctic Current, which moves water eastward around Antarctica continuously. There is no equivalent of the Circum-Antarctic Current in the Northern Hemisphere's Arctic because the oceans there are not connected seaways as they are in the Southern Hemisphere.

Equatorial countercurrents

The anti-clockwise circulation in the Southern Hemisphere and the clockwise circulation in the Northern Hemisphere lead to the oceans having certain similarities. In all ocean basins there are two major ocean currents on either side of the equator, a north equatorial current and a south equatorial current, each of which flows west at approximately 6 km each day. Between the two currents is an equatorial counter-current, which flows east and partially returns some of the water moved by the equatorial currents to the eastern side of the ocean basins.

Boundary currents

Each ocean has a western **boundary current**, which transports warm water from the equator to higher latitudes on the western margin of all the ocean basins. These currents tend to be narrow and flow at high speed; up to 120 km per day. They are

boundary current
a special current that forms when an oceanic gyre is blocked by the presence of a large continental landmass

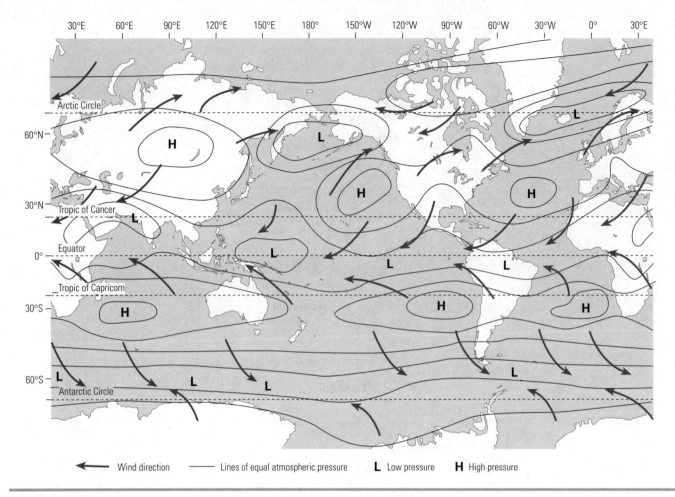

Figure 7.3.3 Air pressure and prevailing winds.

Legend: ← Wind direction — Lines of equal atmospheric pressure **L** Low pressure **H** High pressure

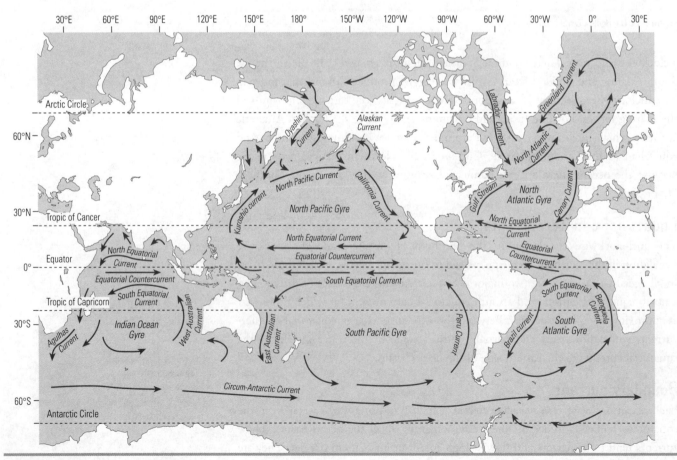

Figure 7.3.4 Surface ocean currents and gyres.

also deeper currents and extend down to depths of 1000 m. The western boundary currents include the Gulf Stream in the North Atlantic, the Kuroshio Current in the North Pacific, the Brazil Current in the South Atlantic, the East Australian Current in the South Pacific and the Agulhas Current in the Indian Ocean.

Moving along the western side of each continent are the eastern boundary currents. These move cold water from high latitudes towards the tropics along the eastern margin of the ocean basins. They tend to be shallow, broad and slow moving; only moving 3–7 km per day. The West Australian Current is an example of an eastern boundary current.

REVIEW ACTIVITIES

1
Name three different types of current that occur in the seas and oceans.

2
Describe or sketch the rotations of the Hadley, Ferrel and Polar cells.

3
Draw a simple sketch that shows the main ocean gyres. On the sketch, indicate the way each gyre rotates.

4
What is a countercurrent?

5
What is meant by the term 'boundary current'?

6
How fast do western boundary currents flow? Do they flow faster or slower than eastern boundary currents?

7
Do western boundary currents flow towards the poles or towards the equator? Why?

EXTENSION ACTIVITIES

8
Explain the general relationship between surface winds and surface ocean currents.

9
Investigate the formation of countercurrents. Draw a sketch map that shows the flow of the Equatorial Countercurrent in the Pacific Ocean and explain how it forms.

10
Research the western boundary currents in the world's oceans so that you can describe and explain the similarities between them.

11
Hold a debate in class where you discuss the disposal of sewage and other wastes by pumping them into the ocean (that is, ocean outfall disposal). Focus on the dispersion and dilution of these waste materials by ocean currents.

DEEP OCEAN CURRENTS

Most movement of water through the deep ocean is driven by density differences caused by variations in the salinity and temperature of the water. This is because the density of sea water depends on its temperature and salinity. Denser water sinks, while less dense water rises; in the same way that a drop of cooking oil rises to float on the water's surface. The density contrast between water of different salinities is not the same as the difference in density between cooking oil and water, but the resulting motions are caused in the same way.

Forming dense sea water

The density of sea water increases in a way that is directly dependent on an increasing concentration of salt over the normal range of seawater salinities. The variation of density with temperature is much more complex, however.

This variation occurs because of the bipolar structure of the water molecule and the way water molecules begin to line up with one another at temperatures near freezing point. The water molecules pack as tightly together as they can at about 4 °C above freezing point. Further cooling leads to a slight decrease in density as the water molecules become organised into rings and pack less tightly. Consequently, sea

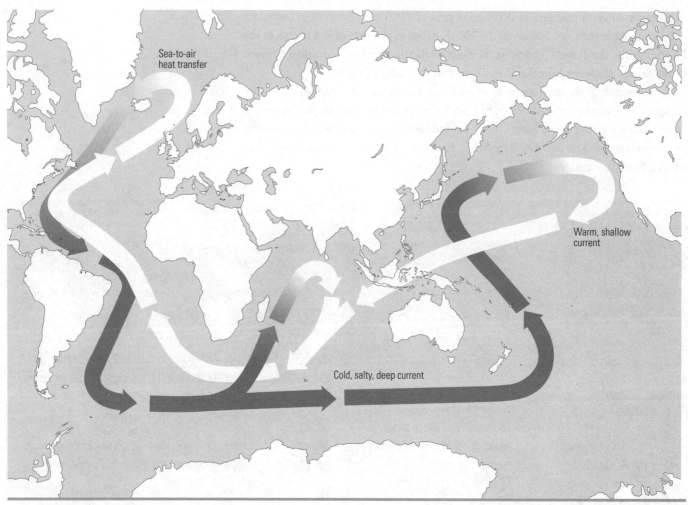

Figure 7.3.5 The global conveyer.

water is densest when its temperature approaches 2 °C. Cooling freshwater behaves in the same way but it is densest at around 4 °C. When ice forms, it has a density about 10% less dense than liquid water, which is why ice floats. The contraction and expansion of water near freezing point is a behaviour that is rarely observed in chemical compounds and is one of the properties that makes water a special substance.

Deep circulation and the global conveyor

Warming sea water or lowering its salt content by mixing it with freshwater makes it less dense and causes it to rise or float. Knowing these facts, we would expect that water that cools near the poles would become dense and sink. About two-thirds of the salt present in sea water is expelled from it when sea ice forms. Therefore, the cooling and freezing of sea water near the poles creates masses of sea water near the surface that are really quite dense relative to other sea water. The cold, saline waters formed at the poles sink and move away from the poles, drawing in warmer surface water towards the poles. This process drives deep ocean currents in an enormous circulating system called the **global conveyor**. (See Figure 7.3.5.)

The global conveyor begins in the North Atlantic, where a water mass called the North Atlantic Deep Water (NADW) forms as surface water cools and sinks in the polar seas around Greenland. This sinking water mass drives a southward flow of water along the bottom of the Atlantic Ocean, where it is thought that the NADW is joined by a similarly formed mass of water called the Antarctic Bottom Water (ABW). The ABW forms in the Weddel Sea, which is located due south-east of the

South Atlantic Ocean. The combined water masses then flow east and split, with one current flowing north into the Indian Ocean and another flowing north into the Pacific Ocean. Here the two bodies of water warm up and rise back to the surface. The large surface gyres then sweep the water through the Indian Ocean and back up into the North Atlantic. One complete circuit of this flow of sea water is estimated to take about 1000 years.

REVIEW ACTIVITIES

1
What is the global conveyor?

2
How long does it take a sample of sea water to make a complete circuit of the global conveyor.

3
Describe the processes that drive deep oceanic circulation and name the physical phenomenon responsible.

4
Why does ice float?

5
Who do the abbreviations NADW and ABW stand for?

EXTENSION ACTIVITIES

6
Outline the reason or reasons why the oceans are thought to be well mixed and name the processes that cause this mixing to occur.

7
Investigate how the global conveyer keeps deep oceanic water well oxygenated and draw some labelled sketches that outline your findings.

TIDES AND TIDAL CURRENTS

Everyone who lives near the sea is familiar with tides: the almost, but not quite, twice-daily cycle of the rise and fall in sea level. In harbours or estuaries it is easy to measure the variation in sea level by chalking measuring marks on one of a jetty's piles or making similar marks on a seawall.

In addition to the near-daily fluctuations, the **tidal range** follows a lunar cycle that is twenty-eight days long. Twice a month, when the Moon and the Earth are in line with the Sun, the tidal range peaks at a maximum value, which is called a **spring tide**. A week after each of these peaks, when the Moon lies in the Earth's orbital path and is perpendicular to the line between the Earth and the Sun, the tidal range drops to a minimum value, which is called a **neap tide**.

This relationship led Isaac Newton to work out that the tides are caused by the gravitational attraction of the Sun and the Moon on the world's oceans. Figure 7.3.6 shows that when the Earth, the Moon and the Sun are aligned, the combined pull of the Sun and Moon is at a maximum and the tidal range is large. When the Moon lies in a position perpendicular to the line between the Earth and the Sun, the combined pull of the Sun and Moon is at a minimum and the tidal range is low.

When the tide rises near the shore or in a harbour, the sea's rise is obstructed by the coast and, as a consequence, the water piles up. This causes the sea water to flow towards the land, producing an ingoing current called a flood tide. After the tidal peak, when the sea level falls, the water piled up on the coast or in the harbour flows back out, producing a flow called an ebb tide. When the tide peaks or troughs there is a brief period lasting about twenty minutes when the sea level effectively stops moving up or down. When this happens, the water stops moving, producing a period called slack water. When the flow direction of the tidal current reverses after a peak or trough, we say the tide has turned.

tidal range
the difference in height between high and low tide

spring tide
highest-range tide of the lunar tide cycle

neap tide
lowest-range tide of the lunar tide cycle

definitions

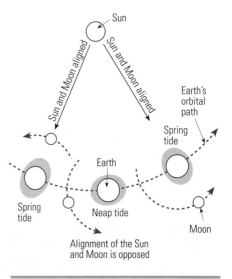

Figure 7.3.6 The influence of the Sun and Moon on tides.

1
Does a neap tide have a larger or smaller tidal range than a spring tide?

2
Outline the relationship between the lunar cycle and the spring and neap tides.

3
Research and then explain why a larger tidal range generally produces faster tidal currents.

4
Find out which parts of the Australian continent have the largest tidal ranges.

---

tsunami
unusual and infrequent surging coastal waves that inundate coastal areas; generally caused by earthquake activity or submarine landslides

definition

TSUNAMI

Tsunami are surging coastal waves that are generally produced by unusual movements of the sea floor resulting from earthquakes or submarine landslides. They can also be caused by the energetic impact of a body of material with the surface of the sea, for example, a meteorite or a comet, collapsing or exploding volcanic islands, and landslides that start on land but finish in the ocean. Small, localised tsunami occur when large blocks of ice collapse off glaciers into the sea in places such as Greenland and Antarctica.

Tsunami are typically described by survivors as a sudden, inrushing wall of water. These survivors often say that before the tsunami struck, the sea level dropped, exposing the harbour bottom and leaving floundering and bewildered fish and other marine life wriggling about in the air. Tsunami produce a type of wave and a sequence of events quite different from the breaking, wind-generated, ocean-swell waves that most of us are used to catching at surf beaches. Figure 7.3.7 shows a tsunami surging up a tidal inlet in Hilo, Hawaii in 1946.

Generally, tsunami are much more severe in areas where the continental shelf is relatively narrow, such as the coasts of Eastern Japan and South-eastern Australia. The way that tsunami waves are generated means that they move across the deep oceans almost unnoticed. In deep ocean water these waves travel at speeds up to 900 km per hour, but the crest of the passing wave is commonly only a metre or so high. When they reach coastal areas, and the water depth beneath the wave decreases, they slow down to about 35 km/h, converting their energy from wave speed to wave height. This is why areas where the shelf is narrow are so badly affected. Wider coast shelves dissipate some of the tsunami's energy as the wave interacts with the sea floor, but this doesn't happen on a narrow shelf.

Figure 7.3.7 A tsunami strikes.

1
Describe the characteristics of tsunami.

2
Explain why tsunami are different from large surf waves.

3
Find an eyewitness account of a tsunami and relate the specific events recounted to the general behaviour of a tsunami.

SUMMARY

Currents can transport water, floating objects and suspended sediment large distances across the oceans.

There are many different types of ocean current, including wind-driven surface currents, deep ocean currents, and tidal currents.

Global wind patterns, the Coriolis effect and the distribution of continental land all contribute to forming the pattern of the ocean gyres.

Gyres are the large-scale, continuously flowing surface currents of the ocean basins.

Dense, cold saline water forms at the poles, and gravity forces it to sink and flow. This process drives a deep ocean current called the global conveyor.

The global conveyor continually mixes together the surface and deep ocean waters.

Tsunami are surging coastal waves that are commonly produced by earthquakes.

Measuring tidal currents

This exercise will improve your understanding of how the tidal currents vary. Due to the difference in the lengths of the tidal cycle and the school day you will probably only be able to measure the variations across the change from the ebb tide to the flood tide, or vice versa.

Materials

- A light fishing float
- A measured length of light fishing line
- A watch
- A stopwatch
- A tall (3 m) measuring pole marked in 20 cm intervals

Caution: This is a field-based exercise that must be supervised by a teacher and undertaken during school time. It is expected that your teacher will choose a suitable, safe site to conduct the fieldwork and modify the procedure according to local conditions.

Procedure

1

Construct a 'tethered' drifter from the fishing float and fishing line. Be careful to work out how much line is needed to deploy the float into the water and how much line will move with the float.

2

Find a suitable, safe location in a tidal inlet or at the downstream end of a stormwater channel that flows into a harbour or estuary. Ensure that it is within the tidal range. This investigation will be easier if there is a bridge that goes over the channel from which the drifter can be deployed.

3

Place the marked measuring pole at a suitable place so that you can use it to measure the position of the tide and the tidal range.

4

Measure the position of the tide and the tidal velocity with the drifter at different times during the tidal cycle by deploying it in the tidal current. The drifter will move with the tidal current. The drifter's velocity can be determined by timing how long it takes to travel a certain distance.

5

Record the height of the tide, the drifter's velocity and the current direction.

6

If it is possible to do so, repeat these observations for a neap tide and a spring tide.

ACTIVITIES

1

How well did your drifter measure the tidal currents?

2

Produce a graph that shows how the position of the tide and the velocity of the tidal current change both varied with time during the part of the tidal cycle you observed.

3

What was the maximum tidal current velocity and when did it occur in relation to the position of the tide?

4

What was the minimum tidal current velocity and when did it occur in relation to the position of the tide?

7.4 Life in the oceans

OUTCOMES

At the end of this chapter you should be able to:

describe the attenuation with depth of light in oceanic waters, and the order in which the different wavelengths of light disappear with depth in oceans

discuss the implications of limited light for the distribution of marine plants in near-shore environments and photosynthetic plankton in the open oceans

describe what is meant by a community of organisms

review the range of abiotic characteristics of an environment that determines the nature of a community within that environment

describe and compare examples of food chains that occur in the top layers of the oceans and those found at great depth, explaining the differences

explain, using examples, why organisms living on the ocean floor will be different from organisms living in the top 30 m of the ocean

explain how increased understanding of ocean currents and sea-floor topography can change the utilisation of ocean resources by society.

AQUATIC DESERTS

There are a few things that you notice about the sea water that you see out in the open ocean. It is a beautiful deep-blue colour quite unlike the blue-green water you see at the beach or in harbours. Sea water is also very clear and, if the sea is calm and you can look straight down into it from high up on a ship's bridge, you can see things that seem to be far below the surface. If you toss a bucket over the side and pull it back in, which is something you can only do safely if the ship is stopped or moving very slowly, the water in the bucket is very clean. These three characteristics (blueness, clearness and cleanness) contrast strongly with the turbid, murky sea water you usually see close to shore. They actually indicate that the deep, blue oceans are quite unproductive and desolate environments.

The fact that sea water is so clear and clean means that not much is living in it. Despite the availability of abundant sunlight, the concentration of nutrients in sea water is too low to sustain any more than a small population of photosynthetic plankton. This means that only small amounts of food are produced for herbivores and omnivores to eat in the open ocean. Consequently, there aren't many of them either. If there are few omnivores and herbivores to eat, then, in turn, there will not be many larger carnivores to eat them. In fact, relatively speaking, the deep oceans are so unproductive that some marine biologists call them 'aquatic deserts'.

VARIATIONS IN LIGHT INTENSITY AND OXYGEN CONTENT WITH DEPTH

We have seen how temperature and salinity change with depth and geographical position in the oceans. Polar water is cold, while tropical water is very warm. Places such as the Red Sea are very saline because there is little precipitation and evaporation rates are high. The Amazon River discharges so much freshwater into the Atlantic Ocean that it is possible to drink the water tens of kilometres off shore. The amount of dissolved oxygen, as well as other gases, and the amount of light present in sea water also change with water depth in the ocean.

Sea water is an effective filterer of sunlight, so the deeper you go down into the ocean the less light there is. The graph given in Figure 7.4.1 shows how the intensity of light present in the oceans decreases with depth. The filtering of light with depth is particularly important because it places a limit on the depth at which photosynthesis can occur. This, in turn, limits the total amount of food that is produced by **phytoplankton** for consumption by the various levels of the food chain.

Oxygen concentrations with depth vary in a different way. An example of the way the amount of oxygen dissolved in sea water varies at different depths is also shown in Figure 7.4.1. The graph shows that the amount of oxygen in sea water decreases from the surface level until about 1000 m, where it reaches a minimum level. Below this point in the water column, the oxygen content increases. This means that the oxygen content of the deep bottom waters approximates that of the surface water.

This pattern is explained by the continual oxygenation of surface waters that results from photosynthesis in the waters above 200 m. Consumption of oxygen by **zooplankton** and the nekton removes progressively more oxygen from sea water down to about 1000 m below the surface. The oxygen content then increases with depth because of the mixing of deoxygenated middle water with oxygenated bottom waters. Remember that these bottom waters begin as oxygenated surface polar waters that sink when they join the great global conveyor.

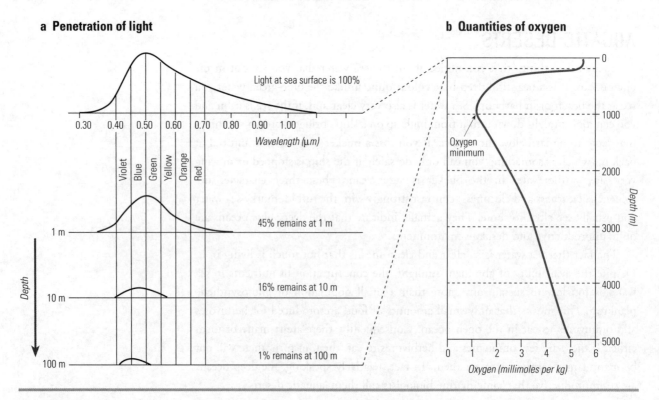

Figure 7.4.1 Light and oxygen variation in the deep ocean.

1

What general differences would you observe if you compared a bucket of water drawn from a river estuary with a bucket of water drawn from the middle of the Pacific Ocean?

2

Explain why some marine biologists call the deep oceans 'aquatic deserts'.

3

What is the source of most of the oxygen dissolved in sea water?

4

Sketch a graph that shows how the intensity of light and the concentration of oxygen vary with depth in the ocean.

5

Outline the reason or reasons why there is a minimum in oxygen concentration at oceanic depths of around 1000 m.

6

Investigate the two main types of plankton (phytoplankton and zooplankton) and locate pictures of them. Outline their similarities and differences.

OCEANIC COMMUNITIES

The combination of the changes in all the factors discussed earlier will obviously create several different physical environments beneath a given spot in the ocean. You can think of the ocean as a stack of a few different layers: the warm, brightly lit surface water; the cool, dimly lit middle water; the cold, dark deep water; and the cold, dark sea floor.

Unsurprisingly, each of these environments is a unique habitat and supports its own community of organisms. In this sense, a community of organisms consists of a group of plants and animals that live in the same environment and are inter-dependent upon one another. The relationships between them are often described by means of a **food web**. (See Figure 7.4.2, page 374.)

Open-ocean communities are generally divided into two groups:
• pelagic communities, which consist of free-swimming organisms (the nekton) and floating organisms (the plankton)
• benthic communities, which consist of organisms that live on the sea floor.

The oceanic animals that we usually eat are mostly taken from the nekton, for example, fish and squid.

definition

food web
a diagram that indicates how the plants, herbivores, omnivores and carnivores that make up a community interact with each other

Surface-water and mid-water oceanic communities

Even though the deep oceans are relatively unproductive places, they are so large that many organisms live in them and their total biomass and diversity is still very impressive. The majority of the nekton and zooplankton live in both the surface and middle waters, while the photosynthetic plankton (phytoplankton) that provides the nekton and the zooplankton with food lives in the surface water. So we will treat this group of organisms as a single community.

The surface-water layer of the ocean is a 200 m zone where the water is warm and well oxygenated. Most of the Sun's light only penetrates to about 100 m and this is the zone in which all the photosynthetic food production takes place.

The mid-water layer of the ocean is the layer of cold water beneath the warm surface layer. It also receives a little light from the surface and is the zone within the ocean where nutrients are at a maximum. Its base, which is located at around 1000 m below the surface, is defined by a minimum in the dissolved oxygen content of the water and by the final disappearance of the remaining light. These two zones are the

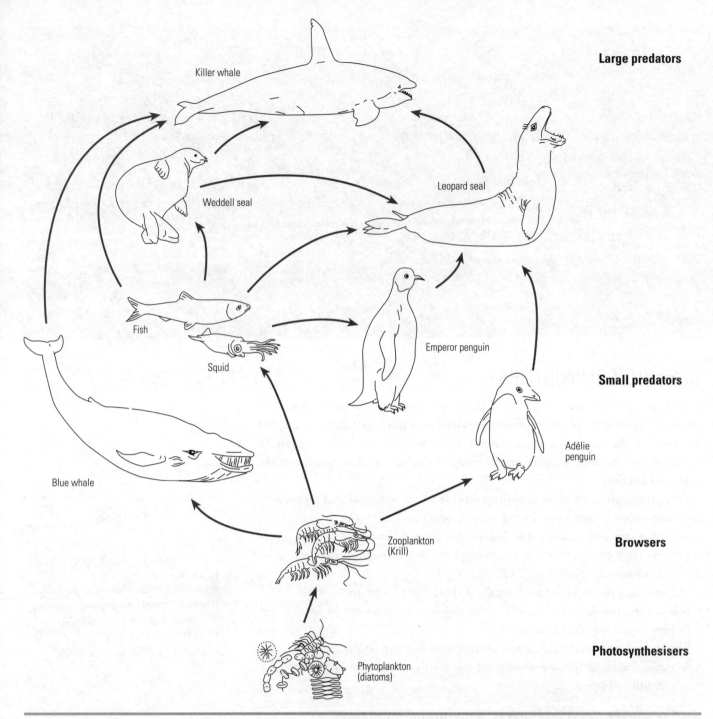

Large predators

Killer whale

Weddell seal

Leopard seal

Fish

Squid

Emperor penguin

Blue whale

Small predators

Adélie penguin

Zooplankton (Krill)

Browsers

Phytoplankton (diatoms)

Photosynthesisers

Figure 7.4.2 A simplified food web for the Southern Ocean. Note how some of the animals feed on a variety of others from different levels of the food web.

part of the open ocean where we find the sorts of fish we are used to eating. Beneath these depths, the pressure generated by the overlying mass of water becomes too intense for fish's swim bladders to work. Similarly, the lungs and endurance of deep-diving mammals are too greatly stressed if they venture far below this depth.

The organisms that perform the photosynthesis that takes place in the surface layer are called phytoplankton (see Figure 7.4.3) and consist of blue-green algae, calcareous algae, diatoms and dinoflagellates. The phytoplankton generally live in the top 150 m of the ocean. These organisms form the base of the food chain and are mostly eaten by tiny (5 mm long) arthropods called copepods. These animals dominate the zooplankton (see Figure 7.4.4) and are almost solely responsible for converting the phytoplankton into the food (mainly protein and fats) that passes on up the food chain to feed the nekton.

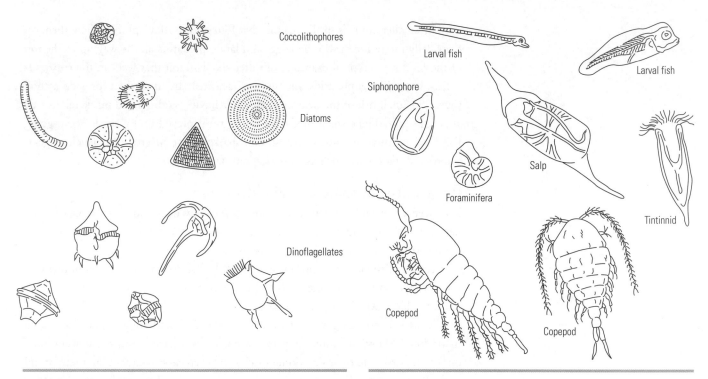

Figure 7.4.3 Phytoplankton.

Figure 7.4.4 Zooplankton.

DAILY MIGRATION OF THE ZOOPLANKTON AND NEKTON

The zooplankton migrates up to the surface to feed on the living phytoplankton at night and then sinks back down to the lower reaches of the middle water during the day. (See Figure 7.4.5.) The obvious explanation for this daily cycle of migration is that the copepods and other zooplankton wish to feed in the more productive surface layer of the ocean without being eaten by predators themselves. The zooplankton are better at hiding in the dark, so they migrate to the surface at night to feed in the dark and then migrate back down into the depths during the day, where they continue to feed on detritus and dead phytoplankton.

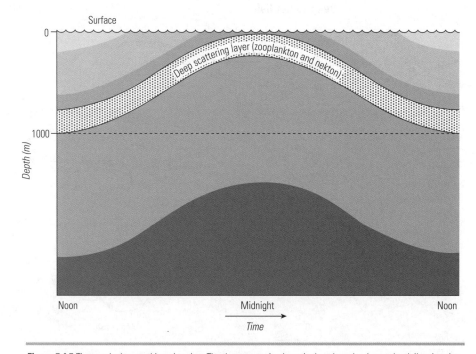

Figure 7.4.5 The zooplankton and its migration. The deep scattering layer is thought to be due to the daily migration of copepods and the fish and crustaceans that feed on them.

The smaller animals of the nekton that feed on the zooplankton follow them on their daily migration, as do the larger and larger predators all the way up to the top of the food web. Typical examples of other zooplankton that feed on the copepods are krill, mysids, euphausids and shrimp (prawns). In turn, these provide a food source for small fish of the nekton (such as anchovies, sardines and mackeral) as well as small squid. These small predators are, in turn, preyed on by larger ocean-going fish (such as tuna, marlin, swordfish and sharks), large invertebrates (such as giant squid) and mammals (such as seals, dolphins and whales).

Deep-water oceanic communities

The deep-water environment is cold and completely dark and is the layer of water between 1000 m and 4000 m. The nutrients that occur there are the remains of material that has drifted down from the surface.

Because of the lack of light, most of the fish that do live in this environment do not have eyes. Although some, such as the lantern fish, have evolved luminous organs that emit light, which they use to track or lure prey.

Fish that live in deep water tend to be quite small and are generally less than a metre long. Many, including gulpers and angler fish, have mouths that seem quite out of proportion to their size. Some species will even eat fish larger than they are and have special expanding stomachs to contain them. This is because catching something is a relatively rare event. All the species that live in this environment have adapted to the great pressures found at these depths, and when they are caught and brought to the surface the poor things often blow up like balloons and burst. You can see for yourself the anatomical differences between the fish of the nekton and deep-water fish in Figure 7.4.6.

Bottom-water oceanic communities

The bottom-water environment lies between 4000 m below sea level and the sea floor, which generally lies at depths between about 4500 m and 5500 m; except for the relatively small areas of ocean occupied by the deep ocean trenches. This is where

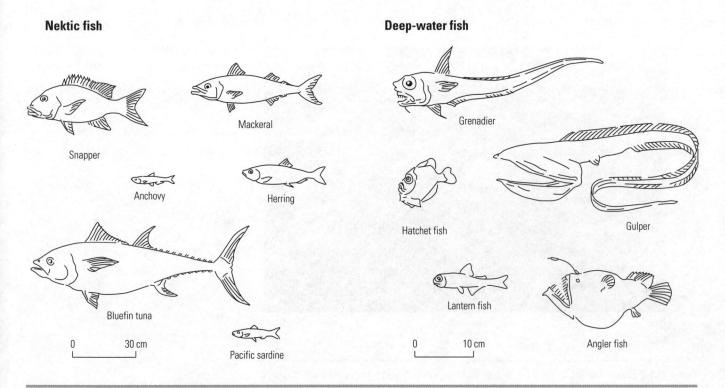

Figure 7.4.6 Typical nektic and deep-water fish.

ram that
ants,
s and
 up a
vith

Zooplankton eat the phytoplankton and each other and
provide a food source for
animals higher in the food
web.

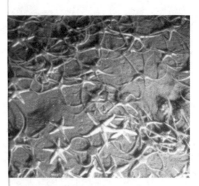

icro-
t the base
veb.

The nekton consists of actively
swimming animals (such as
fish, squid and whales) that
feed on the plankton and
each other.

iyto-
ynthesise
w the sea

Bottom-dwelling animals
include worms, crabs, fish and
echinoderms that scavenge
detritus or feed on each other.

tionships between zooplankton and a

ES

s of bottom-living fish that is important
essional fishers. Also research the
ion of this fish. Report your findings to

utlines the
h. Name
ne the
ch as its
organism
h it lives,
it feeds
ially

2

Compare and contrast the
appearance and lifestyle of a
deep-ocean organism and a
shallow-water organism.

3

Outline how government
organisations collect
information that can be used to
manage or regulate commercial
and recreational fishing.

ep-water fish and bottom-dwelling
em. List some characteristics of these
are different from nektic fish.

4

Why is it necessary to manage
fisheries?

SUMMARY

- Light only penetrates a short distance down into sea water and its intensity attenuates quickly with depth.

- Oxygen concentration in sea water is at a minimum at a depth of 1000 m.

- A community of organisms consists of a group of plants and animals that live in the same environment and are interdependent upon one another.

- A food web is a dia indicates how the herbivores, omnivo carnivores that ma community interact each other.

- Phytoplankton are organisms that are of the marine food

- The limit at which plankton can photo is about 150 m bel surface.

PRACTICAL EXERCISE
Life of the seas

This exercise will improve your understanding of fisheries as a resource and the differences between shallow-water and deep-water ocean life.

Procedure

1

Use the material in this chapter, along with resources you locate on the Internet or in the library, to investigate how fisheries are managed.

2

In particular, research a few of the organisms that form the basis of some of Australia's major fisheries, for example, orange roughy, snapper, bream, Pacific sardine, pilchards, abalone and oysters.

3

Locate pictures of several marine organisms that live in Australian waters. At least one should be a shallow-water species and at least one should be a deep-water species.

ACTIVITI

1

Produce a table that results of your resea each species and ou details about each, s habitat, which type it is, the depth at wh what it feeds on, ho and how it is comme harvested or caught.

SUMMARY

Light only penetrates a short distance down into sea water and its intensity attenuates quickly with depth.

Oxygen concentration in sea water is at a minimum at a depth of 1000 m.

A community of organisms consists of a group of plants and animals that live in the same environment and are interdependent upon one another.

A food web is a diagram that indicates how the plants, herbivores, omnivores and carnivores that make up a community interact with each other.

Phytoplankton are micro-organisms that are at the base of the marine food web.

The limit at which phyto-plankton can photosynthesise is about 150 m below the sea surface.

Zooplankton eat the phyto-plankton and each other and provide a food source for animals higher in the food web.

The nekton consists of actively swimming animals (such as fish, squid and whales) that feed on the plankton and each other.

Bottom-dwelling animals include worms, crabs, fish and echinoderms that scavenge detritus or feed on each other.

PRACTICAL EXERCISE
Life of the seas

This exercise will improve your understanding of fisheries as a resource and the differences between shallow-water and deep-water ocean life.

Procedure

1

Use the material in this chapter, along with resources you locate on the Internet or in the library, to investigate how fisheries are managed.

2

In particular, research a few of the organisms that form the basis of some of Australia's major fisheries, for example, orange roughy, snapper, bream, Pacific sardine, pilchards, abalone and oysters.

3

Locate pictures of several marine organisms that live in Australian waters. At least one should be a shallow-water species and at least one should be a deep-water species.

ACTIVITIES

1

Produce a table that outlines the results of your research. Name each species and outline the details about each, such as its habitat, which type of organism it is, the depth at which it lives, what it feeds on, how it feeds and how it is commercially harvested or caught.

2

Compare and contrast the appearance and lifestyle of a deep-ocean organism and a shallow-water organism.

3

Outline how government organisations collect information that can be used to manage or regulate commercial and recreational fishing.

4

Why is it necessary to manage fisheries?

the remains of organisms that have died and sunk to the bottom come to rest. The large bodies provide rich pickings for scavengers, but most remains are of smaller organisms. Currents sweep this detritus along the sea floor, where bottom-attached creatures consume it. Worms and sediment feeders of many types chomp through the sea-floor muds, extracting from it what food they can. In turn, they provide a food source for crabs, echinoderms and bottom-dwelling fish. Some typical bottom dwellers are shown in Figure 7.4.7.

  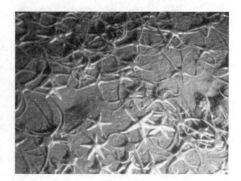

Figure 7.4.7 Typical bottom dwellers: starfish, spider crab and brittle stars.

REVIEW ACTIVITIES

1
Define the term 'nekton'.

2
Outline the differences between phytoplankton and zooplankton.

3
What is a copepod and why are these creatures so important in the marine food web?

4
Describe the daily migration of the zooplankton and nekton and outline why this migration occurs.

5
If you are trawling for fish, at what depth should you set your nets:
a during the day
b at night?
Explain your answers.

6
List some characteristics of deep-water fish that make them different from nektic fish.

7
Name some of the animals that live on the sea floor and describe their food sources.

8
Explain why nektic animals tend to be found in water depths of less than 1000 m.

EXTENSION ACTIVITIES

9
Investigate and outline the relationships between zooplankton and a variety of whale species.

10
Investigate a particular species of bottom-living fish that is important as a commercial target for professional fishers. Also research the issues involved in the exploitation of this fish. Report your findings to the class.

11
Find a variety of pictures of deep-water fish and bottom-dwelling marine animals and examine them. List some characteristics of these creatures and explain why they are different from nektic fish.

At the end of this chapter you should be able to:

- describe the way in which sea water is heated by circulation within newly formed ocean crust

- explain the ability of hydrothermal waters (brines) to scavenge elements from rocks

- describe examples of the unique bacteria and invertebrate species that live around hydrothermal vents

- relate the heating of the water to the cooling of the newly formed crust

- outline and describe the products and process of hydrothermal fluid discharge from deep-sea vents

- identify possible resources from sites where oceans previously existed.

WHAT ARE HYDROTHERMAL VENTS AND HOW DO THEY WORK?

Hydrothermal vents are continuously flowing sea-floor geysers. They are found at active mid-ocean ridges where new basaltic sea floor is being formed and on active underwater volcanoes, which we call sea mounts. The deep-sea vents form when dense, cold sea water flows down small fractures and cracks in the sea floor and seeps through hot, newly formed rocks on the sea floor. Here, the water heats up, and is it heats it becomes less dense. This causes the water to rise and seep back up through the basaltic crust towards the sea floor, where it is ejected as a continuously flowing stream, or jet, of hot water. Once one of these flows starts, it tends to become stronger. The hot water flowing out of the vent causes cold water away from the site of the vent to be sucked down into the ocean crust. This sets up a hydrothermal convection cell that is similar to the cells of moving water that can be seen in a pot of boiling water.

As the hot, salty water (often called brines) seeps through the basalts of ocean crust, it leaches, or scavenges, many elements out of the rocks. These elements include sulfur and metals (such as copper, zinc and gold) and they are carried up to the sea floor with the water. When the hot water jet encounters the freezing-cold water, the hot water quickly mixes with the cold, deep sea water, causing the jet's temperature to drop from around 300–400 °C to almost freezing in less than a minute. This forces the dissolved metals and sulfur in the water to precipitate as fine particles of sulfide and sulfate mineral, such as galena, sphalerite and chalcopyrite. Small amounts of pure gold can also precipitate in this process. The build-up of minerals above the vent forms elaborate chimney structures, often referred to as 'black smokers'. (See Figure 7.5.1.)

The 'smoke' that is visible is actually the precipitating minerals. Even though the water escaping from the vents can reach temperatures up to 400 °C it does not boil. This is because these sea-floor vents generally form at great depth and the pressure of

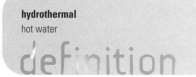

hydrothermal
hot water

definition

Figure 7.5.1 A black smoker.

the surrounding sea water keeps the water from boiling. The intense heat is limited to a very small area; within about 15 cm of the vent's opening. The surrounding water is only 2 °C. Hydrothermal vents play an important role in the geochemical cycles and heat balance of the oceans.

MINERAL DEPOSITS PRODUCED BY HYDROTHERMAL VENTS

Mineral deposits on the bottom of the ocean first attracted attention in the 1960s, when people started to think about mining zinc from sites 2 km deep in the Red Sea. The high cost of recovery of these fairly low-grade deposits has so far made deep-sea mining uneconomical.

Enormous sheets of sediment rich in fine-grained lead, copper and zinc sulfide minerals ejected from hydrothermal vents were discovered in the Bismarck Sea in 1985. More recently, a group of scientists discovered that the eastern Manus Basin contains three active hydrothermal zones. In January 2001, Leg 193 of the Ocean Drilling Program drilled and explored the subsurface parts of these active, mineralised hydrothermal systems. Unlike the deep-sea deposits that had previously been considered for mining, these massive sulfide deposits are high grade and occur at relatively shallow depths (less than 2 km). The average composition of samples recovered from these sites is 10–15% copper, 26% zinc and 2–3% lead, respectively. They also contain up to 21 grams of gold and 200 grams of silver to the tonne.

Ancient hydrothermal vent sites are often exploited for the sulfide minerals they contain. In Australia, they are known as 'sedex deposits', which stands for fine-grained sediment-hosted silver–lead–zinc deposits of mid-Proterozoic age. Examples include the giant Broken Hill, Mount Lyell and Mount Morgan deposits.

HYDROTHERMAL VENT COMMUNITIES

In 1977, scientists in the submersible Alvin discovered a bizarre collection of organisms living at a vent off the Galapagos Islands. Similar communities have since been found at several hundred hydrothermal sites around the world.

Organisms are attracted to the vents by the chemical cocktail that exists there. No sunlight penetrates to the deep ocean, so there is no chance of photosynthesis. Instead, a similar form of energy production, called chemosynthesis, occurs around the vents. Many different species of bacteria and archaea, sometimes called extremophiles, make up the primary producers. They utilise the sulfur, hydrogen, methane and other compounds that are produced when sea water reacts with magma in the crust. Many of these microbes can survive at temperatures above 100 °C. The most abundant chemical dissolved in vent water is hydrogen sulfide. Microbes utilise the energy released from the oxidation of hydrogen sulfide to produce food. These sulfur microbes form mat-like coatings on the surrounding rocks or live within the tissues of other animals.

Worms, clams, crabs, shrimps and octopuses also live around the vents. (See Figure 7.5.2.) The most spectacular organisms discovered so far are probably the giant tube worms. These worms can be over 3 m tall, have fabulous red tops and grow in groups that sway in the current. They have no mouth or gut, instead relying on symbiotic microbes for their food. The red tip contains haemoglobin, which binds oxygen and hydrogen sulfide for the microbes. Giant clams have similar relationships with microbes.

Figure 7.5.2 Animals that inhabit hydrothermal vents: spider crab, tube worms and mussels.

1

Outline the changes that occur to sea water as it cycles through a hydrothermal convection cell.

2

How do hydrothermal brines become concentrated in sulfur and metals?

3

Why are hydrothermal vents often called 'black smokers'?

4

Name the minerals found in the sediments that form beneath the plumes ejected by hydrothermal vents.

5

What is an extremophile and what is special about its metabolism?

6

What is a tube worm?

7

A number of biologists have confidently suggested that some quite complex groups of animals may survive even if all terrestrial and oceanic photosynthesis ceased forever. Why are these biologists so confident of the survival of animal life in spite of such a catastrophe?

8

Use Internet resources to investigate how underwater mining is done and research the findings of the Ocean Drilling Program from the ODP website <http://www-odp.tamu.edu/publications/prelim/193_prel.>.

9

Investigate the special communities that inhabit the sites of hydrothermal vents in the deep ocean and present your findings in the form of a talk or a poster.

10

Research an example of an ancient hydrothermal vent (sedex) deposit. Present your findings in a poster or short talk.

SUMMARY

The heat present in newly formed oceanic crust causes circulating cells of sea water to form in the young crust.

There is an exchange that transfers heat from the crust to the sea water, which cools the crust and heats the sea water.

Hydrothermal vents form on the sea floor at sites where circulating water is expelled from the crust back into the sea.

The brines expelled by hydrothermal vents have often been heated to 300–400 °C and contain large amounts of dissolved lead and zinc sulfide.

Special microbes that metabolise hydrogen sulfide present in expelled hydrothermal vent fluids support unique communities of invertebrates that live in and around hydrothermal vents.

PRACTICAL EXERCISE
Hydrothermal vents

This exercise will improve your understanding of the formation of hydrothermal vents, their sedimentary deposits and the communities they support.

Procedure

1

Use the material in this chapter, along with resources you locate on the Internet or in the library, to investigate hydrothermal vents, their sedimentary deposits and their communities.

2

Locate a variety of photographs showing the structure of hydrothermal vents and the animals that live around them.

ACTIVITIES

1

Outline the structural characteristics of a hydro-thermal vent and explain how these features form.

2

Explain why it is that a number of hydrothermal vents often occur in the same general area, forming a feature that is often called a hydrothermal vent field.

3

What organisms form the basis of the ecosystems that live around hydrothermal vents?

4

List and briefly describe the animals that commonly live around hydrothermal vents and draw a food web that indicates the relationships between these species.

5

Name the commercially important minerals that are formed by the activity of hydrothermal vents and indicate the metals that are extracted from them.

6

Describe the processes that form black smokers and explain why these features are of interest to minerals exploration companies.

At the end of this chapter you should be able to:

outline the origin, characteristics and distribution of different deep-sea sediments in the Pacific Ocean Basin, including calcareous ooze sediments; siliceous ooze sediments; deep-sea clays; manganese nodules; glacial marine sediments; and continental margin sediments

discuss the different circumstances required for the deposition of different deep-sea sediments in the Pacific Ocean Basin.

TYPES OF DEEP-SEA SEDIMENT

Sediments deposited near the coasts of the continents tend to be mostly made up of rocks, sand and mud that have been eroded from the continents and transported to the coast by rivers or glacial ice. These terrestrially derived gravels, sands and silts are called terrigenous sediment. They are usually deposited on the continental shelf and slope and occasionally on the deep sea floor just next to the continents. This material rarely gets deposited far from land out in the deep ocean basins. Instead, most of the sediment that is deposited onto the deep floors of the ocean basins is generated by biological activity in the water column above the site of deposition or consists of very fine-grained sediment that is brought in on the wind. The wind-blown material usually consists of fine ash erupted by volcanoes or continental dusts that have been eroded from the landmasses during wind storms. The sedimentary particles formed in these two ways are generally very small and most are usually less than one hundredth of a millimetre across. This means that the vast majority of deep-sea sediments are the finest-grained sediments that exist.

There are four main groups of sediment that are commonly deposited in the world's oceans:

• **calcareous ooze**, which is generated in the warm sea water and occupies about 35% of the sea floor
• **deep-sea clays**, which occupy the deeper parts of the ocean basins and cover approximately 30% of the sea floor
• **terrigenous sediments**, which collect near the landmasses and occupy about 20% of the sea floor
• **siliceous ooze**, which is generated in colder sea water and covers about 15% of the sea floor.

The four dominant types of sediment deposited in the deep ocean basins are generated in the water column above the site of deposition.

An interesting aspect of both types of ooze is that both silica and calcite start to dissolve back into the sea water once the algae or protozoan that secreted it dies. This

calcareous ooze
fine-grained deep-sea sediment dominated by calcareous skeletal remains of plankton

terrigenous sediments
sediments derived from sources on land

siliceous ooze
fine-grained deep-sea sediment dominated by skeletal remains of plankton made of opaline silica

definitions

means that these materials only collect on the sea floor if the amount produced in the overlying water column is greater than the amount that is dissolved while they sink to the bottom of the sea. The oozes generally dissolve back into sea water if the sea-floor depth is greater than 5000 m.

Deep-sea, or pelagic, clays are generally fairly featureless, reddish-brown or beige clays that are mostly composed of fine-grained volcanic and continental dusts. They commonly occur in the deeper parts of the ocean and collect in water depths of over 4000 m. These dusts are deposited everywhere across the ocean basins, but the amount of pelagic clays that collects on the sea floor is so small that the biological material generated in the overlying column of ocean overwhelms or swamps the amount of clay deposited. Siliceous oozes are mostly made up of disks and plates of opaline silica formed by unicellular algae called diatoms as well as the tiny skeletons of unicellular animals called radiolaria. Calcareous oozes are mostly formed from the tiny skeletons of unicellular animals called foraminifera and calcite plates and disks formed by calcareous algae called coccoliths.

DISTRIBUTION OF DEEP-SEA SEDIMENTS

The map given in Figure 7.6.1 shows the distribution of the main types of sediment in the world's seas and oceans. As you would expect from the preceding descriptions of how these materials form, the sea floor around the continents is covered by continental margin or terrigenous sediments. Shallower, warmer parts of the deep oceans basins are covered by calcareous ooze. Shallower, cooler parts of the ocean are

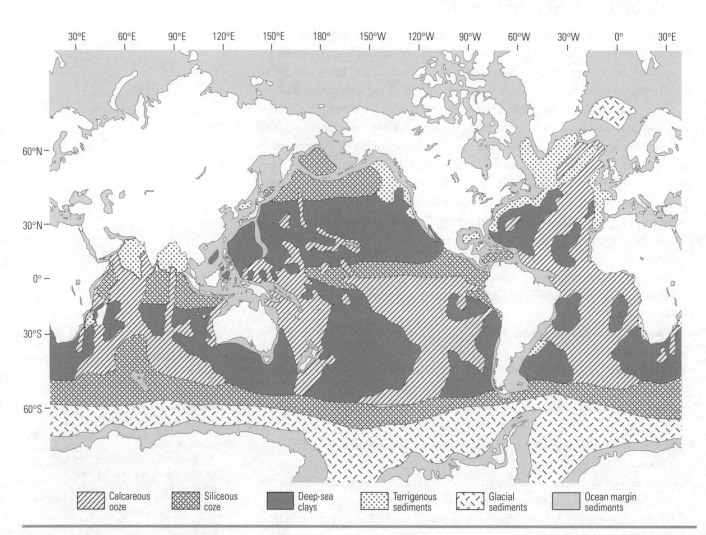

Legend:
- Calcareous ooze
- Siliceous ooze
- Deep-sea clays
- Terrigenous sediments
- Glacial sediments
- Ocean margin sediments

Figure 7.6.1 The distribution of oceanic sediments.

covered by siliceous ooze. The deepest parts of the ocean basins are covered by deep-sea clays.

The map shows that a narrow band of siliceous ooze forms just to the north of the equator in the eastern part of the Pacific Ocean Basin. The presence of this siliceous ooze in this location seems a little odd because we would expect the sea water to be fairly warm in this location. This ooze forms beneath a part of the ocean where the Pacific Ocean's water circulation patterns force deep, cold sea water to rise to the surface (that is, upwell), where it mixes with the surface water. This lowers the water temperature. The upwelling water also brings with it nutrients that favour the production of the radiolaria and diatoms that make up the ooze.

It is possible to further subdivide the groups of ocean sediment shown. For instance, the dominant type of terrigenous sediments deposited near Antarctica are glacial in origin, while the dominant source for terrigenous sediment near Australia is material delivered to the coasts by rivers. Areas of the sea floor that are swept clean of all sediment accumulate manganese nodules or manganese pavements when manganese precipitates out of the sea water. These areas are relatively restricted and cannot be shown on a map of this scale.

REVIEW ACTIVITIES

1
Name the four main groups of oceanic sediment.

2
Describe the materials that make up siliceous and calcareous ooze.

3
Describe the distribution of siliceous ooze in the Pacific Ocean Basin and account for the occurrence of a small patch of this material near the equator.

4
Discuss the different circumstances required for the deposition of different deep-sea sediments in the Pacific Ocean Basin.

EXTENSION ACTIVITIES

5
What sorts of material would you expect to make up the terrigenous sediment currently being deposited off Antarctica?

6
Research and explain the distribution and formation of pelagic clays in the Tasman Sea.

7
Research the formation of manganese nodules and pavements.

SUMMARY

Deep-sea sediments are mostly composed of calcareous ooze, siliceous ooze, and clay that have settled through the column of sea water.

The type of sediment found on a part of the deep sea floor tends to vary according to the depth of the sea floor.

Terrestrial rocks and sediment are eroded from the continents by rivers and glaciers and deposited on the shelves and slopes located at the margins of the continents.

PRACTICAL EXERCISE
Deep-sea sediments

This exercise will improve your understanding of deep-sea sediments.

Procedure

1

Use the material in this chapter, along with resources you locate on the Internet or in the library, to investigate the different types of deep-sea sediments: calcareous ooze, siliceous ooze, deep-sea clays, manganese nodules, glacial marine sediments and continental margin sediments.

2

Locate photographs of as many of these different materials as you can.

ACTIVITIES

1

Produce a poster or a talk that outlines what you found out about the appearance and composition of deep-sea sediments. You should describe the differences in the composition and appearance of these materials.

2

Which deep-sea sediments have alternative names and what are those names?

3

How do we know about deep-sea sediments?

4

Outline how manganese nodules and crusts form and explain their restricted occurrence on the sea floor.

5

Do any of these groups of materials have a potential commercial value?

6

Why do different types of deep-sea sediments form?

Resources

SECTION 1: TECTONIC IMPACTS

Books and other publications

Bureau of Mineral Resources Palaeogeographic Group, *Australia: evolution of a continent*, BMR, Canberra, 1990

Clark, IF & Cook, BF, *Perspectives of the Earth*, Australian Academy of Science, Canberra, 1983

Francis, P, *Volcanoes: a planetary perspective*, Oxford University Press, Oxford, 1993

Hamblin, WK & Christiansen, EH, *Earth's dynamic systems*, 8th edn, Prentice Hall, New Jersey, 1995

Johnson, RW, *Volcanic eruptions and atmospheric change*, Australian Geological Survey Organisation issues paper no. 1, AGSO, Canberra, 1993

Johnstone, AC & Kanter LR, 'Earthquakes in stable continental crust', *Scientific American*, March 1990, pp. 42-9

Kearney, P & Vine, FJ, *Global tectonics*, Blackwell Scientific Publications, Oxford, 1990

Kious, WJ & Tilling, RI, *This dynamic Earth: the story of plate tectonics*, US Geological Survey, Washington, 1999 <http://pubs.usgs.gov/publications/text/dynamic.html>

Scheibner, E, *The geological evolution of New South Wales: a brief review*, NSW Department of Mineral Resources, Sydney, 1999

Smith, DG (ed.), *The Cambridge encyclopedia of Earth sciences*, Cambridge University Press, Cambridge, 1981

Stanley, SM, *Earth system history*, (Chapter 13 The Early Palaeozoic world), WH Freeman and Company, New York, 1999

Veevers, JJ, *Atlas of billion-year history of Australia and neighbours in Gondwanaland*, Gemoc Press, Sydney, 2001

Websites

Alaska Volcano Observatory <http://www.avo.alaska.edu>

Geoscience Australia <http://www.agso.gov.au>

Magma table<http://www.mms.trinity.edu/gck1304/Notes/magmas.htm>

Plate positions and movements <http://sideshow.jpl.nasa.gov/mbh/series.html>

University of British Columbia: the rock cycle <http://www.science.ubc.ca/%7Eeosc221/rock_cycle/rockcycle.html>

US Geological Survey earthquake information <http://quake.usgs.gov/>

Volcanoes and global climate change <http://eospso.gsfc.nasa.gov/eos_homepage/misc_html/nasa_facts.html>

SECTION 2: ENVIRONMENTS THROUGH TIME

Books

Encyclopedia of dinosaurs and prehistoric life, Dorling Kindersley, London, 2000

Attenborough, D, *Life on Earth*, Collins, London, 1979

Briggs, DEG, Erwin DH & Collier, FJ, *The fossils of the Burgess Shale*, Smithsonian Institution Press, Washington, 1994

Conway Morris, S, *The crucible of creation: the Burgess Shale and the rise of animals*, Oxford University Press, Oxford, 1998

Cowen, R, *History of life*, 2nd edn, Blackwell Scientific Publications, Boston, 1995

Darwin, C, *The illustrated origin of species*, Faber and Faber, London, 1979

Gould, SG, *Wonderful life: the Burgess Shale and the nature of history*, Hutchinson Radius, London, 1990

Haines, T, *Walking with beasts*, BBC, London, 2002

Haines, T, *Walking with dinosaurs*, BBC, London, 1999

Levin, HL, *Ancient invertebrates and their living relatives*, Prentice Hall, New Jersey, 1999

Levin, HL, *The Earth through time*, Saunders College Publishing, Fort Worth, 1996

White, ME, *The greening of Gondwana*, 3rd edn, Reed, Sydney, 1986

White, ME, *The nature of hidden worlds*, Reed, Sydney, 1990

Wicander, R & Monroe, JS, *Historical geology: evolution of Earth and life through time*, BrooksCole &Thomson Learning, Pacific Grove, 2000

Zimmer, C, *Evolution: the triumph of an idea*, HarperCollins, New York, 2001

Videos and DVDs

The ballad of Big Al, BBC Natural History Unit, 2000

Evolution: clear blue sky, WGBH/SBS, 2001

Life on Earth, BBC Natural History Unit, 1979

Walking with beasts, BBC Natural History Unit, 2002

Walking with dinosaurs, BBC Natural History Unit, 1999

Websites

Australian Museum <http://www.austmus.gov.au/>

Evolution <http://www.pbs.org/wgbh/evolution/>

Museum of Paleontology, University of California, Berkeley <http://www.ucmp.berkeley.edu/>

National Museum of Natural History, Smithsonian Institute <http://www.mnh.si.edu/>

The Origin of Species <http://www.literature.org/authors/ darwin-charles/the-origin-of-species/>

Scientific American <http://www.sciam.com>

South Australian Museum <http://www.samuseum.sa.gov.au/>

Walking with beasts <http://www.bbc.co.uk/beasts/>

Walking with dinosaurs <http://www.bbc.co.uk/dinosaurs/>

SECTION 3: CARING FOR THE COUNTRY

Books

Botkin, D & Keller, E, *Environmental science*, John Wiley & Sons, New York, 2000

Fogarty, M (ed.), *Senior environments and communities: book 3*, McGraw-Hill, Sydney, 1995

Manuel, M, McElroy, B & Smith, R, *Environmental issues*, Cambridge University Press, Cambridge, 1999

White, ME, *Listen…our land is crying*, Kangaroo Press, Sydney, 1997

White, ME, *Running down*, Kangaroo Press, Sydney, 2000

Winfield, A, *Environmental chemistry*, Cambridge University Press, Cambridge, 2000

Journals

NSW Department of Agriculture and Fisheries, 'Farm planning for tree establishment', *Farm Trees,* no. 3, October 1989

NSW Department of Agriculture and Fisheries, 'Tree planting for gully erosion control', *Farm Trees,* no. 4, February 1989

Videos

Australia, Kyoto and global warming, VEA, 1998

Continent in crisis, VEA, 1990

Darling disaster, VEA, 1994

The greenhouse effect, Learning Essentials, 1994

The making of Australia, VEA, 1998

Pollution, VEA, 1990

Recycling, Learning Essentials, 1994

The salty country, VEA, 1999

The search for the never-never, VEA, 2000

Sewage treatment, Learning Essentials, 1999

Websites

Australian Academy of Science: monitoring the white death—soil salinity <http://www.science.org.au/nova/032/032key.htm>

Department of Agriculture, Fisheries and Forestry <http://www.affa.gov.au/index.cfm>

Environment Australia: Australia State of the Environment 2001 Report <http://www.ea.gov.au/soe/2001>

Family Education Network's Education Please: major air pollutants <http://www.infoplease.com./agi-bin/id/ A0004695.html>

NSW Environment Protection Authority <http://www.epa.nsw.gov.au>

Salinity.org.au <http://www.salinity.org.au>

SECTION 4: INTRODUCED SPECIES AND THE AUSTRALIAN ENVIRONMENT

Books

Complete book of Australian mammals: the national photographic index of Australian wildlife, Cornstalk Publishing, Sydney, 1992

Encyclopedia of Australian animals: frogs—the national photographic index of Australian wildlife, Angus & Robertson, Sydney, 1992

Low, T, *Feral future*, Penguin Books, Sydney, 2000

Parsons, WT & Cuthbertson, EG, *Noxious weeds of Australia*, 2nd edn, CSIRO Publishing, Melbourne, 2001

White, ME, *Listen…our land is crying*, Kangaroo Press, Sydney, 1997

Videos

The biological control of insects, VEA, 1995

The control of native and introduced pests, VEA, 1955

The enemy's enemy: the biological control of insects, VEA, 1994

The fox: Australia's undesirable immigrant, VEA, 1990

Websites

Animal Liberation (South Australia): feral animals <http://www.animalliberation.org.au/feralint.html>

AQIS (Australian Quarantine and Inspection Service) <http://www.aqis.gov.au>

AQIS: shipping <http://www.aqis.gov.au/shipping>

Australian Museum fact sheets: ballast water <http://www.austmus.gov.au/factsheets/ballast.htm>

Department of Agriculture, Fisheries and Forestry <http://www.affa.gov.au/index.cfm>

Environment Australia: Australia State of the Environment 2001 Report <http://www.ea.gov.au/soe/2001>

Feral Cat Coalition: 'The great Australian cat dilemma' <http://www.feralcat.com/sarah1.html>

Natural Heritage Trust: National Feral Animal Control Program <http://www.nht.gov.au/programs/ferals.html>

Weeds Australia <http://www.weeds.org.au>

SECTION 5: ORGANIC GEOLOGY: A NON-RENEWABLE RESOURCE

Books

Allen, PA & Allen, JR, *Basin analysis: principles and applications*, Blackwell Scientific Publications, Oxford, 1990

BMR Palaeogeographic Group, *Australia: evolution of a continent*, Bureau of Mineral Resources, Canberra, 1990

Clark, IF &Cook, BJ, *Geological science: perspectives of the Earth*, Australian Academy of Science, Canberra, 1983

Committee on the Science of Climate Change, *Climate change science: an analysis of some key questions*, National Academy Press, Washington, 2001

Evans, A, *An introduction to economic geology and its environmental impact*, Blackwell Science, Oxford, 1997

Goudie, A, *Environmental change*, 3rd edn, Clarendon Press, Oxford, 1992

IPCC, *Climate change 2001: the scientific basis*, Contribution of Working Group I to the Third Assessment Report of the Intergovernmental Panel on Climate Change, Houghton, JT, Ding, Y, Griggs, DJ, Noguer, M, van der Linden, P, Dai, X & Maskell K (eds), University Press, Cambridge, 2001

Lomborg, B, *The sceptical environmentalist: measuring the real state of the world*, Cambridge University Press, Cambridge, 2001

Prime Minister's Science Council (PMSC), *Global climatic change: issues for Australia*, Papers presented at the first meeting of PMSC, 6 October 1989, Australian Government Publishing Service, Canberra, 1989

Steering Committee of the Climate Change Study, Australian Academy of Technological Sciences and Engineering (AATSE), *Climate change science: current understandings and uncertainties*, AATSE, Parkville, 1995

Stoneley, R, *Introduction to petroleum exploration for non-geologists*, Oxford University Press, London, 1995

Ward, CR, *Coal geology in encyclopedia of physical science and technology*, vol. 3, Academic Press, New York, 1992

Ward, CR, Harrison, HJ, Mallett, CW & Beeston, JW (eds), *Geology of Australian coal basins*, Geological Society of Australia Coal Geology Group. Special Publication 1, 1995

Woodward, J, Place, C & Arbeit, K, 2000 Energy Resources and the Environment. In Ernst, WG (ed.), *Earth systems: processes and issues*, Cambridge University Press, Cambridge, 2000

Yencken, D & Wilkinson, D, *Resetting the compass: Australia's journey towards sustainability*, CSIRO Publishing, Melbourne, 2000

Websites

APPEA (Australian Petroleum Production and Exploration Association): discover the world of oil and gas <http://www.appea.com.au/edusite/index.html>

APPEA: home page <http://www.appea.com.au>

Australian Cooperative Research Centre for Renewable Energy <http://www.acre.murdoch.edu.au>

Australian Greenhouse Office: home page <http://www.greenhouse.gov.au>

Australian Greenhouse Office: renewable energy <http://www.greenhouse.gov.au/renewable/>

BP statistical review of world energy 2002 <http://www.bpamoco.com>

Global change: a review of climate change and ozone depletion <http://www.globalchange.org/>

NSW Department of Mineral Resources <http://www.minerals.nsw.gov.au>

NSW Minerals Council <http://www.nswmin.com.au>

Petrographic atlas <http://mccoy.lib.siu.edu/projects/crelling2/atlas/>

Petroleum Club of Western Australia: School's information program <http://www.petroleumclub.q-net.au>

Rocky Mountains Institute <http://www.rmi.org/>

Sustainable Energy Authority <http://www.seav.vic.gov.au>

United Nations Environmental Programme: introduction to climate change <http://www.grida.no/climate/vital/intro.htm>

University of British Columbia: case histories in applied geophysics <http://www.science.ubc.ca/~eoswr/geop/appgeop/ch-list.html>

US Department of Energy: fossil fuels <http://www.fe.doe.gov/education/>

World Coal Institute <http://www.wci-coal.com/>

SECTION 6: MINING AND THE AUSTRALIAN ENVIRONMENT

Books

Blainey, G, *The rush that never ended*, Melbourne University Press, Melbourne, 1993

Clark, IF & Cook, BJ (eds), *Geological science: perspectives of the Earth*, Australian Academy of Science, Canberra, 1983

Evans, AM, *An introduction to economic geology and its environmental impact*, Blackwell Scientific, Oxford, 1997

Evans, AM, *Ore geology and industrial minerals: an introduction*, Blackwell Scientific Publications, Oxford, 1993

Kesler, SE, *Mineral resources, economics and the environment*, Macmillan College Publishing, New York, 1994

Manning, I, *Native title, mining and mineral exploration: the impact of native title and the right to negotiate on mining and mineral exploration in Australia*, Office of Public Affairs, ATSIC for the Wik Team, ATSIC, 1997

Journals

NSW Department of Mineral Resources, 'Access to land for mineral exploration', *Minfact*, vol. 20, 1994

Roy, PS, Whitehouse, J, Cowell, PJ & Oakes, G, 'Mineral sands occurrences in the Murray Basin, south-eastern Australia', *Economic Geology*, vol. 95, 2000, pp. 1107–28

Websites

BC & Yukon Chamber of Mines: Mineral Exploration Primer <http://www.chamberofmines.bc.ca/educ/primer/explsumm. htm>

BUBL LINK Catalogue of Selected Internet Resources: Minerology <http://link.bubl.ac.uk/mineralogy/>

Discovery of the Cadia Ridgeway Gold-Copper Porphyry Deposit <http://www.smedg.org.au/Sym99cadia.htm>

Environmental Auditing of Operating Mine Sites <http://www.minerals.nsw.gov.au/minfo/59_audit.htm>

Exploration Methods: Mineral Exploration <http://www.bc-mining-house.com/educ/primer/explsumm2.htm>

Landform Design for Rehabilitation <http://www.ea.gov.au/industry/sustainable/mining/booklets/lan dform/land1.html#1>

Tectonic Environments of Ore Deposition <http://www.agcrc.csiro.au/projects/3048MO/>

Virtual Atlas of Opaque and Ore Minerals in their Associations <http://www.smenet.org/opaque-ore/>

Search also for information on the websites of mining companies.

SECTION 7: OCEANOGRAPHY

Books

Byatt, A, Fothergill, A & Holmes, M, *The blue planet: a natural history of the oceans*, BBC, London, 2001

Garrison, T, *Oceanography: an invitation to marine science*, 4th edn, Brooks/Cole-Thomson Learning, Belmont, 2002

Ingmanson, DE & Wallace, WJ, *Oceanography: an introduction*, 5th edn, Wadsworth, Belmont, 1995

Pinet, PR, *Invitation to oceanography*, 2nd edn, Jones and Bartlett, Sudbury, 2000

Skinner, BJ, Porter, SC & Botkin, DB, *The blue planet: an introduction to Earth system science*, 2nd edn, Wiley, New York, 1999

Thurman, HV & Burton, EA, *Introductory oceanography*, 9th edn, Prentice Hall, Upper Saddle River, 2000

Videos and DVDs

The blue planet, BBC, 2001

The living planet: a portrait of the Earth, BBC Video in association with Time-Life Video, 1987

Websites

ABC: Catalyst <http://www.abc.net.au/catalyst/>

ABC: The Lab <http://www.abc.net.au/science/>

ABC: Quantum <http://www.abc.net.au/quantum/archive.htm>

Australian Institute of Marine Science <http://www.aims.gov.au/index-ie2.html>

Centre for Research on Ecological Impacts of Coastal Cities <http://www.eicc.bio.usyd.edu.au/eicc.html>

Marine Science at the University of Sydney <http://www.usyd.edu.au/su/marine/>

Mining the deep <http://www.abc.net.au/quantum/stories/s56129.htm>

NASA (National Aeronautics and Space Administration) <http://www.nasa.gov/>

NASA: destination—Earth <http://www.earth.nasa.gov/>

NOAA (National Oceanic and Atmospheric Administration): home page <http://www.noaa.gov>

NOAA: coral reef <http://www.coralreef.noaa.gov/>

NOAA: Ocean Drilling Program <http://www.oceandrilling.org/>

NOAA: ocean explorer <http://oceanexplorer.noaa.gov/>

NOAA: photo library <http://www.photolib.noaa.gov/>

Tropical Marine Network <http://www.tmnonline.net/>

Index

Page references followed by *f* indicate figures; those followed by *t* indicate tables.